METHODS IN MOLECULAR BIOLOGY

Series Editor
John M. Walker
School of Life and Medical Sciences
University of Hertfordshire
Hatfield, Hertfordshire, UK

For further volumes:
http://www.springer.com/series/7651

For over 35 years, biological scientists have come to rely on the research protocols and methodologies in the critically acclaimed *Methods in Molecular Biology* series. The series was the first to introduce the step-by-step protocols approach that has become the standard in all biomedical protocol publishing. Each protocol is provided in readily-reproducible step-by step fashion, opening with an introductory overview, a list of the materials and reagents needed to complete the experiment, and followed by a detailed procedure that is supported with a helpful notes section offering tips and tricks of the trade as well as troubleshooting advice. These hallmark features were introduced by series editor Dr. John Walker and constitute the key ingredient in each and every volume of the *Methods in Molecular Biology* series. Tested and trusted, comprehensive and reliable, all protocols from the series are indexed in PubMed.

SH2 Domains

Functional Modules and Evolving Tools in Biology

Edited by

Teresa Carlomagno

School of Biosciences, College of Life and Environmental Sciences, University of Birmingham, Birmingham, United Kingdom

Maja Köhn

Signalling Research Centres BIOSS & CIBSS, University of Freiburg, Freiburg, Germany

Humana Press

Editors
Teresa Carlomagno
School of Biosciences, College of Life and Environmental Sciences
University of Birmingham
Birmingham, United Kingdom

Maja Köhn
Signalling Research Centres BIOSS & CIBSS
University of Freiburg
Freiburg, Germany

ISSN 1064-3745 ISSN 1940-6029 (electronic)
Methods in Molecular Biology
ISBN 978-1-0716-3392-2 ISBN 978-1-0716-3393-9 (eBook)
https://doi.org/10.1007/978-1-0716-3393-9

This Humana imprint is published by the registered company Springer Science+Business Media, LLC, part of Springer Nature.
The registered company address is: 1 New York Plaza, New York, NY 10004, U.S.A.

Preface

SH2 domains are essential mediators of phosphotyrosine signaling pathways: they bind to disordered protein stretches containing phosphorylated tyrosine residues, thereby regulating the activity of several proteins, including kinases and phosphatases. Being involved in numerous cellular functions, SH2 domains have evolved to accommodate selectivity and specificity for diverse peptide sequences. Despite their small sizes and decades of structural and biophysical studies, the relationships between sequence, structure, dynamics and function of SH2 domains remain nebulous and are thus subject of active research.

In this book, we compile recent methods to study a variety of aspects related to understanding and modulating the biological function and mechanism of SH2 domains. In Part I, we present methodology aimed at determining the structures and dynamics of SH2 domains and their complex with phosphopeptides both in solution and in the crystal. These studies are indispensable to understand how structure and dynamic properties determine binding selectivity and specificity.

In Part II, we present methods to understand the biophysical properties and predict interactions of SH2 domains by either measuring or calculating the affinity of phosphopeptides to SH2 domains, by detecting transient binding events, and by interrogating protein databases to discover new SH2 domains binding motifs.

In Part III, we turn our attention to inhibitors of SH2 domains, leading the way for chemical tool development and drug discovery. The chapter presents methods to block SH2 domains by covalent linkage to phosphopeptides, to screen for small-molecule binders to SH2 domains and to determine their selectivity, as well as methods to synthesize a specific class of SH2 domain inhibitors with a macrocyclic peptide scaffold. In addition, we describe methods to study lipid recognition by SH2 domains, a property that is fundamental for the activity of SH2 domains at the plasma membrane.

In Part IV, we describe how to evolve and/or engineer SH2 domains with specific binding properties. These methods are important to develop SH2 domains into tools for cell and molecular biology, drug discovery, and even diagnostics.

Finally, SH2 domains are part of larger proteins, whose function they regulate through a plethora of mechanisms, including localization and enzymatic site blockage. In Part V, we describe how to measure the regulation of protein tyrosine phosphatase activity through allosteric binding of peptides to SH2 domains, a biochemical assay to study the interaction of SH2 domain–containing proteins with phosphopeptides at the membrane, and how to use phosphopeptides as baits to discover new SH2 domain–containing proteins.

We hope that this collection of methods will facilitate biophysical and biochemical research in this exciting, varied, and versatile class of regulatory and signaling domains.

Birmingham, United Kingdom — *Teresa Carlomagno*
Freiburg, Germany — *Maja Köhn*

Contents

Contributors

FABIO C. L. ALMEIDA • *National Center for Structural Biology and Bioimaging (CENABIO)/National Center for Nuclear Magnetic Resonance (CNRMN), Federal University of Rio de Janeiro, Rio de Janeiro, Brazil; Institute of Medical Biochemistry – IBqM, Federal University of Rio de Janeiro, Rio de Janeiro, Brazil*

MASSIMILIANO ANSELMI • *Theoretical Physics and Center for Biophysics, Saarland University, Saarbrücken, Germany*

QIRAT F. ASHRAF • *Department of Chemical and Physical Sciences, University of Toronto Mississauga, Mississauga, ON, Canada; Department of Chemistry, University of Toronto, Toronto, ON, Canada*

YISHU BAO • *Department of Chemistry, The Chinese University of Hong Kong, Hong Kong, SAR, China*

ANGELA BERG • *Institute of Organic Chemistry, Leipzig University, Leipzig, Germany*

THORSTEN BERG • *Institute of Organic Chemistry, Leipzig University, Leipzig, Germany*

KYLI BERKLEY • *Department of Chemistry, University of Illinois at Chicago, Chicago, IL, USA*

SARA BOBONE • *Department of Chemical Science and Technologies, University of Rome Tor Vergata, Rome, Italy*

GIANFRANCO BOCCHINFUSO • *Department of Chemical Science and Technologies, University of Rome Tor Vergata, Rome, Italy*

TITUS J. BOGGON • *Department of Pharmacology, Yale University, New Haven, CT, USA; Department of Molecular Biophysics and Biochemistry, Yale University, New Haven, CT, USA; Yale Cancer Center, Yale University, New Haven, CT, USA*

PAOLO CALLIGARI • *Department of Chemical Science and Technologies, University of Rome Tor Vergata, Rome, Italy*

TERESA CARLOMAGNO • *School of Biosciences, University of Birmingham, Birmingham, UK; Institute of Cancer and Genomic Sciences, University of Birmingham, Birmingham, UK*

ICARO P. CARUSO • *National Center for Structural Biology and Bioimaging (CENABIO)/ National Center for Nuclear Magnetic Resonance (CNRMN), Federal University of Rio de Janeiro, Rio de Janeiro, Brazil; Multiuser Center for Biomolecular Innovation (CMIB), Department of Physics, São Paulo State University (UNESP), Sao Jose do Rio Preto, Sao Paulo, Brazil*

LUCÍA B. CHEMES • *Structural and Computational Biology Unit, European Molecular Biology Laboratory, Heidelberg, Germany; Instituto de Investigaciones Biotecnológicas, Universidad Nacional de San Martín (UNSAM) – Consejo Nacional de Investigaciones Científicas y Técnicas (CONICET), San Martín, Argentina; Escuela de Bio y Nanotecnologías (EByN), Universidad Nacional de San Martín, San Martín, Argentina*

WONHWA CHO • *Department of Chemistry, University of Illinois at Chicago, Chicago, IL, USA*

ELVIN D. DE ARAUJO • *Department of Chemical and Physical Sciences, University of Toronto Mississauga, Mississauga, ON, Canada*

TOBY J. GIBSON • *Structural and Computational Biology Unit, European Molecular Biology Laboratory, Heidelberg, Germany*

JULIAN GRÄB • *Institute of Organic Chemistry, Leipzig University, Leipzig, Germany*

PATRICK T. GUNNING • *Department of Chemical and Physical Sciences, University of Toronto Mississauga, Mississauga, ON, Canada; Department of Chemistry, University of Toronto, Toronto, ON, Canada*
MENACHEM J. GUNZBURG • *Monash Institute of Pharmaceutical Sciences, Monash University, Melbourne, VIC, Australia*
JIANTAO GUO • *Department of Chemistry, University of Nebraska-Lincoln, Lincoln, NE, USA; The Nebraska Center for Integrated Biomolecular Communication (NCIBC), University of Nebraska-Lincoln, Lincoln, NE, USA*
JOCHEN S. HUB • *Theoretical Physics and Center for Biophysics, Saarland University, Saarbrücken, Germany*
HAITAO JI • *Drug Discovery Department, H. Lee Moffitt Cancer Center & Research Institute, Tampa, FL, USA; Departments of Oncologic Sciences and Chemistry, University of South Florida, Tampa, FL, USA*
AZIN KIANI • *Faculty of Biology, Institute of Biology III, University of Freiburg, Freiburg, Germany; Signalling Research Centres BIOSS and CIBSS, University of Freiburg, Freiburg, Germany; Faculty of Chemistry and Pharmacy, University of Freiburg, Freiburg, Germany*
JOHN P. KIRKPATRICK • *School of Biosciences, University of Birmingham, Birmingham, UK*
BARBARA KLÜVER • *Institute of Organic Chemistry, Leipzig University, Leipzig, Germany*
MAJA KÖHN • *Faculty of Biology, Institute of Biology III, University of Freiburg, Freiburg, Germany; Signalling Research Centres BIOSS and CIBSS, University of Freiburg, Freiburg, Germany*
MICHELANGELO MARASCO • *Molecular Pharmacology Program, Sloan Kettering Institute for Cancer Research, Memorial Sloan Kettering Cancer Center, New York, NY, USA*
GREGORY D. MARTYN • *Donnelly Centre for Cellular and Biomolecular Research, University of Toronto, Toronto, ON, Canada; Department of Molecular Genetics, University of Toronto, Toronto, ON, Canada*
FERNANDO A. MELO • *Multiuser Center for Biomolecular Innovation (CMIB), Department of Physics, São Paulo State University (UNESP), Sao Jose do Rio Preto, Sao Paulo, Brazil*
RICHARD MORIGGL • *Institute of Animal Breeding and Genetics, University of Veterinary Medicine, Vienna, Austria*
VITTORIA NANNA • *BMWZ and Institute of Organic Chemistry, Leibniz University Hannover, Hannover, Germany; School of Biosciences, University of Birmingham, Birmingham, UK*
WEI NIU • *Department of Chemical & Biomolecular Engineering, University of Nebraska-Lincoln, Lincoln, NE, USA; The Nebraska Center for Integrated Biomolecular Communication (NCIBC), University of Nebraska-Lincoln, Lincoln, NE, USA*
ANNA ORLOVA • *Institute of Animal Breeding and Genetics, University of Veterinary Medicine, Vienna, Austria*
PABLO RIOS • *Faculty of Biology, Institute of Biology III, University of Freiburg, Freiburg, Germany; Signalling Research Centres BIOSS and CIBSS, University of Freiburg, Freiburg, Germany*
HUGO SÁMANO-SÁNCHEZ • *Structural and Computational Biology Unit, European Molecular Biology Laboratory, Heidelberg, Germany; Zhejiang University School of Medicine, International Campus, Zhejiang University, Haining, China; Biomedical Sciences, Edinburgh Medical School, The University of Edinburgh, Edinburgh, UK*

KAROLINE SANCHES • *Medicinal Chemistry, Monash Institute of Pharmaceutical Sciences, Monash University, Parkville, VIC, Australia; ARC Centre for Fragment-Based Design, Monash University, Parkville, VIC, Australia*
MIKI SENDA • *Structural Biology Research Center, Institute of Materials Structure Science, High Energy Accelerator Research Organization (KEK), Tsukuba, Ibaraki, Japan*
TOSHIYA SENDA • *Structural Biology Research Center, Institute of Materials Structure Science, High Energy Accelerator Research Organization (KEK), Tsukuba, Ibaraki, Japan*
ASHUTOSH SHARMA • *Department of Chemistry, University of Illinois at Chicago, Chicago, IL, USA*
SACHDEV S. SIDHU • *Donnelly Centre for Cellular and Biomolecular Research, University of Toronto, Toronto, ON, Canada; Department of Molecular Genetics, University of Toronto, Toronto, ON, Canada*
LORENZO STELLA • *Department of Chemical Science and Technologies, University of Rome Tor Vergata, Rome, Italy*
AMY L. STIEGLER • *Department of Pharmacology, Yale University, New Haven, CT, USA*
CLAUDIA STORTI • *Department of Chemical Science and Technologies, University of Rome Tor Vergata, Rome, Italy*
XIAOLEI SU • *Department of Cell Biology, Yale School of Medicine, New Haven, CT, USA; Yale Cancer Center, Yale University, New Haven, CT, USA*
GIANLUCA VEGGIANI • *Donnelly Centre for Cellular and Biomolecular Research, University of Toronto, Toronto, ON, Canada*
SEAN PAUL WALDROP • *Department of Chemistry, University of Nebraska-Lincoln, Lincoln, NE, USA*
RUI WANG • *Department of Chemistry, The Chinese University of Hong Kong, Hong Kong, SAR, China*
GABRIELLE M. WATSON • *The Walter and Eliza Hall Institute of Medical Research, Parkville, VIC, Australia*
JACQUELINE A. WILCE • *Department of Biochemistry and Molecular Biology, Biomedicine Discovery Institute, Monash University, Clayton, VIC, Australia*
JIANG XIA • *Department of Chemistry, The Chinese University of Hong Kong, Hong Kong, SAR, China*
TAO XIAO • *Drug Discovery Department, H. Lee Moffitt Cancer Center & Research Institute, Tampa, FL, USA; Departments of Oncologic Sciences and Chemistry, University of South Florida, Tampa, FL, USA*
LONGHUI ZENG • *Department of Cell Biology, Yale School of Medicine, New Haven, CT, USA*
MIN ZHANG • *Drug Discovery Department, H. Lee Moffitt Cancer Center & Research Institute, Tampa, FL, USA; Departments of Oncologic Sciences and Chemistry, University of South Florida, Tampa, FL, USA*

Part I

Structure and Dynamics of SH2 Domains

Chapter 1

Methods for Structure Determination of SH2 Domain–Phosphopeptide Complexes by NMR

Vittoria Nanna, Michelangelo Marasco, John P. Kirkpatrick, and Teresa Carlomagno

Abstract

Nuclear magnetic resonance (NMR) spectroscopy is a powerful technique to solve the structure of biomolecular complexes at atomic resolution in solution. Small proteins such as Src-homology 2 (SH2) domains have fast tumbling rates and long-lived NMR signals, making them particularly suited to be studied by standard NMR methods. SH2 domains are modular proteins whose function is the recognition of sequences containing phosphotyrosines. In this chapter, we describe the application of NMR to assess the interaction between SH2 domains and phosphopeptides and determine the structure of the resulting complexes.

Key words Nuclear magnetic resonance (NMR) spectroscopy, Src-homology 2 (SH2) domain, Phosphopeptides, Phosphotyrosine, Structure determination, Intermolecular interactions

1 Introduction

Numerous eukaryotic cellular processes such as proliferation, metabolic homeostasis and differentiation are regulated by the phosphorylation of tyrosine residues [1]. Src-homology 2 (SH2) domains are conserved protein modules of approximately 100 residues, whose function is the recognition of peptide sequences containing a phosphotyrosine. SH2 domains can discriminate between different phosphorylated motifs, providing the specificity to mediate phosphotyrosine signaling in cellular processes [2]. Due to their small size, SH2 domains are well suited to characterization via standard nuclear magnetic resonance (NMR) spectroscopy methods, which benefit from fast tumbling rates and hence slow relaxation of the NMR signals. To date, a large number of structures of SH2 domains, either in their apo form or bound to phosphopeptides or other ligands, have been solved by either NMR spectroscopy or X-ray crystallography. [2] Solution NMR has the advantage

Teresa Carlomagno and Maja Köhn (eds.), *SH2 Domains: Functional Modules and Evolving Tools in Biology*, Methods in Molecular Biology, vol. 2705, https://doi.org/10.1007/978-1-0716-3393-9_1,

of studying the protein at room temperature under near-native conditions in solution, as well as allowing studies of site-specific dynamics (*see* Chapter 2).

Structural studies of SH2 domain–phosphopeptide complexes by NMR require isotope-labeled protein (usually ^{15}N,^{13}C-labeled) and unlabeled ligand peptide. The very first step is an NMR-monitored titration to assess both binding kinetics and the stoichiometry of the interaction. Increasing amounts of peptide are added to a buffered solution of the protein of interest, and, after each titration point, a 2D ^{15}N-HSQC (heteronuclear single quantum coherence) spectrum of the protein is recorded. The ^{15}N-HSQC spectrum correlates the amide hydrogen resonances $^{1}H_{N}$ to the respective ^{15}N resonances, and is often called the "fingerprint" of the protein. This spectrum can be recorded in a short time and requires only ^{15}N-labeling, which is relatively inexpensive compared to ^{13}C-labeling. Except for the prolines, every N–H group in the protein, including amide groups of asparagines and glutamines, Hε–Nε from arginines and Hε–Nε from tryptophans, yields a peak. The binding of the ligand to the protein results in changes of the position of the peaks in the spectrum (also called *chemical shift*). Once the binding kinetics and stoichiometry have been identified, the experiments required for the determination of the structure of the complex can be designed and measured. NMR-based structure determination relies on the inter-proton distances derived from multidimensional NOESY (Nuclear Overhauser Enhancement SpectroscopY) spectra. The intensity of the cross peaks between two hydrogen nuclei in the NOESY experiment is dependent on the time-averaged distance between the two nuclei. Programs such as ARIA (Ambiguous Restraints for Iterative Assignment) convert the peak intensities into interatomic distances and, through an iterative protocol, calculate a structure ensemble that satisfies these distances [3].

2 Materials

The authors assume that the readers approaching this chapter are already familiar with the sub-cloning of a new protein construct in an expression vector and with the expression and purification of unlabeled proteins. Here, we list materials required for the expression of ^{13}C,^{15}N-labeled proteins that are necessary for performing the NMR experiments, as well as the materials required for the NMR experiments and their analysis. All the solutions should be prepared using ultrapure water.

2.1 Expression of SH2 Domains

1. *E. coli* strains suitable for recombinant protein expression (e.g., BL21(DE3) or Tuner (DE3) competent cells).
2. Expression vector encoding the protein of interest.
3. Luria-Bertani (LB) medium.
4. Trace-element mix (1000×): 50 mL of 0.1 M $FeCl_3$ in 0.12 M HCl, 2 mL of 1 M $CaCl_2$, 1 mL of 1 M $MnCl_2$, 1 mL of 1 M ZnSO4, 1 mL of 0.2 M $CoCl_2$, 2 mL of 0.1 M $CuCl_2$, 1 mL of 0.2 M $NiCl_2$, 2 mL of 0.1 M Na_2MoO_4, 2 mL of 0.1 M Na_2SeO_4, 2 mL of 0.1 M H_3BO_4 and adjust to a final volume of 100 mL with water. Mix overnight (*see* **Note 1**).
5. 10× M9 salt solution: dissolve 37 g of Na_2HPO_4, 30 g KH_2PO_4, and 5 g NaCl in 800 mL of water and adjust the pH to 7.2 with NaOH. Add water to the final volume of 1 L. Either autoclave or sterile filter.
6. M9 minimal medium: to 860 mL of milliQ water, add 2 mL of 1 M $MgSO_4$, 30 mg of thiamine hydrochloride, 4 g of glucose, 1 g of NH_4Cl, 100 mL of 10× M9 salts solution, and the appropriate antibiotic based on the selection marker of the plasmid (e.g., ampicillin, kanamycin or chloramphenicol). Use ^{13}C-D-glucose and/or ^{15}N-ammonium chloride for labeled protein expression. Sterilize by filtration. Then, add 100 μL of 1 M $CaCl_2$ and 1 mL of trace-element mix (1000×) (*see* **Note 2**).
7. 1 M stock solution of isopropyl-β-D-thiogalactoside (IPTG) or appropriate inducing agent as required by the plasmid.
8. A spectrophotometer to measure the optical density of the bacterial culture at 600 nm (OD_{600}).
9. A centrifuge capable of generating at least 3500 × g of centrifugal force.

2.2 Purification of SH2 Domains

The materials required for the protein purification might differ based on the protocol adopted. In general, there exists no protocol best fitted to all SH2 domains; the purification protocol should be optimized case-by-case in order to obtain the best protein yield and purity:

1. Lysis buffer.
2. Sonicator or French press.
3. Centrifuge capable of a least 18,500 × g centrifugal force.
4. Ni-NTA column or appropriate affinity column based on the construct.
5. Wash buffer: typically 1 M NaCl, 50 mM Tris–HCl, 5% glycerol, 10 mM imidazole, 5 mM β-mercaptoethanol, pH 7.6.

6. Elution buffer: typically 1 M NaCl, 50 mM Tris–HCl, 5% glycerol, 500 mM imidazole, 5 mM β-mercaptoethanol, pH 7.6.
7. Fast protein liquid chromatography (FPLC) system.
8. Size exclusion chromatography (SEC) column (HiLoad Superdex75 or equivalent).
9. SEC buffer: typically 100 mM MES (2-(*N*-morpholino)ethanesulfonic acid), 150 mM NaCl, 3 mM TCEP (tris(2-carboxyethyl)phosphine), pH 6.5.
10. Concentrators with an appropriate molecular weight cutoff.

2.3 NMR Sample Preparation and Spectra Analysis

The materials required for the NMR sample preparation are as follows:

1. Buffered solution: 100 mM MES, 150 mM NaCl, 3 mM TCEP, pH 6.5, and 0.05% NaN_3 containing the labeled protein.
2. Phosphopeptides. The stock solution of the peptide should be resuspended in a buffer matching the one chosen for the protein to a concentration of 5 mM or higher. Adjust the pH of the stock solution appropriately. Customized synthetic phosphopeptides are either purchased from specialized companies or obtained from in-house facilities by solid-phase synthesis. The resulting peptides are usually purified by reverse-phase HPLC chromatography, and they should be more than 95% pure (*see* **Note 3**).
3. Deuterium oxide (D_2O).
4. 50 μM of 4,4-dimethyl-4-silapentane-1-sulfonic acid (DSS).
5. NMR tube (*see* **Note 4**).

To perform the NMR experiments, one requires the following:

1. NMR spectrometer, equipped with a triple-resonance HCN probe, which enables the simultaneous application of radiofrequency pulses at ^1H, ^{13}C and ^{15}N frequencies.
2. Spectrometer control software (we use Bruker TopSpin, but will be typically be dictated by the manufacturer of the spectrometer).
3. Data processing software (we use nmrPipe [4]).
4. NMR data visualization and assignment software (we use CcpNmr Analysis [5]).
5. Structure calculation software (we use ARIA).
6. Structure validation software or web-server.

3 Methods

3.1 Expression of the Protein

^{13}C- and ^{15}N-edited NMR experiments of $^{13}C,^{15}N$-labeled biomolecules are sufficient to study systems up to 25 kDa, such as SH2 domains. Isotopic enrichment is achieved by growing cell cultures in minimal medium, where $^{15}NH_4Cl$ and ^{13}C-glucose are the only source of nitrogen and carbon, respectively. We will describe a general expression protocol for *Escherichia coli*, which is the most common expression organism to produce isotope-labeled proteins. Most SH2 domains are expressed in bacterial hosts as soluble protein [6]. Whenever recombinant proteins expressed in bacteria aggregate, have low expression, or are insoluble or misfolded, other expression systems should be considered as an alternative, such as insect cells [7], yeast [8], or mammalian cells [9]. A standard expression protocol in *E. coli* is as follows:

1. Transform the expression plasmid into the appropriate *E. coli* expression line. The expression plasmid should encode for the protein tagged with a 6× His or other affinity tag to allow for target protein isolation. Some fusion proteins such as SUMO, thioredoxin, or glutathione S-transferase (GST) can improve solubility, help folding, and prevent the precipitation of the target protein in inclusion bodies.
2. Prepare a 50 mL overnight preculture by picking a colony from the plate and adding it to LB medium supplemented with the appropriate antibiotics, and shake it overnight at 37 °C.
3. The following day, spin down the appropriate amount of preculture according to the volume of main culture to grow. The dilution factor between the preculture and the main culture is 1:100. Resuspend the pellet in the M9 minimal medium supplemented with antibiotics and add it to 500 mL of medium in a 2 L flask. Grow the cells to OD_{600} of 0.6 at 37 °C with shaking at 250 rpm. Induce protein expression with the appropriate induction agent (i.e., 1 mM IPTG), and express according to pre-established optimal conditions of temperature and time. To save costs, the optimization of the expression protocol can be done with non-isotopically labeled material.
4. Harvest the cells by centrifugation at 3500 × g for 20 min and store the pellets at −20 °C.

3.2 Protein Purification

1. Thaw the cell pellet on ice and resuspend it in lysis buffer.
2. Disrupt the cells by sonication or French press.
3. Centrifugate the lysate for 1 h at 18,500 × g and 4 °C to remove the cell debris. The target protein should be contained in the supernatant.

4. Using an FPLC system, His-tagged proteins can be purified by immobilized metal ions affinity chromatography (IMAC) columns, like HisTrap column (we use columns from Cytiva). Load the supernatant onto the column pre-equilibrated with wash buffer. Wash the column until the absorbance at 280 nm reaches the baseline, and elute the His-tagged protein with the elution buffer.
5. We advise to remove the affinity tag before NMR analysis. For this purpose, a cleavage site should be designed between the affinity tag and the protein sequence. Common proteases are human rhinovirus 3C, tobacco etch virus (TEV), and thrombin. The protease should contain the same tag as the target protein for the following purification step. If the protease needs specific buffer conditions for the cleavage, then add a desalting step before the overnight incubation with the protease. Otherwise, directly dialyze overnight against 2 L of wash buffer.
6. After the cleavage, run a second IMAC-based purification to separate the tag and the protease from the target protein.
7. Perform a size exclusion chromatography (SEC) run as the last purification step. Pool the fractions from the second IMAC purification and concentrate to a final volume of 1–2 mL. Load the sample on a gel filtration column (we use the Superdex 75 pg column from Cytiva). NMR experiments require high purity. If the protein is still not pure after gel filtration, alternative purification steps should be considered, such as ion exchange chromatography (*see* **Note 5**).

3.3 NMR Spectroscopy

3.3.1 NMR Sample Preparation

1. Identification of the optimal buffer. The identification of the optimal sample buffer for the NMR experiments is critical to ensure protein stability throughout the long measurements. Thermal shift assays (TSA) are an efficient way to screen different buffer conditions and additives by monitoring protein unfolding during thermal denaturation [10]. In TSAs protein unfolding is detected by monitoring the change in fluorescence intensity of a fluorescent, hydrophobic dye, which binds to the denatured protein. Although phosphate-containing buffer might yield optimal stability of SH2 domains, it is not recommended to use these buffers for binding assays, as high concentrations of phosphate ions in the buffer might compete with phosphopeptide binding. Whenever possible, a slightly acidic pH is preferred for NMR experiments to reduce the chemical exchange rate of the amide hydrogens with the aqueous solvent.
2. Add to the protein in the optimized buffer 10% of D_2O and 50 μM DSS (sodium trimethylsilylpropanesulfonate).

3. The NMR sample concentration for measuring 3D experiments should be as high as possible (ideally between 300 μM and 1 mM). The volume required differs based on which type of NMR tube is used (*see* **Note 4**).
4. Before starting any relevant experiments for structure determination, test the protein sample stability. NMR experiments such as 3D NOESY have long acquisition times (from two or three days to a week). The protein must be stable at room temperature during the entire course of the NMR experiment. The best way to monitor the protein stability is to record a 1D ^{1}H spectrum and a ^{15}N-HSQC spectrum at regular time intervals. If the spectra remain identical, the sample does not change with time. If the protein undergoes degradation, sharp, poorly dispersed peaks appear in the ^{15}N-HSQC spectrum.
5. Characterize the protein–peptide complex by NMR titrations. Follow the changes induced by the binding of the phosphopeptide to the protein by measuring a ^{15}N-HSQC spectrum for each titration point. The endpoint of an NMR titration is reached when, upon addition of further ligand, the positions of the peaks no longer change (*see* **Note 6**).
6. Evaluate the kinetics of the binding event. In general terms, the binding of the protein (P) and the ligand (L) to form the complex (PL) is described by the equation:

$$\mathrm{P} + \mathrm{L} \underset{k_{\mathrm{on}}}{\overset{k_{\mathrm{off}}}{\leftrightarrows}} \mathrm{PL} \tag{1}$$

where k_{on} and k_{off} are the association and dissociation rate constants, respectively.

The binding equilibrium is described by the dissociation constant K_{D}, defined as

$$K_{\mathrm{D}} = \frac{[\mathrm{P}][\mathrm{L}]}{[\mathrm{PL}]} \tag{2}$$

where [P], [L], and [PL] are the molar concentrations of, respectively, protein, ligand, and complex at the equilibrium.

The dissociation constant K_{D} can also be written as the ratio of the dissociation rate and association rate constants (Eq. 3):

$$K_{\mathrm{D}} = \frac{k_{\mathrm{off}}}{k_{\mathrm{on}}} \tag{3}$$

NMR experiments do not report directly on k_{on} and k_{off} rates but rather on the exchange rate k_{ex}:

$$k_{\mathrm{ex}} = k_{\mathrm{on}}[\mathrm{L}] + k_{\mathrm{off}} \tag{4}$$

The term *chemical exchange* refers to all phenomena observed in NMR spectroscopy when a system interconverts between at least two distinct states over time. An example of chemical exchange is the equilibrium between the free and bound states of a protein in the presence of a ligand. Three exchange regimes are distinguishable depending on the magnitude of the exchange rate relative to the difference in NMR frequencies between the free and bound states ($\Delta\nu$):

Slow exchange regime:

$$k_{\mathrm{ex}} \ll |\Delta\nu| \tag{5}$$

Intermediate exchange regime:

$$k_{\mathrm{ex}} \approx |\Delta\nu| \tag{6}$$

Fast exchange regime:

$$k_{\mathrm{ex}} \gg |\Delta\nu| \tag{7}$$

By monitoring the changes in the ^{15}N-HSQC spectrum of the protein upon titration of the ligand, determine the exchange regime of the protein–ligand complex under investigation (Fig. 1). If during the titration the intensities of the peaks of the apo protein species diminish and new peaks appear elsewhere in the spectrum, with their intensity increasing with the ligand concentration, the complex is in the slow exchange regime. The new peaks represent the ligand-bound state of the protein. If the peaks of the protein species move across the spectrum without significantly changing in intensity or linewidth as the ligand concentration increases, the complex is in the fast exchange regime. In this case, a generic NMR property A measured for the protein peaks at each titration point is a population-weighted average of A in the free and ligand-bound protein species:

$$A_{\mathrm{avg}} = A_{\mathrm{free}}\tilde{n}p_{\mathrm{free}} + A_{\mathrm{bound}}\tilde{n}p_{\mathrm{bound}} \tag{8}$$

Finally, if the protein peaks broaden or completely disappear at low ligand concentrations and reappear at higher ligand concentrations, the complex is in the intermediate exchange regime.

The stoichiometry and K_{D} of the protein–peptide complex determine how the NMR sample should be prepared (*see* Subheading 3.3.2). Studying systems in the intermediate exchange regime can be challenging due to line-broadening.

3.3.2 NMR Experiments

NMR structure determination is based on the measurement of interatomic distances between hydrogens; thus, it is essential to assign all hydrogen resonances of both the protein and the peptide. The most common assignment approach relies on the combination of experiments that correlate resonances via through-bond or through-space interactions between the respective nuclei.

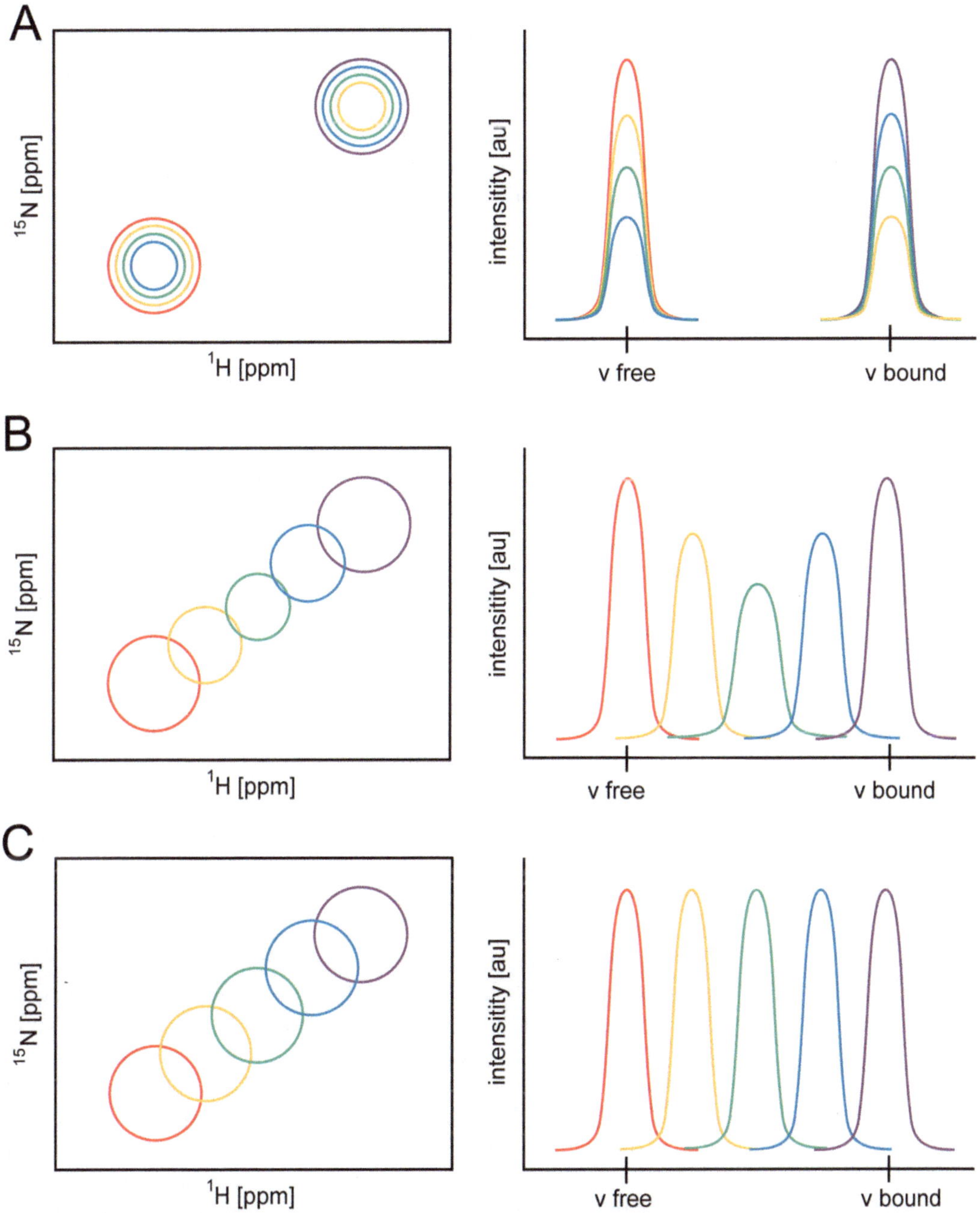

Fig. 1 Exchange regimes and corresponding ^{15}N-HSQC spectra. Changes in the appearance of the protein peaks in a ^{15}N-HSQC spectrum upon addition of increasing quantities of the peptide depend on the exchange regime of the complex. In each panel, a schematic peak is shown over a series of ^{15}N-HSQC spectra measured after each peptide addition (left). The intensities of the peaks are shown on the (right). The peaks of the free and bound protein species are in red and purple, respectively. The protein peaks at peptide–protein ratios below the protein saturation value are in yellow, green, and blue. (**a**) Slow exchange: the peak corresponding to the free protein (red) diminishes in intensity as the peak corresponding to the bound protein (purple) appears elsewhere in the spectrum and increases in intensity with the ligand concentration. (**b**) Intermediate exchange: upon addition of the peptide ligand, the protein peak moves continuously from the position corresponding to the free state towards the position corresponding to the bound protein. At the same time, the peak either

1. Assign the resonances of the protein backbone atoms (^1H_N, ^{15}N, Cα, Cβ, and C′). This step relies on the 2D ^{15}N-HSQC and three-dimensional (3D) triple-resonance experiments HNCACB, HN(CO)CACB, and HNCO (Fig. 2) [11–13]. In the NMR sample used for the assignment of protein resonances, the phosphopeptide should be added in amounts that guarantee protein saturation. The HNCACB experiment correlates the amide ^1H_N and ^{15}N resonances to the carbon resonances of $^{13}C\alpha$ and $^{13}C\beta$ of both the same amino acid and the preceding one. The HN(CO)CACB experiment correlates the amide ^1H_N and ^{15}N resonances to the carbon resonances of the $^{13}C\alpha$ and $^{13}C\beta$ of only the preceding amino acid. A sequential link can be made when a pair of peaks appear at the same $^{13}C\alpha$ and $^{13}C\beta$ frequencies in both the HNCACB and HN(CO)-CACB experiments. Certain amino acids, such as glycine, serine, threonine, and alanine, have characteristic Cα and Cβ chemical shifts making them easy to identify. If a sequential stretch of amino acids is identified that represents a unique motif in the protein sequence, the amide resonances can be unambiguously assigned to the corresponding residues, thus providing starting points for the sequential assignment. The HNCO spectrum facilitates the identification of overlapped peaks in the ^{15}N-HSQC spectrum and allows the assignment of the carbonyl carbon C′ in the backbone, which is useful to predict the dihedral angles φ and ψ of the corresponding peptide bond.
2. Assign the chemical shifts of protein aliphatic side-chain hydrogens. This assignment step is performed through a combination of HN(CO)CACB, H(CCCO)NH-TOCSY, and HCCH-TOCSY experiments (Fig. 3). The H(CCCO)NH-TOCSY correlates ^1H_N and ^{15}N amide resonances of one residue (residue i) with all side-chain hydrogens of the preceding residue (residue $i-1$) [14–16]. As mentioned before, the HN(CO)-CACB provides the Cα and Cβ frequencies of residue $i-1$. Using the combination of H(CCCO)NH-TOCSY and HN (CO)CACB, the Hα-Cα peaks of each residue can be assigned in the ^{13}C-HSQC spectrum. The HCCH-TOCSY experiment is a double resonance experiment that correlates the ^{13}C and ^1H frequencies of C-H groups to all other hydrogens in the side chain of the same residue.

Fig. 1 (continued) broadens or completely disappears at low ligand concentrations and reappears at higher ligand concentrations. (**c**) Fast exchange: upon addition of the peptide ligand, the protein peak moves continuously from the position corresponding to the free state towards the position corresponding to the bound protein without changes in the linewidth other than those expected from the increased rotational correlation time of the complex

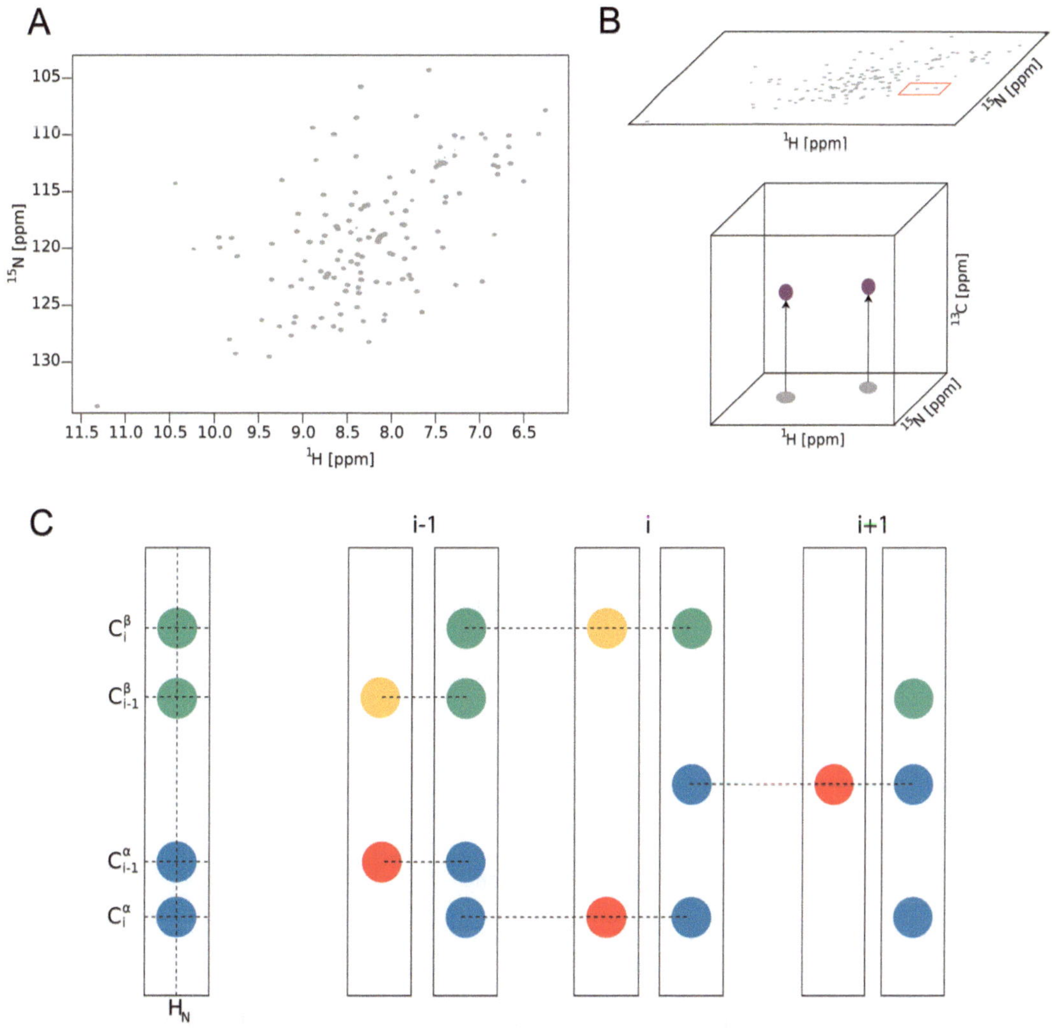

Fig. 2 2D and 3D spectra required for the assignment of the protein backbone resonances. (**a**) ^{15}N-HSQC spectrum of the ^{13}C,^{15}N-labeled N-SH2 domain of PLCγ1 in complex with a phosphopeptide. Each amide group in the protein, with the exception of proline, yields a peak in the spectrum. Well-dispersed peaks in this spectrum indicate that the protein is folded. (**b**) 3D experiments for protein backbone assignment can be depicted as a 2D ^{15}N-HSQC spectrum whose peaks are spread (in some experiments multiple times) along a third axis representing the ^{13}C dimension. (**c**) Peaks in 3D spectra are commonly visualized as strips. For each pair of ^{1}H and ^{15}N resonances in the ^{15}N-HSQC spectrum, the strip shows the ^{13}C frequencies in the full vertical dimension and the ^{1}H frequency of the pair in the horizontal dimension. The ^{15}N dimension is perpendicular to the plane of the strip and determines from which ^{1}H,^{13}C plane of the 3D spectrum the strip originates. Left, example strip from an HNCACB experiment. Note the presence of two Cα peaks (blue) and two Cβ peaks (green). Right, example of sequential assignment based on the HNCACB and HN(CO)CACB experiments. In the strips generated from the HN(CO)CACB spectrum, the Cα and Cβ peaks are in red and in yellow, respectively. Residue *i* can be connected to residue *i* − 1 because a pair of peaks appears at the same Cα and Cβ frequencies in both experiments. Residue *i* can be identified as a glycine since it yields only a Cα peak (no Cβ) at a characteristic frequency

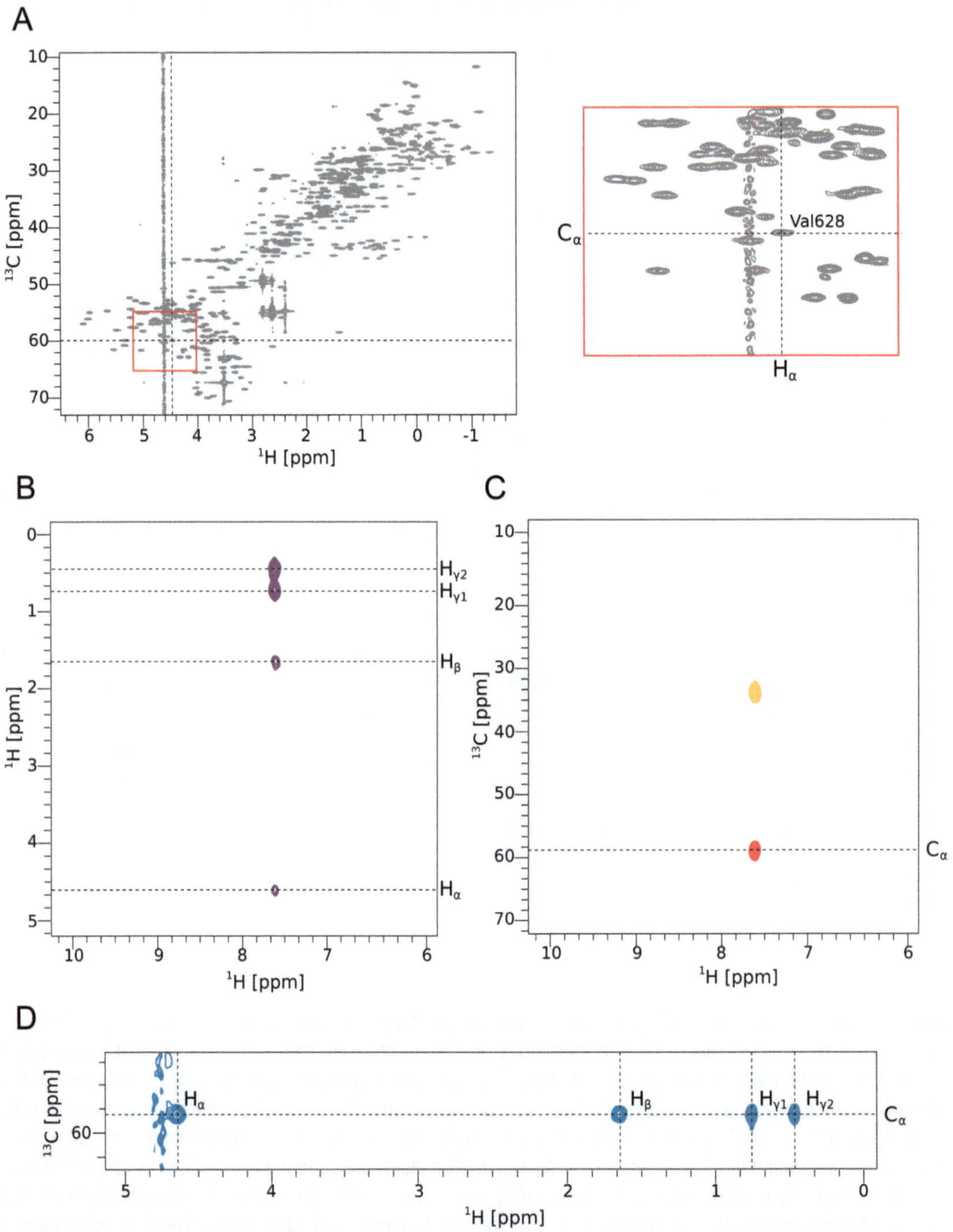

Fig. 3 2D and 3D spectra required for the assignment of the protein side-chain resonances. (**a**) ^{13}C-HSQC spectrum of the ^{13}C,^{15}N-labeled N-SH2 domain of PLCγ1 in complex with a phosphopeptide (left). Expansion of one region of the ^{13}C-HSQC (right). The Hα resonance indicated by the dashed line is found in the H(CCCO)NH-TOCSY experiment (panel **b**), correlated to the ^{15}N and ^{1}H resonances of the amide group of the following residue: the same ^{15}N and ^{1}H resonances show a peak with the Cα frequency indicated by the dashed line in the HN(CO)CACB spectrum (panel **c**). (**d**) Ideally, the HCCH-TOCSY experiment yields strips for all side-chain carbon resonances, connecting all side-chain hydrogen resonances. In the example shown, the diagonal peak is centered on the Hα and Cα resonances of V628, and the cross peaks correlate this Hα and Cα pair with all side-chain hydrogens of the same residue (Hα, Hβ, Hγ1, Hγ2)

3. Assign the chemical shifts of protein aromatic side-chain hydrogens. This assignment is accomplished through 2D (HB)CB(CGCD)HD and 2D (HB)CB(CGCDCE)HE experiments, which correlate the Cβ resonance of an aromatic residue with the resonances of either hydrogen Hδ or hydrogens Hδ and Hε of the same residue, respectively [17]. Once the chemical shifts of the aromatic side-chain hydrogens are known, the corresponding carbon resonances can be assigned from a ^{13}C-HSQC spectrum.
4. Use 2D double-^{13}C,^{15}N-filtered-NOESY and 2D double-^{13}C,^{15}N-filtered-TOCSY experiments to assign the resonances of the bound unlabeled peptide [18]. With the term *filtered*, we indicate the rejection of the signals from hydrogens attached to an isotope-labeled heteroatom (^{13}C or ^{15}N). Hence, the filter elements in the double-^{13}C,^{15}N-filtered-NOESY and TOCSY select signals arising from protons attached to ^{12}C or ^{14}N nuclei. Since the protein is ^{15}N,^{13}C-labeled, the double-filtered experiments yield only signals from the unlabeled peptide. Other unlabeled species in solution containing hydrogens, such as the buffer components, are best replaced by the corresponding deuterated species (*see* **Note 7**). The double-^{13}C,^{15}N-filtered-TOCSY experiment correlates all hydrogens located within the same amino acid of the peptide. The assignment is based on pattern recognition of the cross peaks. The identified amino acids can be linked sequentially using the double-^{13}C,^{15}N-filtered-NOESY, which correlates each hydrogen with all nearby hydrogens in the peptide. For complexes with a low K_D, use a slight excess of protein to maximize the amount of bound peptide in solution. For complexes with a K_D in the μM–mM range, prepare a sample that contains as much protein in excess as necessary to saturate the peptide.
5. Measure interatomic distances from multidimensional NOESY spectra. The most widely used set of structural restraints in structure determination by NMR consists of inter-hydrogen distances derived from multidimensional NOESY spectra. The intensities of peaks correlating the frequencies of two hydrogen atoms in these NOESY spectra (the NOE peaks) can be converted into internuclear distances of the correlated atom pairs (*see* **Note 8**).

   In the structure calculation of a protein–peptide complex, we distinguish three sets of distance restraints: intramolecular protein NOEs, intermolecular protein–peptide NOEs, and intramolecular peptide NOEs.

   To obtain these distances, record the following experiments:

   Intramolecular protein NOEs: 3D NOESY-^{15}N-HSQC and 3D NOESY-^{13}C-HSQC, which yield both intramolecular

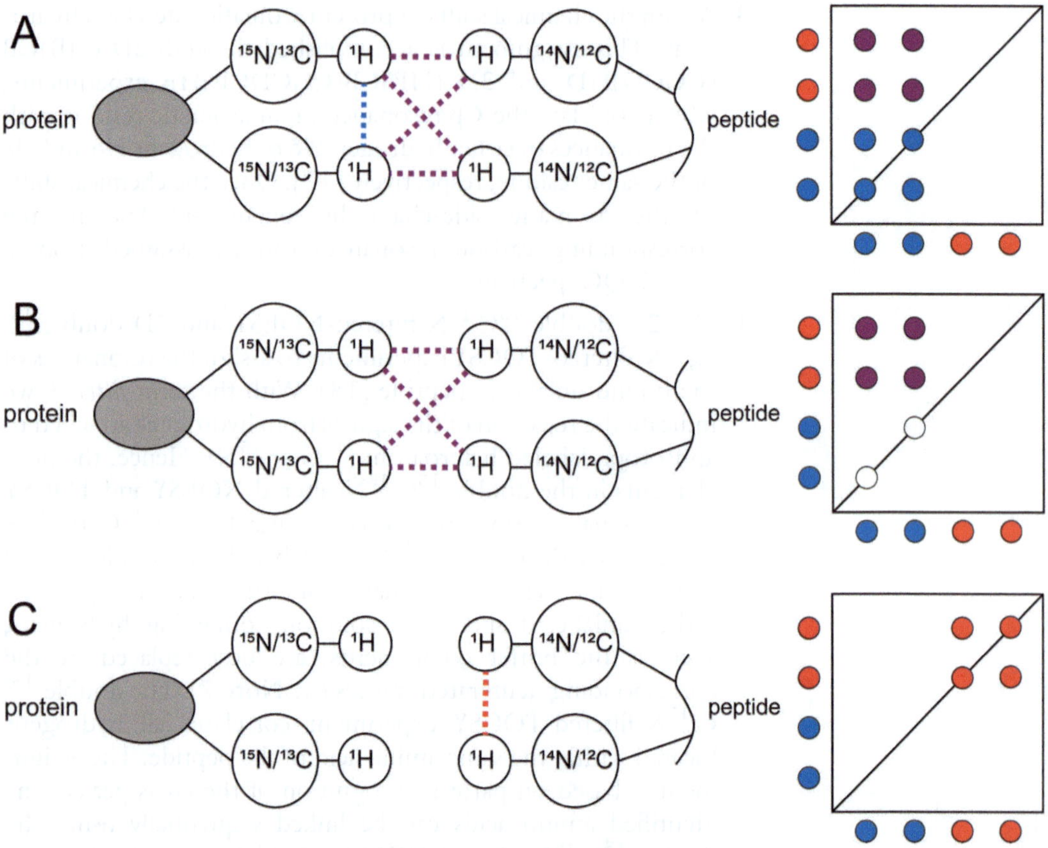

Fig. 4 The protein–peptide complex is depicted on the left-hand side with its isotope labeling pattern; ^{1}H atoms are shown bound to the corresponding heteroatom, either NMR active (^{13}C and ^{15}N) in the protein or NMR inactive (^{12}C and ^{14}N) in the peptide. A schematic representation of different NOESY experiments is given on the right-hand side. Diagonal peaks and intramolecular NOE cross peaks from the protein and the peptide are colored in blue and red, respectively. Intermolecular NOE cross peaks are colored in purple. (**a**) Scheme of a plane from a NOESY-$^{15}N/^{13}C$-HSQC spectrum; it contains both intramolecular protein (blue) and intermolecular protein–peptide (purple) NOE peaks. (**b**) Scheme of a plane from a $^{13}C,^{15}N$-filtered-NOESY-$^{13}C/^{15}N$-HSQC spectrum; it contains only intermolecular protein–peptide NOEs (purple). (**c**) Scheme of 2D double-$^{13}C,^{15}N$-filtered-NOESY spectrum; it contains exclusively intramolecular peptide NOEs

protein and intermolecular protein–peptide NOE peaks and hence distances; [19–21] (*see* **Note 9**).

Intermolecular protein–peptide NOEs: 3D $^{13}C,^{15}N$-filtered-NOESY-^{13}C-HSQC and 3D $^{13}C,^{15}N$-filtered-NOESY-^{15}N-HSQC [18], which yield intermolecular protein–peptide NOEs (Fig. 4); (*see* **Note 10**).

Intramolecular peptide NOEs: 2D double-$^{13}C,^{15}N$-filtered-NOESY, which yields exclusively intramolecular peptide NOEs [18]. (*see* **Note 11**).

6. Perform structure calculations. Supply the chemical-shift assignments and the list of NOE cross-peaks (positions and intensities) from the NOESY spectra to the software ARIA, a program for automated NOE assignment and structure calculation (*see* **Note 12**). The first step of the iterative protocol in an ARIA run is the calibration; it consists in the calculation of a scale factor C (named calibration factor) that is applied in the conversion of the peak intensities into distances. We have described that for complexes in fast exchange regime the conversion of the NOEs into distances is not accurate. To reduce the errors in the calculation of the scale factor, the intermolecular NOEs require a different calibration procedure (*see* **Note 13**). In addition to the distance restraints derived from NOESY spectra, ARIA can integrate dihedral angle restraints for the protein backbone that can be predicted from the backbone chemical shifts using software such as TALOS-N [22] or DANGLE [23]. Informative resources on ARIA are available elsewhere [3, 24]. At the end of the ARIA run, inspect the list of violated restraints, and reanalyze the spectra to correct erroneous assignments or wrongly selected NOE cross peaks. Run ARIA again with the corrected input data, and repeat this cycle until the structure ensemble and the restraint violation list pass standard quality control criteria established for NMR structure calculations [25].
7. Assess the quality of the structure with independent structure validation tools. These tools detect possible distortions in the geometry of the protein by comparing it to parameters derived from high-resolution protein structures. A useful validation resource is the web server PSVS (protein structure validation software suite), which integrates several widely used structure quality evaluation tools [26].
8. Deposit the final structural data in publicly available databases. After validation of the structure ensemble, deposit chemical-shift information in the BioMagResBank (BMRB) and the coordinates of the structure ensemble as well as the NOE data in the Protein Data Bank (PDB).

4 Notes

1. The final trace-element solution turns a golden color.
2. The $CaCl_2$ and the trace-element mix should be added after the filtration to avoid the precipitation of salt on the filtering membrane.

3. It is possible to protect the peptides from proteolysis by protecting the N- and C-termini with acetyl and amide groups, respectively.
4. NMR tubes are available in various diameters and shapes. The most common diameters are 5 and 3 mm: these tubes require a sample volume of 500–650 μL and 170–180 μL, respectively. Shigemi tubes are usually chosen when a limited amount of sample is available since they require smaller volumes (around 300 μL for a 5-mm diameter Shigemi tube). In principle, the signal-to-noise ratio (S/N) decreases with the diameter of the tube (for a solution at constant concentration). However, the concomitant decrease in volume allows for the preparation of more concentrated solution (for the same amount of protein, providing that the protein is soluble at the higher concentration) and performs better in the presence of salty buffers (the loss of S/N with salty buffers being less severe than would be expected simply from the relative sample volumes). It is important to add the proper volume in each NMR tube since insufficient volume may cause sub-optimal magnetic field homogeneity in the active volume and consequently spectra with poor resolution and line shape. For samples in salty buffers, 3 mm tubes should be preferred.
5. To reach the concentration required for the NMR sample, proteins are often concentrated after the SEC run using centrifugal concentrators. The filters of new concentrators are protected with glycerol, which must be removed by at least three washes with ultrapure water or buffer.
6. The protein spectrum is sensitive to pH, salt concentration, volume, and solvent of the ligand stock (e.g., DMSO). Appropriate controls should be measured. The concentration of the peptide stock solution should be high to minimize the volume change during the titration (do not exceed 10%).
7. Since deuterated buffers are expensive, it is preferable to prepare only the final sample in deuterated NMR buffer. We usually perform a buffer exchange by ultrafiltration. The protein sample is diluted tenfold in deuterated NMR buffer and reconcentrated to the initial volume using a centrifugal concentrator. The dilution–concentration process is repeated multiple times (at least three) to completely remove the protonated buffer. The phosphopeptide stock solution can be directly prepared using the deuterated buffer.
8. The inter-hydrogen distances obtained from the NOE intensities should be corrected for spin diffusion. This is done through integration of the relaxation matrix in the ARIA protocol.

9. In the protein–peptide complex used for structure determination by NMR, the protein is uniformly ^{15}N,^{13}C-labeled and the peptide is unlabeled. The NOESY-^{15}N-HSQC and NOESY-^{13}C-HSQC correlate ^{1}H-^{15}N and the ^{1}H-^{13}C groups with the resonances of nearby hydrogen atoms, respectively. Hence, the NOESY-^{15}N-HSQC and 3D NOESY-^{13}C-HSQC experiments show cross peaks arising from both intramolecular protein NOEs and protein–peptide intermolecular NOEs (wherein the ^{1}H-^{15}N and the ^{1}H-^{13}C groups belong to the protein). In other words, the two experiments detect NOEs between proximal hydrogen atoms as long as *at least* one of them is directly bound to a ^{13}C or ^{15}N atom. Since the peptide is unlabeled and the amount of peptide with ^{13}C and ^{15}N in natural abundance can be neglected, no intra-peptide NOEs are detected in these experiments. The vast majority of peaks stem from intramolecular protein NOEs. For these experiments, it is fundamental to maximize the peptide-bound population of the protein. For protein–peptide complexes with a low K_D (~100 times lower than the concentration of the complex), full saturation is achieved by adding peptide in slight excess with respect to the stoichiometric ratio of the two species in the complex. Complexes with low K_D have slow k_{off} values and thus the NMR peaks of the protein are usually in the slow exchange regime. If the peaks of the protein are found to be in the fast exchange regime during peptide titration experiments, the K_D is usually in the μM–mM range. The relative amount of peptide–protein necessary to reach protein saturation (i.e., no further movement of the protein peaks upon addition of more quantity of the peptide) depends on the total concentration of the species and must be determined in the preliminary NMR titration experiments. For K_D values in the mM range, saturation of the protein might require as much as a 100-fold excess of peptide.

10. Isotope filtering approaches come in useful to distinguish between the intramolecular protein NOEs and the protein–peptide intermolecular NOEs as well as to detect the intra-peptide NOEs. In ^{13}C,^{15}N-filtered-NOESY-^{13}C-HSQC and ^{13}C,^{15}N-filtered-NOESY-^{15}N-HSQC experiments, a filter element selects NOEs arising between hydrogen pairs provided that *only* one of the hydrogen atoms in the pair is directly attached to a ^{13}C or ^{15}N atom, thus enabling selective detection of intermolecular protein–peptide NOEs [18]. For tightly binding peptides, which usually have a K_D much lower than the concentration of the species in solution, the sample should be prepared with a slight excess of peptide to ensure that the protein is fully bound. The chemical-shift positions of the intermolecular NOEs measured in such conditions correspond

to those of the fully bound states for both peptide and protein. For weakly binding peptides—where either the protein or the peptide (or both) exists as a mixture of free and bound forms and where exchange between the different forms is likely to be fast on the chemical-shift timescale—the positions of the NOE peaks correspond to the population-weighted chemical shifts of the free and bound forms for both the protein and the peptide:

$$\delta_{\text{obs}} = \delta_{\text{free}} \tilde{n} p_{\text{free}} + \delta_{\text{bound}} \tilde{n} p_{\text{bound}} \quad (9)$$

Thus, the chemical shifts of both the protein and the peptide resonances depend on the peptide–protein concentration ratio (and the K_D). In this case of a weakly binding peptide, it is recommended to measure the $^{13}C,^{15}N$-filtered-NOESY-^{13}C-HSQC and $^{13}C,^{15}N$-filtered-NOESY-^{15}N-HSQC experiments with a sample that contains as much peptide as necessary to completely saturate the protein (as for the NOESY-^{15}N-HSQC and NOESY-^{13}C-HSQC experiments). With such a sample, the chemical-shift positions of the protein resonances of the NOE peaks correspond to those of the bound protein, while the positions of the peptide resonances are averaged between the free and bound forms according to Eq. 9. In order to assign the peptide resonances of the intermolecular NOE peaks, it is necessary to do a complete ^1H chemical-shift assignment of the peptide under these conditions (i.e., with the same concentrations of protein and peptide). This entails recording the double-filtered NOESY and TOCSY spectra under the same conditions; here, the NOESY spectrum is solely used to aid the chemical-shift assignment but is not used to extract intra-peptide NOEs for structure calculations. In the case of very weak binding, where a large excess of peptide is required to saturate the protein, it may be feasible to simply use the ^1H chemical-shift assignment of the free peptide in lieu of that for the partially (very slightly) bound state.

11. This set of NOEs is measured from a double-$^{13}C,^{15}N$-filtered 2D NOESY spectrum acquired on a sample in which the protein is in excess (the degree of excess protein required being dependent on the strength of the binding). This NOESY spectrum is used to provide distance restraints for the structure calculation and also to assist the ^1H chemical-shift assignment of the peptide in the fully bound state. For the latter purpose, one also records a corresponding a double-$^{13}C,^{15}N$-filtered 2D TOCSY spectrum.

12. ARIA uses the structure generation engine CNS (Crystallography and NMR system) to generate a molecular topology file (MTF). The SH2 domain–phosphopeptide complex contains

the non-canonical amino acid phosphotyrosine and thus requires modifications to the CNS protocols and ARIA dictionaries. In particular, the CNS topology, linkage, and parameter files must be changed. Definition of the non-canonical residues should be added to the files atomnames.xml and iupac.xml of the ARIA dictionary.

13. The treatment of the intermolecular protein–peptide NOEs in the case of a weakly binding peptide may require some additional considerations with respect to their calibration (conversion of NOE intensities into distance restraints). This is because these NOEs are necessarily measured under sample conditions where the peptide is in excess and therefore exists in an exchanging mixture of free and bound forms. If we accept the usual assumption that weak binding corresponds to fast exchange on the chemical-shift timescale, then it follows that the unbinding rate-constant k_{off} is sufficiently fast that the lifetime of the complex ($1/k_{off}$) is short compared to the NOE mixing time. In other words, fast exchange on the chemical-shift timescale usually implies fast exchange on the NOE mixing timescale. The net result is that any particular peptide molecule spends only a fraction of the experimental mixing time bound to the protein. On average, this "effective" mixing time is the experimental mixing time scaled by the mole-fraction of bound peptide (fraction of bound to total peptide), and hence the overall intensity of the intermolecular NOEs is reduced compared to those that would be observed for a tight-binding peptide and would require a different calibration factor for conversion into distance restraints.

    For the intermolecular NOEs derived from the edited/filtered NOESY spectra, the adjustment of the calibration factor is automatically and silently achieved during the standard ARIA calibration procedure; this is simply because the calibration factors are always peak list specific and the edited/filtered NOESY spectra have their own peak lists that contain only intermolecular NOEs. However, the NOEs measured in the edited NOESY spectra are a mixture of protein intramolecular and protein–peptide intermolecular NOEs. In the weak-binding case, it is recommended to split the NOEs from the edited NOESY spectra into two peak lists per spectrum, one containing the intramolecular NOEs and the other the intermolecular NOEs. The intermolecular NOEs from these edited NOESY spectra can then be calibrated independently, as for those from the edited/filtered spectra.

    A final point concerns adjustments to the parameters used for the spin-diffusion correction to the calibration factors for the intermolecular NOE peak lists that may also be advisable in the case of weak binding. The spin-diffusion correction relies on simulation of the theoretical NOEs, for which is required—in addition to a template structure—parameters related to the relaxation properties of the protein (the rotational correlation time) and also the experimental setup (spectrometer frequency and length of mixing time). The accuracy of the spin-diffusion correction may therefore be slightly compromised in the case of weak binding, because the effective mixing time defined earlier will be shorter than the experimental mixing time. While by no means a rigorous solution, the most straightforward adjustment to improve the accuracy of the spin-diffusion correction is simply to set the mixing time used by ARIA for the spin-diffusion-corrected calibration of the intermolecular NOE peak lists as the estimated effective mixing time.

Acknowledgments

This work was supported by a Leverhulme International Professorship to T.C.

References

1. Hunter T (2009) Tyrosine phosphorylation: thirty years and counting. Curr Opin Cell Biol 21:140–146. https://doi.org/10.1016/j.ceb.2009.01.028
2. Liu BA, Jablonowski K, Raina M et al (2006) The human and mouse complement of SH2 domain proteins-establishing the boundaries of phosphotyrosine signaling. Mol Cell 22: 851–868. https://doi.org/10.1016/j.molcel.2006.06.001
3. Rieping W, Habeck M, Bardiaux B et al (2007) ARIA2: automated NOE assignment and data integration in NMR structure calculation. Bioinformatics 23:381–382. https://doi.org/10.1093/bioinformatics/btl589
4. Delaglio F, Grzesiek S, Vuister GW et al (1995) NMRPipe: a multidimensional spectral processing system based on UNIX pipes. J Biomol NMR 6:277–293. https://doi.org/10.1007/BF00197809
5. Vranken WF, Boucher W, Stevens TJ et al (2005) The CCPN data model for NMR spectroscopy: development of a software pipeline. Proteins 59:687–696. https://doi.org/10.1002/prot.20449
6. Machida K, Thompson CM, Dierck K et al (2007) High-throughput phosphotyrosine profiling using SH2 domains. Mol Cell 26: 899–915. https://doi.org/10.1016/j.molcel.2007.05.031
7. Skora L, Shrestha B, Gossert AD (2015) Isotope labeling of proteins in insect cells. Methods Enzymol 565:245–288. https://doi.org/10.1016/bs.mie.2015.05.013
8. Sugiki T, Ichikawa O, Miyazawa-Onami M et al (2012) Isotopic labeling of heterologous proteins in the yeast Pichia pastoris and Kluyveromyces lactis. Methods Mol Biol 831:19–36. https://doi.org/10.1007/978-1-61779-480-3_2
9. Hansen AP, Petros AM, Mazar AP et al (1992) A practical method for uniform isotopic labeling of recombinant proteins in mammalian cells. Biochemistry 31:12713–12718. https://doi.org/10.1021/bi00166a001
10. Kozak S, Lercher L, Karanth MN et al (2016) Optimization of protein samples for NMR using thermal shift assays. J Biomol NMR 64: 281–289. https://doi.org/10.1007/s10858-016-0027-z
11. Grzesiek S, Bax A (1992) An efficient experiment for sequential backbone assignment of medium-sized isotopically enriched proteins. J Magn Reson (1969) 99:201–207. https://doi.org/10.1016/0022-2364(92)90169-8
12. Grzesiek S, Bax A (1992) Improved 3D triple-resonance NMR techniques applied to a 31 kDa protein. J Magn Reson (1969) 96: 432–440. https://doi.org/10.1016/0022-2364(92)90099-S

13. Wittekind M, Mueller L (1993) HNCACB, a high-sensitivity 3D NMR experiment to correlate amide-proton and nitrogen resonances with the alpha- and beta-carbon resonances in proteins. J Magn Reson B 101:201–205. https://doi.org/10.1006/jmrb.1993.1033
14. Grzesiek S, Anglister J, Bax A (1993) Correlation of backbone amide and aliphatic side-chain resonances in 13C/15N-enriched proteins by isotropic mixing of 13C magnetization. J Magn Reson B 101:114–119. https://doi.org/10.1006/jmrb.1993.1019
15. Logan TM, Olejniczak ET, Xu RX et al (1992) Side chain and backbone assignments in isotopically labeled proteins from two heteronuclear triple resonance experiments. FEBS Lett 314: 413–418. https://doi.org/10.1016/0014-5793(92)81517-P
16. Montelione GT, Lyons BA, Emerson SD et al (1992) An efficient triple resonance experiment using carbon-13 isotropic mixing for determining sequence-specific resonance assignments of isotopically-enriched proteins. J Am Chem Soc 114:10974–10975. https://doi.org/10.1021/ja00053a051
17. Yamazaki T, Forman-Kay JD, Kay LE (1993) Two-dimensional NMR experiments for correlating carbon-13.beta. and proton.delta./.epsilon. chemical shifts of aromatic residues in 13C-labeled proteins via scalar couplings. J Am Chem Soc 115:11054–11055. https://doi.org/10.1021/ja00076a099
18. Zwahlen C, Legault P, Vincent SJF et al (1997) Methods for measurement of intermolecular NOEs by multinuclear NMR spectroscopy: application to a bacteriophage λ N-peptide/boxB RNA complex. J Am Chem Soc 119: 6711–6721. https://doi.org/10.1021/ja970224q
19. Marion D, Driscoll PC, Kay LE et al (1989) Overcoming the overlap problem in the assignment of 1H NMR spectra of larger proteins by use of three-dimensional heteronuclear 1H-15N Hartmann-Hahn-multiple quantum coherence and nuclear Overhauser-multiple quantum coherence spectroscopy: application to interleukin 1 beta. Biochemistry 28:6150–6156. https://doi.org/10.1021/bi00441a004
20. Zuiderweg ER, Fesik SW (1989) Heteronuclear three-dimensional NMR spectroscopy of the inflammatory protein C5a. Biochemistry 28:2387–2391. https://doi.org/10.1021/bi00432a008
21. Marion D, Kay LE, Sparks SW et al (1989) Three-dimensional heteronuclear NMR of nitrogen-15 labeled proteins. J Am Chem Soc 111:1515–1517. https://doi.org/10.1021/ja00186a066
22. Shen Y, Bax A (2013) Protein backbone and sidechain torsion angles predicted from NMR chemical shifts using artificial neural networks. J Biomol NMR 56:227–241. https://doi.org/10.1007/s10858-013-9741-y
23. Cheung M-S, Maguire ML, Stevens TJ et al (2010) DANGLE: a Bayesian inferential method for predicting protein backbone dihedral angles and secondary structure. J Magn Reson 202:223–233. https://doi.org/10.1016/j.jmr.2009.11.008
24. Bardiaux B, Malliavin T, Nilges M (2012) ARIA for solution and solid-state NMR. Methods Mol Biol 831:453–483. https://doi.org/10.1007/978-1-61779-480-3_23
25. Vuister GW, Fogh RH, Hendrickx PMS et al (2014) An overview of tools for the validation of protein NMR structures. J Biomol NMR 58: 259–285. https://doi.org/10.1007/s10858-013-9750-x
26. Bhattacharya A, Tejero R, Montelione GT (2007) Evaluating protein structures determined by structural genomics consortia. Proteins 66:778–795. https://doi.org/10.1002/prot.21165

Chapter 2

NMR Methods to Study the Dynamics of SH2 Domain–Phosphopeptide Complexes

Michelangelo Marasco, John P. Kirkpatrick, Vittoria Nanna, and Teresa Carlomagno

Abstract

Nuclear magnetic resonance (NMR) spectroscopy is the method of choice for studying the dynamics of biological macromolecules in solution. By exploiting the intricate interplay between the effects of protein motion (both overall rotational diffusion and internal mobility) and nuclear spin relaxation, NMR allows molecular motion to be probed at atomic resolution over a wide range of timescales, including picosecond (bond vibrations and methyl-group rotations), nanosecond (loop motions and rotational diffusion), and microsecond–millisecond (ligand binding, allostery). In this chapter, we describe different NMR pulse schemes (R_1, $R_{1\rho}$, heteronuclear NOE, and CPMG relaxation dispersion) to characterize the dynamics of SH2 domains. As an example, we use the N-SH2 domain of protein tyrosine phosphatase SHP2 in complex with two phosphopeptides derived from immune checkpoint receptor PD-1 (ITIM and ITSM).

Key words Nuclear magnetic resonance (NMR) spectroscopy, Src-homology 2 (SH2) domain, Phosphopeptides, Protein dynamics, Allostery, Spin relaxation, Chemical exchange

1 Introduction

Tyrosine phosphorylation is a fundamental mechanism whereby cells regulate a diverse family of processes including signal transduction, cytoskeletal remodeling, cell growth, and proliferation [1]. In this context, specificity is conferred by protein modules that recognize polypeptides only when they carry a phosphotyrosine (pY) mark and that can discriminate between different phosphopeptides based on the amino acids adjacent to pY. Of all the protein domains capable of accomplishing this feat, Src-homology 2 (SH2) domains are by far the most prevalent and well characterized. Traditionally, phosphopeptide binding to SH2 domains has been explained as a two-pronged interaction, with the pY phosphate group stabilized by a network of salt bridges and the amino acids immediately C-terminal to pY hosted in a so-called specificity

Teresa Carlomagno and Maja Köhn (eds.), *SH2 Domains: Functional Modules and Evolving Tools in Biology*, Methods in Molecular Biology, vol. 2705, https://doi.org/10.1007/978-1-0716-3393-9_2,

pocket on the SH2 domain, which is generally hydrophobic in nature [2, 3].

Over time, this description has been recognized as overly simplistic, and it is now accepted that the interaction between an SH2 domain and its cognate phosphopeptide is dependent on many contributing factors, including the influence of the motions of several conserved loops (especially the EF and BG loops) on binding affinity and kinetics [4]. Protein dynamics becomes even more important for those SH2 domains that have evolved the additional ability to regulate the activity of catalytic domains that are part of the same enzyme. For instance, in the case of protein tyrosine phosphatase SHP2, two SH2 domains (termed N-SH2 and C-SH2) couple the binding of phosphopeptides with its activation [5]. Detailed studies of the dynamical responses of SH2 domains to phosphopeptide binding are expected to be critical for developing our understanding of this essential contributor to cellular signaling [6, 7].

Due to its unique ability to explore molecular motions in atomic detail and cover a wide range of timescales, NMR is the method of choice to address such questions. Different pulse schemes have been developed to probe motional processes occurring on different timescales. Motions occurring on the ps–ns timescale (which include bond vibrations, breakage and formation of hydrogen bonds, rotational diffusion, and side-chain motions) are studied by monitoring the longitudinal (R_1) and transverse (R_2 or $R_{1\rho}$) relaxation rates of ^{15}N nuclei in the protein, as well as their heteronuclear NOEs (hetNOE), using a set of well-established experiments performed on a ^{15}N-labeled protein sample [8]. The results of these experiments are typically analyzed according to the model-free approach of Lipari and Szabo [9], which yields easily interpretable parameters describing both the overall rotational diffusion of the protein (the rotational correlation time, τ_c, in the case of isotropic rotational diffusion or the corresponding tensor in the case of anisotropic diffusion) and the local mobility of amide group ^{15}N–^{1}H bond vectors. The latter is characterized by the S^2 order parameter (a measure of the angular amplitude of the bond vector in the molecular frame) and the internal correlation time τ_i (which is related to the timescale of the local motion). The model-free approach relies on a formal decomposition of the autocorrelation function—which is a complete description of the time dependence of the magnetic interactions responsible for nuclear spin relaxation—into a product of two terms, related, respectively, to the motions due to global rotational diffusion and local mobility. Fourier transformation of this autocorrelation function yields the spectral density function, $J(\omega)$. The spectral density function is an alternative description of the motional time dependence that shows the relative contribution of motions at different frequencies to the overall motion of the interaction vector; nuclear spin relaxation rates depend on the spectral densities at a particular set of frequencies that are dictated by the energy separations of the nuclear spin states. In the model-free approach, this spectral density function is called the model-free spectral density and is a sum of two terms:

$$J(\omega) = \frac{2}{5}\left[\left(\frac{S^2\tau_c}{1+(\omega\tau_c)^2}\right)+\left(\frac{(1-S^2)\tau_e}{1+(\omega\tau_e)^2}\right)\right]$$

Where ω is the frequency in rad·s^{-1}, S^2 is the order parameter (squared), τ_c is the rotational correlation time, and τ_e is related to τ_c and the internal correlation time τ_i according to

$$\tau_e = \left(\tau_c^{-1} + \tau_i^{-1}\right)^{-1}$$

Notwithstanding the importance of local mobility for protein function, most biological processes that involve conformational changes, allostery, and folding/unfolding occur on the μs–ms timescale and as such are not accessible using the experiments mentioned above. Such motions are instead revealed by an additional contribution to the peak linewidths, which arises due to time-dependent modulations of the isotropic chemical shifts; the modulations of the chemical shifts are the result of the structural or chemical changes associated with the dynamic process of interest. The magnitudes of the changes in the chemical shift are typically such that when they are modulated on the μs–ms timescale they can induce significant broadening of the NMR peak linewidths, which results from additional contributions to the overall transverse relaxation rates. This phenomenon is known as exchange broadening (and the increase in the transverse relaxation rate is termed the exchange contribution to R_2, usually written as $R_{2,ex}$); the underlying dynamic processes are typically described as chemical exchange processes, even when the associated changes are purely conformational. The simplest description for chemical exchange processes is the two-site model, with two interchanging states: A ⇄ B. By convention, state B is taken to be the less populated state (or minor state), and the rate constants for the forward and backward reactions are k_{+1} and k_{-1}, respectively.

The most common approach for detailed characterization of exchange broadening is via CPMG (Carr–Purcell–Meiboom–Gill) relaxation dispersion experiments, where the transverse relaxation rate is measured in the presence of a train of rapidly repeated CPMG elements [10]. The application of the CPMG train acts to partially suppress the exchange contribution to the transverse relaxation, and the experiment involves repeated measurements of the transverse relaxation rate in the presence of CPMG trains applied at different rates. The resulting relaxation dispersion profile shows the dependence of the effective transverse relaxation rate, $R_{2,eff}$, on the CPMG rate, ν_{CPMG} (also known as the CPMG field strength), and

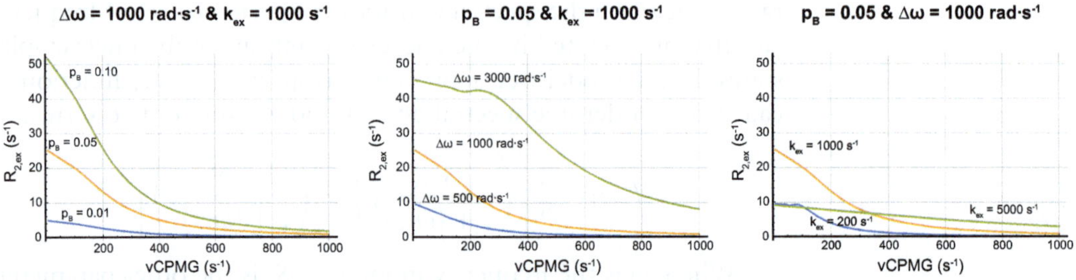

Fig. 1 Simulation of CPMG relaxation dispersion profiles following the Carver–Richards equation [32] for the case where the evolution of the magnetization during the CPMG period is dominated by a single exponential. The parameters are defined according to [33], with the exception that ν_{CPMG} is used instead of τ_{CP}, with $\nu_{CPMG} = 1/(2\tau_{CP})$

this can then be fit to an appropriate function—which describes the theoretical form for the proposed exchange model—to extract physical parameters relating to the exchange process. In the case of the two-site model, these parameters are the exchange rate constant, k_{ex} ($k_{ex} = k_{+1} + k_{-1}$), the fractional population of the minor state, p_B, and the difference between the resonance frequencies in the two states, $\Delta\omega$. The parameters k_{ex} and p_B will be global parameters common to all nuclei participating in the same exchange process, while the parameter $\Delta\omega$ is nucleus specific (see Fig. 1).

Here, we describe how these NMR techniques are applied to study the dynamics of SH2 domains. We use the example of the N-SH2 domain of SHP2 bound to two different phosphopeptides derived from immune checkpoint receptor PD-1, namely, the immune tyrosine-based inhibitory motif (ITIM) and immune tyrosine-based switch motif (ITSM) [11, 12]. One important consideration prior to commencing such measurements concerns the importance of making measurements of the same types of relaxation rates on NMR spectrometers with different magnetic field strengths. For the investigation of the fast timescale (ps–ns) dynamics, acquiring datasets at two field strengths is not essential, but is still recommended. Meaningful interpretation of data acquired at a single field strength via the model-free approach is possible, but in this case, certain situations that can lead to erroneous results from model-free fitting are difficult to recognize. Principal among these is the case where the protein undergoes local motions on ns timescales. The model-free analysis relies on the criterion that the local and global motions are on widely different timescales, i.e., ps and ns timescales, respectively. If this criterion is not fulfilled, then the extracted parameters will not be meaningful. There is a so-called "extended" model-free formalism which can account for local motions on two timescales [13], but there remains a difficulty in accurately determining the global rotational correlation time in the case of slow internal motion. The first stage of the model-free analysis always involves identification of a subset of amides that are assumed to have negligible local flexibility (and are also not participating in any chemical exchange processes); the R_2/R_1 ratios from these amides are then used to determine the global rotational correlation time (or rotational diffusion tensor in the case of diffusion anisotropy). This identification process involves filtering the amides based on their heteronuclear NOEs; the hetNOEs are highly sensitive to local motions on the ps timescale, which allows amides undergoing such motions to be excluded from the set used for deriving the rotational correlation time. However, the hetNOE is much less sensitive to internal motions on slower timescales, and hence amides undergoing these motions will not be identified at this stage; inclusion of these amides in the set used to derive the rotational correlation time will lead to underestimation of τ_c and distort all subsequent analyses. Measurement of the relaxation rates at two or more fields provides a means by which internal motions on slower timescales can be recognized; in such a case, the transverse relaxation rates will show a field-strength dependence, so that the degree by which τ_c is underestimated from the R_2/R_1 ratios will increase with increasing field strength. Therefore, if the values of τ_c calculated using the subset of supposedly rigid amides are not the same for the datasets acquired at different field strengths, some of those amides are likely to be undergoing internal motion on the ns timescale, and a more considered analysis is required.

For characterization of chemical exchange by analysis of CPMG relaxation dispersion data, measurement of dispersion profiles at two (or more) static field strengths is necessary, as the variable parameters in the functions used to fit the relaxation dispersion profiles typically exhibit strong covariance with each other. In practice, this means that a particular experimentally measured dispersion profile can be almost equally well fit by different sets of variable parameters. The problem of covariance is significantly alleviated by simultaneously fitting dispersion profiles acquired at different static field strengths, principally because the nucleus-specific parameter $\Delta\omega$ is field dependent, while the global parameters k_{ex} and p_B are not.

2 Materials

2.1 Expression of ^{15}N-Labeled N-SH2

1. pETM22–N-SH2 (*see* **Note 1**).
2. Luria–Bertani (LB) agar plates with kanamycin.
3. 50 mg/mL of kanamycin stock 1000×. To be prepared in ultrapure water and filtered through a 0.2 μm filter.
4. Tuner (DE3) *E. coli* competent cells.

5. Trace element mix (1000×): 50 mL of 0.1 M $FeCl_3$ in 0.12 M HCl, 2 mL of 1 M $CaCl_2$, 1 mL of 1 M $MnCl_2$, 1 mL of 1 M ZnSO4, 1 mL of 0.2 M $CoCl_2$, 2 mL of 0.1 M $CuCl_2$, 1 mL of 0.2 M $NiCl_2$, 2 mL of 0.1 M Na_2MoO_4, 2 mL of 0.1 M Na_2SeO_4, 2 mL of 0.1 M H_3BO_4 and adjust to a final volume of 100 mL with ultrapure water.
6. 10× M9 salt solution: dissolve 75.2 g of Na_2HPO_4, 30 g KH_2PO_4, and 5 g NaCl in 800 mL of ultrapure water and adjust the pH to 7.2 with NaOH. Add water to a final volume of 1 L. Sterile filter.
7. M9 minimal medium for ^{15}N labeling: to 860 mL of ultrapure water, add 2 mL of 1 M $MgSO_4$, 30 mg of thiamine hydrochloride, 4 g of glucose, 1 g of $^{15}NH_4Cl$, 100 mL of 10× M9 salts solution, and kanamycin. Sterile filter. Then, add 100 μL of 1 M $CaCl_2$ and 1 mL of trace element mix (1000×).
8. 1 M stock solution of isopropyl-β-D-thiogalactoside (IPTG).
9. A spectrophotometer for the measurement of the optical density of the bacterial culture (OD_{600}).
10. A centrifuge capable of generating at least 4000 g of centrifugal force.

2.2 Purification of ^{15}N-Labeled N-SH2

1. Lysis buffer: 1 M NaCl, 50 mM Tris–HCl, 5% glycerol, 10 mM imidazole, 5 mM β-mercaptoethanol, pH 7.6. Add one tablet of EDTA-free protease inhibitors, 100 μg of lysozyme, and 50 μg of DNAse per 20 mL of buffer.
2. Sonicator.
3. A centrifuge capable of generating at least 18000 g of centrifugal force.
4. FPLC system.
5. Ni-NTA column (e.g., HisTrap FF column).
6. Wash buffer: 1 M NaCl, 50 mM Tris–HCl, 5% glycerol, 10 mM imidazole, 5 mM β-mercaptoethanol, pH 7.6.
7. Elution buffer: 1 M NaCl, 50 mM Tris–HCl, 5% glycerol, 500 mM imidazole, 5 mM β-mercaptoethanol, pH 7.6.
8. Recombinant His_6-tagged rhinovirus 3C protease.
9. Dialysis tubing with the appropriate pore size (a typical SH2 domain has a molecular weight of around 10 kDa).
10. Protein concentrators with the appropriate molecular weight cutoff.
11. Size-exclusion chromatography column (e.g., HiLoad 16/600 Superdex 75 pg column).
12. NMR buffer: 100 mM MES (pH 6.8), 150 mM NaCl, 3 mM TCEP (Tris(2-carboxyethyl)phosphine), and 0.05% NaN_3.

2.3 NMR Experiments and Data Analysis

1. Purified N-SH2.
2. Phosphopeptides (synthesized via Fmoc solid-phase peptide synthesis and purified to >95% purity via reverse-phase HPLC):
   (a) ITIM: Ac-FSVD(pY)GELDFQ-NH_2.
   (b) ITSM: Ac-EQTE(pY)ATIVFQ-NH_2.
3. Deuterium oxide (D_2O).
4. 50 μM of 4,4-dimethyl-4-silapentane-1-sulfonic acid (DSS).
5. 5 mm glass tubes for NMR.
6. NMR spectrometers operating at static field strengths between 600 MHz and 1 GHz, equipped with N_2- or He-cooled inverse HCN triple-resonance cryogenic probeheads.
7. Spectrometer control software.
8. Data processing software (NMRPipe) [14].
9. NMR data visualization and assignment software (CcpNmr Analysis) [15].
10. Software for lineshape fitting and for fitting exponential decays of extracted peak intensities (FuDA) [16].
11. Software for model-free analysis (TENSOR2) [17].
12. Software for fitting the relaxation dispersion curves (ChemEx) [18].

3 Methods

3.1 Expression of ^{15}N-Labeled N-SH2

1. Transform pETM22–N-SH2 into Tuner (DE3) cells and plate them on an LB agar plate with kanamycin.
2. Isolate a colony and start an overnight culture with 25 mL of ^{15}N M9 minimal medium and kanamycin. Keep it shaking at >200 rpm at 37 °C.
3. The day after, transfer the culture into a 50 mL falcon tube and pellet the bacteria at 4000 g for 10 min. Then use the pellet to start a large 500 mL culture in ^{15}N M9 minimal medium. Use a 2 L flask and shake at at 37 °C and >250 rpm for proper aeration.
4. Monitor the initial OD_{600} and keep monitoring it every 20 min after the first 1–2 h. When $OD_{600} = 0.6$–0.8, rapidly chill the bacterial culture in an ice–water mix for 5 min, add 0.1 mM IPTG, and transfer to a shaker at 20 °C. Keep it shaking at >250 rpm overnight.
5. The day after, harvest the culture by centrifuging it at 4000 g for 30 min and store the bacterial pellets at −20 °C until further use.

3.2 Purification of ^{15}N-Labeled N-SH2

1. Resuspend the bacterial pellet in 20 mL of ice-cold lysis buffer and gently shake on ice for 20 min (*see* **Note 2**).
2. Lyse the cells by sonication (8 s pulses separated by 16 s of pause, with amplitude of 50%, for 20 min).
3. Clarify the lysate by centrifuging it at 18000 g for 1 h. Collect the supernatant and filter it through a 0.2 μm syringe filter.
4. Prewash the HisTrap column with 8–10 column volumes (CV) of wash buffer on an FPLC system kept at 8–13 °C.
5. Load the supernatant on the column at 1–2 mL/min.
6. Wash the column with 10 CV of wash buffer.
7. Elute the bound protein with 5 CV of elution buffer. Collect and pool the fractions containing the protein of interest.
8. Prepare a dialysis bag containing the eluted protein and 3C protease (1:100 protease:protein ratio). Dialyze against two liters of wash buffer. Let the process proceed overnight in a cold room to remove excess imidazole (*see* **Note 3**) and to allow the protease to cleave the thioredoxin tag.
9. The day after, perform a second HisTrap step. The protein of interest is recovered from the flow-through. Pool the fractions and concentrate to a final volume of maximum 2 mL, depending on the amount of protein present.
10. Spin the concentrated protein on a tabletop centrifuge for 2 mL tubes at full speed and for at least 15 min, to remove any precipitated protein. Use a pipette to retrieve the supernatant.
11. Load the protein on a HiLoad 16/600 Superdex 75 pg column, previously equilibrated with NMR buffer (*see* **Note 4**). Collect the fractions containing the N-SH2, pool them, and concentrate them to the final desired concentration (ideally, protein concentrations of around 0.5–0.75 mM should be achieved).
12. Aliquot the protein in batches of 0.5 mL and flash freeze in liquid nitrogen unless immediately used. Store the frozen aliquots at −80 °C (*see* **Note 5**).

3.3 NMR Experiments

1. Prepare the NMR sample. Resuspend the phosphopeptide in NMR buffer (*see* **Note 6**) to a final concentration of 5 mM. If necessary, adjust the pH by pipetting in small amounts of concentrated NaOH (*see* **Note 7**). Assemble the N-SH2–phosphopeptide complex by mixing the appropriate amounts of N-SH2 and phosphopeptide to achieve full saturation of the protein-binding site (*see* Chapter 1). In this example, five equivalents of ITIM and two equivalents of ITSM are required to fully saturate the binding site of N-SH2. The final

concentration of the protein–peptide complex should be around 0.5 mM, with a volume of at least 0.5 mL (*see* **Note 8**). Mix 0.5 mL of protein–phosphopeptide complex with 50 μL of D_2O + DSS and put it in a 5 mm glass tube for NMR.

2. Acquire the NMR spectra (we assume that the resonances of all amide groups in the N-SH2–ITIM and N-SH2–ITSM complexes have been assigned, as described in the Chapter 1). Carry out the measurement of the R_1 relaxation rates of the backbone ^{15}N atoms at one or two magnetic field strengths, using established experiments based on a gradient-selected, sensitivity-enhanced, refocused $^1H,^{15}N$-HSQC pulse sequence [19–22]. Use relaxation delays with values between 20 ms and 1200 ms. ^1H-^{15}N cross-relaxation can be suppressed by applying ^1H amide-selective IBURP-1 inversion pulses at intervals of 10 ms during the relaxation delay [23]. Carry out the measurement of the ^{15}N $R_{1\rho}$ relaxation rates of the backbone ^{15}N atoms at one or two magnetic field strengths, using established experiments based on a gradient-selected, sensitivity-enhanced, refocused $^1H,^{15}N$-HSQC pulse sequence [19–22]. Use relaxation delays with values between 3 ms and 100 ms. ^1H-^{15}N cross-relaxation can be suppressed by applying between one and four ^1H amide-selective inversion pulses during the ^{15}N spin lock relaxation period [24]. The ^{15}N magnetization is spin-locked using a B_1 field-strength of 2.5 kHz [25]. Measure backbone amide [26] ^{15}N steady-state heteronuclear NOEs according to [27, 28]. Use a five-second train of high-power 180° pulses applied at 10.9 ms intervals to saturate amide proton magnetization [29]. The reference and saturated spectra can be acquired in an interleaved fashion, using a long recycle delay of 15 s. Carry out CPMG relaxation dispersion measurements of backbone ^{15}N atoms according to Hansen et al. [30] at two magnetic field strengths. CPMG datasets are measured using a constant time delay of 50 ms, and the CPMG field strength is varied between 20 and 1200 Hz.

3.4 Data Analysis

1. Process the relaxation spectra using NMRPipe with partial Lorentzian-to-Gaussian apodization in both dimensions and limited linear prediction in the ^{15}N dimension. Inspect the resulting spectra using CcpNmr Analysis.
2. Extract the peak intensities via lineshape fitting with FuDA, and also use FuDA to extract the relaxation rates by fitting the respective intensity-decay profiles (in the case of the hetNOE data, the values of the hetNOE are given by a simple ratio of the peak intensities in the saturated and reference sub-spectra).
3. Fit the R_1, $R_{1\rho}$, and $\{^1H\}^{15}N$ hNOE relaxation data to the model-free spectral density according to the formalism of Lipari and Szabo, using the program TENSOR2. Extract the

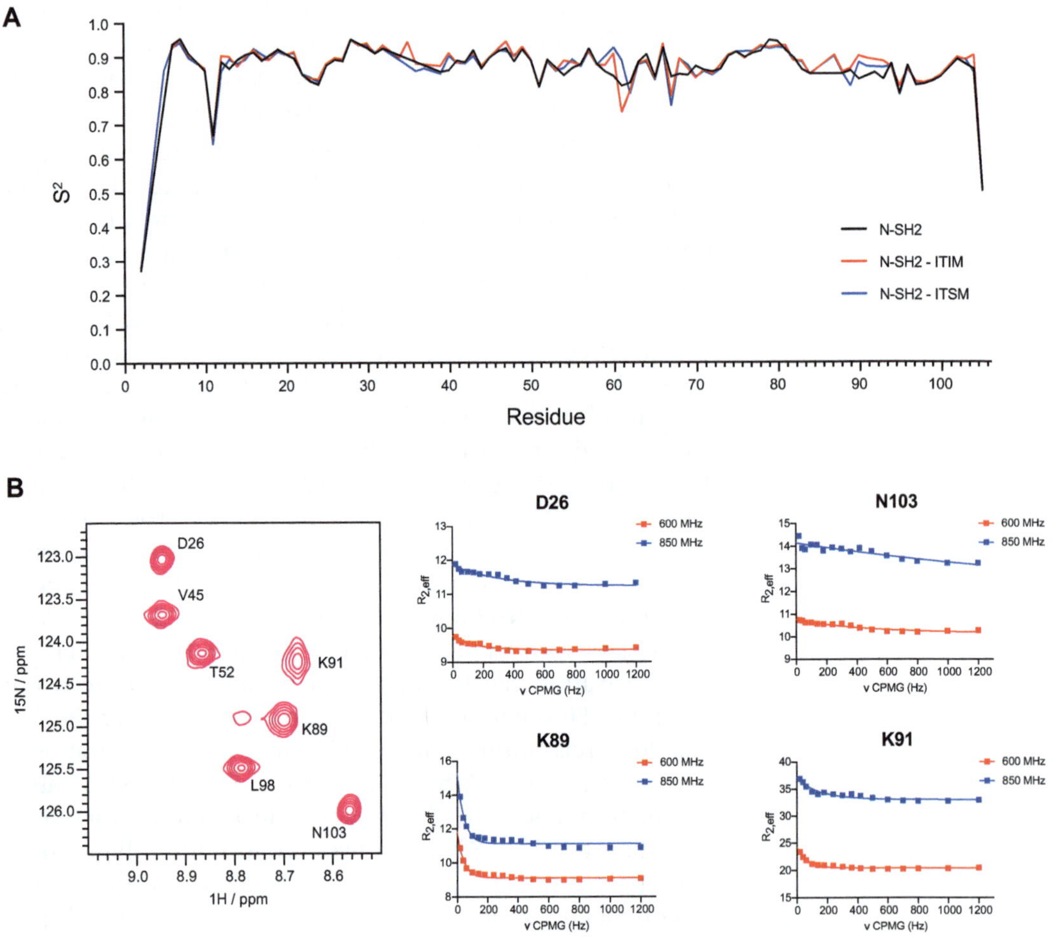

Fig. 2 (**a**) S^2 order parameters of N-SH2 (unbound and bound to ITIM and ITSM) obtained via the model-free analysis of Lipari and Szabo. It can be seen that N-SH2 is a fairly rigid molecule (S^2 close to one for most residues) and that peptide binding does not significantly perturb protein motions happening on the ps–ns timescale. (**b**) This excerpt of the ^{15}N-HSQC spectrum of N-SH2–ITIM includes the resonances of residues whose amide groups show limited (D26 and N103) or more pronounced (K89 and K91) exchange contributions to their transverse relaxation. The CPMG profiles of these residues are shown on the right

global rotational diffusion tensor and the local S^2 parameters and correlation times (Fig. 2a). A detailed guide of how to use the program is available at http://thcgerm.free.fr/IMG/pdf/Relaxation_TD2.pdf.

4. Fit the CPMG relaxation dispersion curves with ChemEx. The curves obtained at both fields must be fitted simultaneously on a per-residue basis to a two-state exchange model (Fig. 2b). The variable parameters that can be obtained are k_{ex}, p_B, $\Delta\omega$, and R_2^0 (the exchange-free transverse relaxation rate). Chemex can be freely downloaded at https://github.com/gbouvignies/ChemEx.

4 Notes

1. This plasmid confers resistance to kanamycin and allows the expression of N-SH2 as a fusion protein with thioredoxin, which can be cleaved using His_6-tagged recombinant rhinovirus 3C protease.
2. Although N-SH2 is a very stable protein, we nevertheless recommend carrying out lysis and the initial purification steps using ice-cold buffers and in a cold environment to minimize proteolysis by contaminating proteases. After cleavage of the thioredoxin tag, all further steps can be performed at room temperature. We have found that concentrating the protein at room temperature reduces the chance of aggregation and precipitation by improving the solubility of N-SH2.
3. Instead of dialysis, excess imidazole can also be removed by running the sample on a desalting column.
4. The yield of N-SH2 produced with the method described can be very high. For this reason, it may be necessary to perform two separate size-exclusion chromatography runs to prevent overloading of the column.
5. Always assess the purity of the sample by SDS-PAGE before acquiring NMR experiments.
6. The choice of the appropriate buffer for NMR experiments is critical, and a great deal of care must be used to optimize its pH and ionic strength. Ideal NMR buffers have low pH (usually in the range 5–7, to minimize the amide proton–water exchange rate and the consequent loss of sensitivity) and low ionic strength (<200 mM, since higher salt concentrations lead to a reduction in sensitivity). However, not all SH2 domains (and proteins in general) are stable under these conditions. Ideally, high-throughput screening of buffer conditions should be performed before attempting lengthy NMR experiments, for example, as described in [31].
7. Synthetic phosphopeptides can give rise to very acidic solutions when resuspended in buffers, because of both the phosphate group and the likely presence of TFA (trifluoroacetic acid). The NMR buffer we use contains 100 mM MES (2-(N-morpholino)ethanesulfonic acid), which is usually sufficient to prevent the pH from becoming too low. Low pH conditions can also prevent complete solubilization of the peptide. Always make sure that the pH of the peptide stock matches that of the NMR buffer by using pH paper or a pH meter.
8. The concentration of the phosphopeptide stock must be determined very carefully, in order to obtain a reliable quantitative analysis of the data. The concentration of solutions of peptides that do not carry tryptophan or other aromatic amino acids

cannot be reliably determined by measuring the absorbance at 280 nm. In this case (as for ITIM and ITSM), we recommend quantifying the stock by integrating the appropriate methyl-group proton peaks in a 1D NMR spectrum and comparing their volume to those of a calibration compound at known concentration.

Acknowledgment

This work was supported by a Leverhulme International Professorship to T.C.

References

1. Stoker AW (2005) Protein tyrosine phosphatases and signalling. J Endocrinol 185(1):19–33. https://doi.org/10.1677/joe.1.06069
2. Eck MJ, Shoelson SE, Harrison SC (1993) Recognition of a high-affinity phosphotyrosyl peptide by the Src homology-2 domain of p56lck. Nature 362(6415):87–91. https://doi.org/10.1038/362087a0
3. Huang H, Li L, Wu C, Schibli D, Colwill K, Ma S, Li C, Roy P, Ho K, Songyang Z, Pawson T, Gao Y, Li SS (2008) Defining the specificity space of the human SRC homology 2 domain. Mol Cell Proteomics 7(4):768–784. https://doi.org/10.1074/mcp.M700312-MCP200
4. Marasco M, Carlomagno T (2020) Specificity and regulation of phosphotyrosine signaling through SH2 domains. J Struct Biol X 4: 100026. https://doi.org/10.1016/j.yjsbx.2020.100026
5. Hof P, Pluskey S, Dhe-Paganon S, Eck MJ, Shoelson SE (1998) Crystal structure of the tyrosine phosphatase SHP-2. Cell 92(4):441–450. https://doi.org/10.1016/s0092-8674(00)80938-1
6. Marasco M, Kirkpatrick J, Nanna V, Sikorska J, Carlomagno T (2021) Phosphotyrosine couples peptide binding and SHP2 activation via a dynamic allosteric network. Comput Struct Biotechnol J 19:2398–2415. https://doi.org/10.1016/j.csbj.2021.04.040
7. Finerty PJ, Muhandiram R, Forman-Kay JD (2002) Side-chain dynamics of the SAP SH2 domain correlate with a binding hot spot and a region with conformational plasticity. J Mol Biol 322(3):605–620. https://doi.org/10.1016/s0022-2836(02)00803-3
8. Kovermann M, Rogne P, Wolf-Watz M (2016) Protein dynamics and function from solution state NMR spectroscopy. Q Rev Biophys 49: e6. https://doi.org/10.1017/S0033583516000019
9. Lipari G, Szabo A (1982) Model-free approach to the interpretation of nuclear magnetic resonance relaxation in macromolecules. 1. Theory and range of validity. J Am Chem Soc 104(17):4546–4559. https://doi.org/10.1021/ja00381a009
10. Loria JP, Rance M, Palmer AG (1999) A relaxation-compensated Carr–Purcell–Meiboom–Gill sequence for characterizing chemical exchange by NMR spectroscopy. J Am Chem Soc 121(10):2331–2332. https://doi.org/10.1021/ja983961a
11. Marasco M, Berteotti A, Weyershaeuser J, Thorausch N, Sikorska J, Krausze J, Brandt HJ, Kirkpatrick J, Rios P, Schamel WW, Köhn M, Carlomagno T (2020) Molecular mechanism of SHP2 activation by PD-1 stimulation. Sci Adv 6(5):eaay4458. https://doi.org/10.1126/sciadv.aay4458
12. Marasco M, Kirkpatrick JP, Carlomagno T (2020) 1H, 13C, 15N chemical shift assignments of SHP2 SH2 domains in complex with PD-1 immune-tyrosine motifs. Biomol NMR Assign 14(2):179–188. https://doi.org/10.1007/s12104-020-09941-y
13. Clore GM, Szabo A, Bax A, Kay LE, Driscoll PC, Gronenborn AM (1990) Deviations from the simple two-parameter model-free approach to the interpretation of nitrogen-15 nuclear magnetic relaxation of proteins. J Am Chem Soc 112(12):4989–4991. https://doi.org/10.1021/ja00168a070
14. Delaglio F, Grzesiek S, Vuister GW, Zhu G, Pfeifer J, Bax A (1995) NMRPipe: a multidimensional spectral processing system based on UNIX pipes. J Biomol NMR 6(3):277–293. https://doi.org/10.1007/BF00197809

15. Vranken WF, Boucher W, Stevens TJ, Fogh RH, Pajon A, Llinas M, Ulrich EL, Markley JL, Ionides J, Laue ED (2005) The CCPN data model for NMR spectroscopy: development of a software pipeline. Proteins 59(4):687–696. https://doi.org/10.1002/prot.20449
16. Hansen DF, Yang D, Feng H, Zhou Z, Wiesner S, Bai Y, Kay LE (2007) An exchange-free measure of 15N transverse relaxation: an NMR spectroscopy application to the study of a folding intermediate with pervasive chemical exchange. J Am Chem Soc 129(37):11468–11479. https://doi.org/10.1021/ja072717t
17. Dosset P, Hus JC, Blackledge M, Marion D (2000) Efficient analysis of macromolecular rotational diffusion from heteronuclear relaxation data. J Biomol NMR 16(1):23–28. https://doi.org/10.1023/a:1008305808620
18. Vallurupalli P, Bouvignies G, Kay LE (2012) Studying "invisible" excited protein states in slow exchange with a major state conformation. J Am Chem Soc 134(19):8148–8161. https://doi.org/10.1021/ja3001419
19. Dayie KT, Wagner G (1994) Relaxation-rate measurements for 15N–1H groups with pulsed-field gradients and preservation of coherence pathways. J Magn Reson Ser A 111(1):121–126. https://doi.org/10.1006/jmra.1994.1236
20. Farrow NA, Muhandiram R, Singer AU, Pascal SM, Kay CM, Gish G, Shoelson SE, Pawson T, Forman-Kay JD, Kay LE (1994) Backbone dynamics of a free and phosphopeptide-complexed Src homology 2 domain studied by 15N NMR relaxation. Biochemistry 33(19):5984–6003. https://doi.org/10.1021/bi00185a040
21. Kay LE, Torchia DA, Bax A (1989) Backbone dynamics of proteins as studied by nitrogen-15 inverse detected heteronuclear NMR spectroscopy: application to staphylococcal nuclease. Biochemistry 28(23):8972–8979. https://doi.org/10.1021/bi00449a003, note = PMID: 2690953
22. Mulder FAA, de Graaf RA, Kaptein R, Boelens R (1998) An off-resonance rotating frame relaxation experiment for the investigation of macromolecular dynamics using adiabatic rotations. J Magn Reson 131(2):351–357. https://doi.org/10.1006/jmre.1998.1380
23. Geen H, Freeman R (1991) Band-selective radiofrequency pulses. J Magn Reson (1969) 93(1):93–141. https://doi.org/10.1016/0022-2364(91)90034-Q
24. Korzhnev DM, Skrynnikov NR, Millet O, Torchia DA, Kay LE (2002) An NMR experiment for the accurate measurement of heteronuclear spin-lock relaxation rates. J Am Chem Soc 124(36):10743–10753. https://doi.org/10.1021/ja0204776
25. Hansen DF, Kay LE (2007) Improved magnetization alignment schemes for spin-lock relaxation experiments. J Biomol NMR 37(4):245–255. https://doi.org/10.1007/s10858-006-9126-6
26. Bax A, Clore GM, Gronenborn AM (1990) 1H-1H correlation via isotropic mixing of 13C magnetization, a new three-dimensional approach for assigning 1H and 13C spectra of 13C-enriched proteins. J Magn Reson (1969) 88(2):425–431. https://doi.org/10.1016/0022-2364(90)90202-K
27. Li YC, Montelione GT (1993) Solvent saturation-transfer effects in pulsed-field-gradient heteronuclear single-quantum-coherence (PFG-HSQC) spectra of polypeptides and proteins. J Magn Reson B 101(3):315–319. https://doi.org/10.1006/jmrb.1993.1049
28. Renner C, Schleicher M, Moroder L, Holak TA (2002) Practical aspects of the 2D 15N-[1h]-NOE experiment. J Biomol NMR 23(1):23–33. https://doi.org/10.1023/a:1015385910220
29. Ferrage F, Cowburn D, Ghose R (2009) Accurate sampling of high-frequency motions in proteins by steady-state (15)N-{(1)H} nuclear Overhauser effect measurements in the presence of cross-correlated relaxation. J Am Chem Soc 131(17):6048–6049. https://doi.org/10.1021/ja809526q
30. Hansen DF, Vallurupalli P, Kay LE (2008) An improved 15N relaxation dispersion experiment for the measurement of millisecond time-scale dynamics in proteins. J Phys Chem B 112(19):5898–5904. https://doi.org/10.1021/jp074793o
31. Kozak S, Lercher L, Karanth MN, Meijers R, Carlomagno T, Boivin S (2016) Optimization of protein samples for NMR using thermal shift assays. J Biomol NMR 64(4):281–289. https://doi.org/10.1007/s10858-016-0027-z
32. Carver JP, Richards RE (1972) A general two-site solution for the chemical exchange produced dependence of T2 upon the carr-Purcell pulse separation. J Magn Reson (1969) 6(1):89–105. https://doi.org/10.1016/0022-2364(72)90090-X
33. Palmer AG 3rd, Kroenke CD, Loria JP (2001) Nuclear magnetic resonance methods for quantifying microsecond-to-millisecond motions in biological macromolecules. Methods Enzymol 339:204–238. https://doi.org/10.1016/s0076-6879(01)39315-1

Chapter 3

Crystal Structure Analysis of SH2 Domains in Complex with Phosphotyrosine Peptides

Miki Senda and Toshiya Senda

Abstract

While the number of tertiary structures solved by cryoelectron microscopy has rapidly increased, X-ray crystallography is still a popular method to determine the tertiary structure of proteins at atomic resolution. However, there are still problems associated with X-ray crystallography, including crystallization and crystal twinning. Indeed, we encountered crystallization and twinning problems in the crystal structure analysis of the SH2 domains complexed with a phosphorylated peptide derived from the oncoprotein CagA. In this chapter, we describe the methods used to overcome these problems. In addition, we provide details of the optimization of the crystallization conditions and cryo-conditions, which are usually not given in published crystal structure analyses.

Key words X-ray crystallography, Phosphotyrosine peptide, Crystallization, Anaerobic crystallization, Microseeding, Cryoprotectant, Twinning, Crystallographic refinement

1 Introduction

CagA is an oncoprotein of *Helicobacter pylori*. *H. pylori* injects the CagA protein into gastric epithelial cells using a type IV secretion system (TFSS) [1]. The injected CagA is localized on the inner surface of the plasma membrane, and Tyr residues in the EPIYA segment of the C-terminal intrinsically disordered region are phosphorylated [2]. The phosphorylated EPIYA (EPIpYA) segment, which includes a Glu-Pro-Ile-Tyr-Ala (EPIYA) sequence motif, binds SHP2 protein phosphatase, activating the SHP2 and disturbing the cellular signaling [3, 4]. Since this molecular process seems to be related to the development of gastric cancer, the molecular mechanisms of the SHP2 activation, particularly the interaction between tandem SH2 domains of SHP2 and the phosphorylated EPIYA (EPIpYA) segment of CagA, have attracted the attention of researchers. Interestingly, a difference in the amino acid sequence of the EPIYA segment affects the activation activity of SHP2 [4]. CagA from *H. pylori* in East Asian countries

Teresa Carlomagno and Maja Köhn (eds.), *SH2 Domains: Functional Modules and Evolving Tools in Biology*, Methods in Molecular Biology, vol. 2705, https://doi.org/10.1007/978-1-0716-3393-9_3,

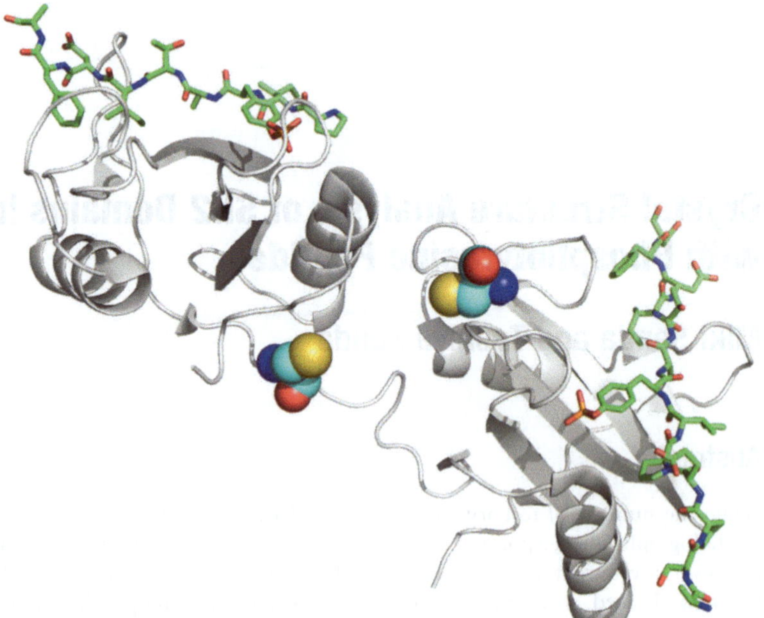

Fig. 1 Crystal structure of the EPIpYA-D complex (PDB ID: 5X94). Carbon atoms in tandem SH2 domains and EPIpYA-D peptides are shown in white and green, respectively. Free Cys residues in the tandem SH2 are shown as spheres. The tandem SH2 has two free Cys residues, which may form random disulfide bridges in solution

has a different EPIYA segment (EPIYA-D) from that found in European countries (EPIYA-C) and shows a higher SHP2 activation activity than the European-type CagA. Our crystallographic analysis of the tandem SH2–EPIpYA-C and tandem SH2–EPIpYA-D complexes (hereafter, EPIpYA-C and EPIpYA-D complexes, respectively) revealed a molecular mechanism of SHP2 activation and explained the geographic difference in gastric cancer incidence between the East Asian and European types of CagA [5].

While there have been many examples of crystal structure analyses of the tandem SH2 domains complexed with a phosphorylated Tyr peptide, crystal structure analysis of the tandem SH2 domains (hereafter tandem SH2) from SHP2 with a phosphorylated CagA peptide encountered several problems. The first problem occurred in the crystallization screening step. Following the tandem SH2 purification, we carried out the crystallization screening using a commercially available crystallization kit. However, no crystals were obtained. Moreover, insoluble precipitations of the tandem SH2 appeared in many droplets. Insoluble precipitations are a bad sign for crystallization. While the actual reason for the insoluble precipitation formation of the tandem SH2 is unknown, one possible explanation is the oxidation of the tandem SH2. Free Cys residues in tandem SH2 domains may form a disulfide bridge randomly (*see* Fig. 1), causing insoluble precipitations. Therefore, we tested an anaerobic crystallization. Comparison of the aerobic crystallization screening with the anaerobic one revealed a significant difference—namely, fewer precipitations were observed in droplets of anaerobic crystallization [6]. Based on this result, we decided to crystallize the EPIpYA-C and -D complexes under anaerobic conditions. Anaerobic crystallization successfully produced crystals of these complexes. While the anaerobic crystallization techniques overcame the first problem, we encountered two additional difficulties: reproducibility of the crystallization and pseudo-merohedral twinning of the crystal. Since these are common problems in macromolecular crystallography, we believe it is worthwhile discussing them in detail, along with our solutions.

This chapter thus focuses on typical problems in crystal structure analysis and useful techniques to overcome them.

2 Materials

2.1 Anaerobic Experiments

1. An anaerobic chamber: we use Anaero Box "HARD" (Hirasawa, Tokyo) (*see* **Note 1**).
2. The chamber requires a gas mixture (96% N_2 and 4% H_2) and a palladium catalyst (AZX-250) to make an anaerobic environment.
3. An oxygen indicator.

2.2 Crystallization Screening

1. Protein concentrators: we use Amicon Ultra-4 (MWCO 3000).
2. Buffer A for the protein and EPIpYA-CD peptides: 20 mM Tris–HCl pH 8.0 and 100 mM NaCl.
3. Purified tandem SH2 [5].
4. Chemically synthesized CagA peptides EPIpYA-C (VSP*EPIpYA*TIDDL) and EPIpYA-D (ASP*EPIpYA*TIDFD) with phosphorylated Tyr (pY).
5. Crystallization plate, thick siliconized circle cover slides, and micro-bridges.
6. Crystallization screening buffers: we use Crystal Screen (Hampton Research) (*see* **Note 2**).
7. Anaerobic chamber with incubator at 20 °C.
8. Stereomicroscope (in the anaerobic chamber).

2.3 Optimization of Crystallization Conditions

1. Crystallization plate: crystallization plate with sealant, thick siliconized circle cover slides, and micro-bridges.
2. Crystallization solution for the EPIpYA-C complex crystal: 27.0–30.5% (w/v) PEG4000 (*see* **Note 3**), 0.1 M Tris–HCl pH 9.0, 0.13 M sodium acetate (*see* **Note 4**).

3. Crystallization solution for the EPIpYA-D complex crystal: 27.3–31.0% (w/v) PEG4000, 0.1 M Tris–HCl pH 8.0, 0.13 M sodium acetate.
4. Artificial mother liquors for the EPIpYA-C complex crystal: 31% (w/v) PEG4000, 0.07 M Tris–HCl pH 9.0, 0.1 M sodium acetate.
5. Artificial mother liquors for the EPIpYA-D complex crystal: 31% (w/v) PEG4000, 0.07 M Tris–HCl pH 8.0, 0.1 M sodium acetate.
6. Anaerobic chamber with incubator at 20 °C.
7. Stereomicroscope (in the anaerobic chamber).

2.4 Microseeding

1. Artificial mother liquors for the EPIpYA-C complex crystal: 31% (w/v) PEG4000, 0.1 M Tris–HCl pH 9.0, 0.13 M sodium acetate.
2. Artificial mother liquors for the EPIpYA-D complex crystal: 31% (w/v) PEG4000, 0.1 M Tris–HCl pH 8.0, 0.13 M sodium acetate.
3. Crystallization plate with sealant, thick siliconized circle cover slides, and micro-bridges.
4. Needles (for crushing crystals).
5. Stereo microscope.
6. Depression glasses (custom-made glassware) (*see* Fig. 2) [7].

2.5 Crystal Scooping and Freezing

1. Cryo-loop.
2. Cryo-pin (*see* **Note 5**).
3. Liquid nitrogen.
4. Dewar (for liquid nitrogen).
5. UniPucks.

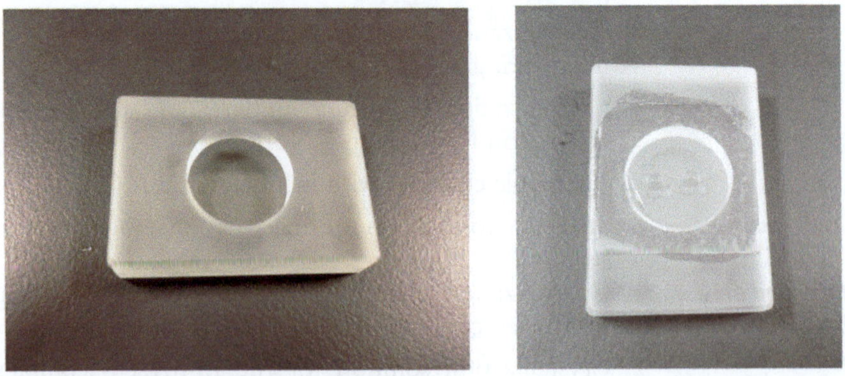

Fig. 2 The custom-made depression glass used in soaking experiments. Photos show the depression glass used in the soaking experiments. The glass has a round depressed well in the middle. The hole can be sealed to avoid the evaporation of liquid

6. Stereo microscope.
7. Dry shipper.
8. Depression glasses (custom-made glassware).

2.6 Cryoprotectant Optimization

1. Cryoprotectant stock solution: we use CryoPro from Hampton Research.
2. Depression glasses (custom-made).
3. Cryo-loop with cryo-pin.
4. Liquid nitrogen.
5. Dewar.
6. UniPucks.
7. Dry shipper.
8. Stereo microscope.
9. Cryoprotectant solution for the EPIpYA-C complex crystal: 26% (w/v) PEG4000, 0.07 M Tris–HCl pH 9.0, 0.1 M sodium acetate, 17% (w/v) diethylene glycol.
10. Cryoprotectant solution for the EPIpYA-D complex crystal: *see* Table 1.
11. Depression glasses (custom-made glassware).

Table 1
Screening of cryoprotectants for the EPIpYA-D complex

No.	Cryoprotectant[a]	Resolution 1[b]	Resolution 2[b]
1	30% (w/v) PEG4000	2.3	–
2	33% (v/v) glycerol	2.05	2.3
3	33% (v/v) ethylene glycol	2.35	2.1
4	16.5% (v/v) diethylene glycol	2.05	Frosted
5	33% (v/v) PEG200	Frosted	3.4
6	33% (v/v) PEG400	2.5	2.7
7	33% (v/v) PEG MME550	2.7	2.6
8	25% (w/v) polyvinylpyrrolidone K15	2.6	–
9	35% (w/v) sucrose	2.6	3.3
10	35% (w/v) sorbitol	3.0	2.5
11	15% (w/v) maltose	3.4	–

[a]The cryoprotectant solutions were prepared by mixing 50–100% cryoprotectant stock solution (Hampton Research) with 39% (w/v) PEG4000, 0.1 M Tris–HCl pH 8.0, and 0.13 M sodium acetate in a 1:1 or 1:2 ratio

[b]These resolution values were obtained by visual inspection of the snapshots. The resolution values in this table tend to be better than those obtained by data processing and scaling

2.7 Data Collection and Processing

1. PC (Windows, Mac, or Linux machine).
2. Software for data processing and scaling: XDS [8] (https://xds.mr.mpg.de/) and CCP4 [9] (https://www.ccp4.ac.uk/) (no Windows version is available for XDS).
3. Software for crystallographic analysis: CCP4 [9] and PHENIX [10, 11] (https://phenix-online.org/).

2.8 Twinning

1. PC (Windows, Mac, or Linux machine).
2. Software for data processing and scaling: XDS [8] and CCP4 [9].
3. Software for crystallographic calculations: CCP4 [9] and PHENIX [10, 11].

2.9 Phasing Using the Molecular Replacement Method

1. PC (Windows, Mac, or Linux machine).
2. Software for molecular replacement: MOLREP [12] in CCP4 or PHASER [13] in PHENIX.
3. Initial model for the molecular replacement: in the example we give, 4DGP of PDB [the coordinates can be obtained from PDB (RCSB PDB (https://www.rcsb.org/), PDBj (https://pdbj.org/), PDBe (https://www.ebi.ac.uk/pdbe/), or other database sites].

2.10 Crystallographic Refinement

1. PC (Windows, Mac, or Linux machine).
2. Software for crystallographic refinement: PHENIX.refine in PHENIX [10, 11].
3. Software for model building: Coot [14].
4. CCP4 (FREERFLAG in CCP4 was used for setting free-R flags) [9].

3 Methods

3.1 Anaerobic Experiments (General Preparation) [6, 15]

1. Keep all plasticwares used in the anaerobic experiments, such as tubes, pipet tips, and crystallization plates, in the anaerobic chamber for more than 1 week to remove oxygen molecules absorbed on their surfaces.
2. Check the oxygen concentration in the chamber using an oxygen indicator before starting the experiments.
3. Prepare all solutions (protein, peptide, crystallization solutions, etc.) under aerobic conditions.
4. Place the prepared crystallization solutions inside the anaerobic chamber for at least 3 days to remove oxygen from the solutions. The caps of the tubes should be loosened to facilitate the exchange of gases.

5. Place the protein solution in the anaerobic chamber, typically for a few hours.
6. Check the oxygen concentration in each solution (including the protein solution) using an oxygen indicator. If the oxygen indicator detects oxygen molecules, wait for one more night.

3.2 Crystallization Screening (see Note 6)

1. Measure the tandem SH2 concentration using the absorbance of 280 nm with a UV-vis spectrometer. The molar extinction coefficient ($\varepsilon = 22.92 \times 10^3\ M^{-1}\ cm^{-1}$ in our example) can be calculated using a web tool (*see* **Note 7**).
2. Dissolve a synthesized CagA peptide in Buffer A. The concentration of the peptide is 7.5 mM.
3. Put the tandem SH2 solution in an Amicon protein concentrator and concentrate by centrifuge to nearly 70–100 mg/m at 4 °C. During the centrifugation, check the protein concentration repeatedly, and stop the centrifugation when the protein concentration reaches the target concentration range. In our example, the concentration of the protein solution was not adjusted even when being overconcentrated (*see* **Note 8**).
4. To establish a molar ratio of 1:3 for tandem SH2/peptide, add 34.8 μL of 7.5 mM peptide solution to 30 μL of 76 mg/mL (2.9 mM) tandem SH2 solution. It is also possible to mix the SH2 and peptide solutions at a molar ratio of 1:4 before concentration. Then, concentrate the solution to nearly 70–100 mg/mL by centrifugation with the Amicon protein concentrator.

The following procedures are carried out inside the anaerobic chamber:

5. Place the protein and crystallization solutions in the anaerobic chamber and remove oxygen from the solutions (*see* above).
6. Add 0.5 mL of each protein crystallization solution to a well of a crystallization plate.
7. Mix the protein solution and each crystallization solution on a coverslip. The volumes of both solutions are typically 1 μL (but as little as 0.5 μL should still work). There is no need to stir the mixed solution.
8. Turn the coverslip around and use it to seal the well containing the corresponding crystallization solution.
9. Observe each droplet just after the crystallization screening to check the formation of precipitation (*see* **Note 9**).
10. Store the crystallization plate in an incubator at 20 °C.
11. Monitor the crystallization plate every week and categorize as follows: (i) clear drop, (ii) precipitant, and (iii) crystal. In the case of the EPIpYA-D complex, thin needle crystals appeared

under the condition with 30% (w/v) PEG4000, 0.1 M Tris–HCl pH 8.5, and 0.2 M sodium acetate (#22 of Crystal Screen). This condition is the starting point for the optimization of the crystallization conditions.

3.3 Optimization of Crystallization Conditions

Before starting the optimization of the crystallization conditions, physicochemical characteristics of the target protein should be recognized from the results of the initial crystallization screening. When many droplets exhibit precipitate, the protein concentration is likely too high. In this case, crystallization screening with a lower protein concentration should be tested. When no precipitation appears, you need to increase the protein concentration. In the case of the EPIpYA complexes, we observed a high level of precipitation that we suspected was due to the oxidation of free Cys residues. Therefore, we tried to perform an anaerobic crystallization:

1. If no crystals are obtained at the previous step, change the molar ratio of SH2/peptide (*see* **Note 10**).
2. If crystals appear in the crystallization screening, before optimizing the crystallization conditions, perform a preliminary diffraction study to verify that the obtained crystals are protein crystals.
3. Optimize the crystallization conditions based on these first conditions. Continue working in the anaerobic chamber. In the case of the EPIpYA-D complex, the crystallization solution of 30% (w/v) PEG4000, 0.1 M Tris–HCl (pH 8.5), and 0.2 M sodium acetate gave crystals. Based on this condition, the crystallization condition was optimized by changing the PEG concentration from 27% to 35% (w/v) in increments of 1%. However, since the crystallization solution tended to show a phase separation at 35% (w/v) PEG4000, the concentration of sodium acetate was reduced to 0.13 M to prevent the phase separation. The optimization revealed that 27.0–30.5% (w/v) of PEG4000 gave good results. The final conditions of the crystallization solution for the EPIpYA-D complex were 0.1 M Tris–HCl (pH 8.5), 0.13 M sodium acetate, and PEG4000 27.3–31% (w/v). Several PEG4000 concentrations, ranging from 27.3% to 31.0% in increments of 0.3%, were used in every crystallization experiment, because the best concentration of the PEG4000 varies depending on the protein concentration. The crystallization conditions of the EPIpYA-C complex were optimized in the same manner. The sitting drop method sometimes yields substantially different crystallization results from the hanging drop method. Indeed, the sitting drop was much better than the hanging drop for both

the EPIpYA-C and EPYpYA-D complexes (*see* Fig. 4 of Ref. [6]).

4. After optimizing the crystallization conditions, determine the conditions of the artificial mother liquor [7]. The artificial mother liquor is used for soaking experiments, such as for optimizing cryoprotectant conditions and preparing ligand complex crystals. The pH of the artificial mother liquor should be the same as that of the droplet generating the crystal. The precipitant concentration in the artificial mother liquor is slightly higher than that of the crystallization solution to avoid crystal dissolution. In the case of the EPIpYA-D complex, the conditions of the artificial mother liquor were 31% (w/v) PEG4000, 0.07 M Tris–HCl pH 8.0, and 0.1 M sodium acetate.

The artificial mother liquor should be able to keep the crystal without dissolution, preferably longer than overnight. Check the artificial mother liquor as follows:

5. Put the artificial mother liquor (3–5 μL) in a depression glass and transfer a crystal into the artificial mother liquor. Then seal the depression glass (*see* Fig. 2) and place it in an incubator.
6. Observe the soaked crystal with a stereo microscope. Check the size of the crystal. When the crystal is dissolved, change the conditions of the artificial mother liquor. Store the depression glass in an incubator (20 °C) and observe the crystal the next day. Optimize the concentration of the precipitant to avoid crystal dissolution.

3.4 Microseeding

Microseeding often improves the reproducibility of crystallization experiments. It is possible to control the number and size of crystals in a droplet by changing the seed concentration. In the case of the SH2–EPIpYA-C and -D complexes, microseeding significantly improved the reproducibility of the crystallization experiments. While crystallization without microseeding gave crystals of the EPIpYA complexes with low reproducibility, microseeding crystallization made the complex crystals in 2 days with high reproducibility. The following steps are carried out inside the anaerobic chamber:

1. Scoop (a) crystals from a droplet (*see* Subheading 3.5) and transfer them to the artificial mother liquor (10 μL) in a depression glass.
2. Crush crystals with a needle.
3. Check the size of the crushed crystals using a stereo microscope.
4. Prepare a dilution series (×1, ×1/10, ×1/100, ×1/1000) of microseeds to optimize the microseeding conditions.

5. Put crystallization solution (0.5 mL) in each well of the crystallization plate.
6. Mix the tandem SH2 protein (1 μL), crystallization (1 μL), and microseed solutions (0.2 μL) on a coverslip.
7. Turn over the coverslip and seal the corresponding well.
8. Store the crystallization plate in an incubator at 20 °C.
9. Observe crystallization droplets using a stereo microscope. Typically, we observe crystallization droplets after 1, 2, 3, 5, 7, and 14 days of crystallization.
10. When too many tiny crystals appear in one droplet, use a diluted microseed solution. The optimal number of crystals in one droplet is about 10–20 in the case of the EPIpYA complexes.
11. The microseed solution can be stored in a freezer at −80 °C. The microseed solution stock increases the reproducibility of the microseeding experiment.

3.5 Crystal Scooping and Freezing

1. Before scooping crystals, check the type of cryo-pin for the beamline you will use. In addition, the height of the cryo-loop from the basement of the cryo-pin is an important parameter for data collection, because a robot typically carries out crystal mounting to the diffractometer. The information is usually provided on a web page of the beamline.
2. Take the coverslip off the crystallization plate, turn it over, and place it under a stereo microscope.
3. Add the corresponding crystallization solution (e.g., 2 μL) to the droplet when the droplet volume is insufficient for scooping.
4. Check the size of the crystal and select a cryo-loop with an appropriate size. If the cryo-loop is too big, crystal scooping is difficult. If the cryo-loop is too small, the crystal will be damaged by scooping.
5. Do not touch the crystal when scooping.
6. If crystals adhere to the coverslip, do not remove the crystal with a needle, as touching the crystal with a needle frequently damages the crystal. Instead, add crystallization solution to the droplet to dissolve the surface of the crystal and wait for several minutes. Then, remove the crystal from the coverslip by generating a gentle flow with the cryo-loop.
7. When crystals are prepared under anaerobic conditions, crystal scooping and freezing should be performed in the anaerobic chamber (*see* **Note 11**). However, when no damages occur

under aerobic conditions, the crystal can be scooped outside the anaerobic chamber. In the case of the EPIpYA-C and -D complexes, no damages were observed on the anaerobically prepared crystals. Therefore, we scooped and froze the crystals under aerobic conditions (outside the anaerobic chamber).

3.6 Cryoprotectant Optimization

Cryoprotectant optimization is carried out under aerobic conditions:

1. Prepare the artificial mother liquor.
2. Prepare the cryoprotectant solution based on the conditions of the artificial mother liquor (*see* Table 1).
3. Put 3 μL of each cryoprotectant solution into a depression glass.
4. Soak a crystal in each cryoprotectant solution for 30 s, scoop the crystal, and freeze it using liquid nitrogen. Soak more than one crystal for each cryoprotectant solution to check the reproducibility. The cryoprotection of the EPIpYA-C and -D complexes was done by soaking the crystals for only 30 s.
5. Store the frozen crystals in a UniPuck cassette in liquid nitrogen.
6. Store the UniPuck cassette in a dry shipper until data collection.
7. Collect preliminary diffraction data from each crystal (*see* **Note 12**). Data collection should be done from more than one crystal for each cryo-condition to confirm reproducibility.
8. Compare the resolution of the cryoprotected crystals and choose an appropriate cryoprotectant(s) (*see* Table 1).
9. Optimize the soaking concentration and time for each selected cryoprotectant. The soaking temperature sometimes affects the result [16]. Furthermore, the multistep soaking method may improve the results [17]. When optimizing the conditions, one should use more than one crystal for each condition and confirm reproducibility.
10. Determine the conditions for the cryoprotectant solution.

3.7 Data Collection and Processing

1. Send your UniPuck cassette to a synchrotron facility with your data collection parameters. Data collection and processing in the advanced synchrotron facilities are typically performed automatically. The data collection strategy should be carefully determined if the crystal is twinned (*see* Subheading 3.8).
2. Select an appropriate wavelength for data collection. The MR (molecular replacement)-SAD method is frequently useful for rapid structure determination (*see* **Note 13**).

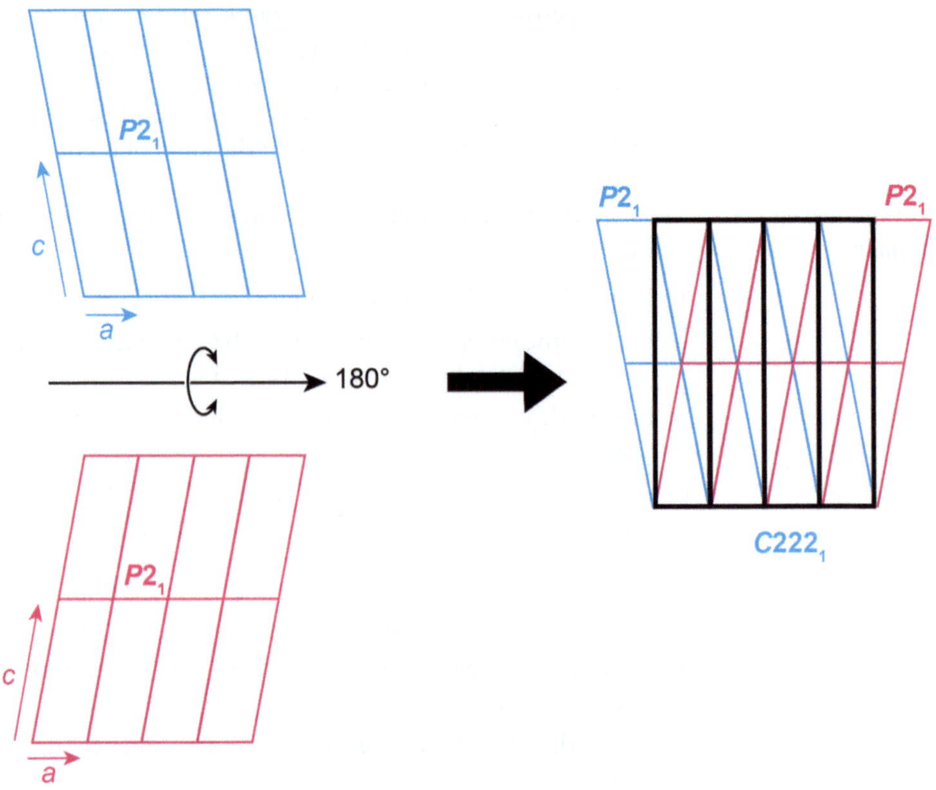

Fig. 3 Pseudo-merohedral twinning of a $P2_1$ crystal. The apparent space group of the EPIpYA-D crystal is $C222_1$. This pseudo (apparent) space group arises from the superposition of the two $P2_1$ lattices as shown in this figure. The two $P2_1$ lattices, which are shown in blue and purple, are related by a twofold axis along the *a*-axis. As a result, a pseudo-$C222_1$ lattice appears

Crystal quality is frequently different, even if crystals were harvested from the same droplet (*see* **Note 14**). Understanding the relationship between crystal characteristics, such as size and shape, and crystal quality is essential to collecting good diffraction data.

3.8 Twinning

The (merohedral or pseudo-merohedral) twinning is a frequent problem in macromolecular crystallography. In the case of the EPIpYA-D complex [5], data processing suggested the apparent space group $C222_1$. However, the molecular replacement (MR) calculation did not work well with this space group despite the relatively good resolution of the crystal. We therefore suspected a pseudo merohedral twinning, and after further examination, we confirmed that the crystal indeed belonged to space group $P2_1$ with pseudo-merohedral twinning (*see* Fig. 3). In many cases, crystal twinning causes problems in phasing and crystallographic refinement. It was thus necessary to reprocess the diffraction data with the right space group and detwin the data to perform appropriate crystallographic analysis. This section explains how to analyze the crystal structure with twinning:

1. Process and scale the diffraction data with the actual (correct) space group.
2. Analyze the processed data with Xtriage in PHENIX and check the twinning section in the Xtriage log file. The comments provide information about twinning, including the type of twinning, twin fraction, and twin operators. In the case of the EPIpYA-D complex, Xtriage in PHENIX suggested pseudo-merohedral twinning of the crystal with $P2_1$ symmetry [18] and provided the twin fraction and twin operators.
3. Collect several sets of diffraction data from multiple crystals. Twin fractions differ from crystal to crystal. Select a crystal with a small twin fraction. Crystals with smaller twin fractions typically give better results in crystallographic refinement. Crystals with a large twin fraction—namely, crystals close to perfect twinning—typically cannot give good results from the crystallographic refinement even if the resolution is good.
4. The twin fraction frequently varies depending on the position of the crystal. Try several crystals to clarify the character of the crystal (*see* **Note 15**). Collect the diffraction data at the position with a small twin fraction.
5. A helical scan with a small beam is useful to mitigate the effect of the radiation damage. However, the helical scan causes trouble when the crystal is twinned (*see* **Note 15**). Choose an appropriate mode of data collection by considering the character of the crystals (*see* **Note 15**).
6. Based on the twin fraction obtained from Xtriage, detwin the data using the program DETWIN in CCP4.
7. Perform the MR calculation with the detwinned data. In the case of the EPIpYA-D complex, it was impossible to solve the structure without detwinning (*see* Subheading 3.9).
8. When refining the crystal structure, input an estimated value of the twin fraction and twin operators. The twin fraction will be optimized during the crystallographic refinement.

3.9 Phasing Using the Molecular Replacement Method

1. The molecular replacement (MR) method requires an initial model. When the crystal structure of the same protein is already solved, choose the same protein as the initial model. When the crystal structure information of your target protein is not available, choose the crystal structure of a homolog protein with the highest sequence homology to your target protein as the initial model. In recent years, it has become common to use a model predicted by AlphaFold [19]. This frequently works quite well. However, regions with low reliability should be removed from the initial model.

2. There are several choices for MR software. Use automated MR software, such as MOLREP [12] in CCP4 and Phaser [13] in PNENIX.
3. After obtaining the initial phases, check the quality of the electron density using software for model building, such as COOT [14]. Check the fitting between the initial model and the initial map and molecular packing. If the fitting and packing are reasonable, you can start the crystallographic refinement.
4. Another critical option for rapid structure determination is the MR-(native) SAD method, which combines experimental phase information with the phases obtained by the MR method (*see* **Note 13**).

In the case of the EPIpYA-D complex, we used detwinned data for the MR calculation. Without detwinning, MR calculations could not provide good results. The diffraction data processed with the space group $P2_1$ were detwinned using the program DETWIN in CCP4. The obtained data were used for the MR calculation with an initial model prepared from 4DGP. Since 4DGP contains a phosphatase domain, only the tandem SH2 portion was isolated from the coordinates and used as an initial model.

3.10 Crystallographic Refinement

Crystallographic refinement is the final but critical step for structure determination. Many books and manuals show a general method for crystallographic refinement. Therefore, we will omit the general method and instead provide some tips for crystallographic refinement:

1. Use Phenix.refine in PHENIX [10] to check both the refinement statistics and the geometry with a Ramachandran plot. The program provides an integrated environment for crystallographic refinement.
2. For crystallographic refinement of (pseudo)-merohedral twin crystals, make sure to set the freeR flags with care. Because of the pseudo-symmetry, random selection is inappropriate. When setting freeR flags, use an appropriate optional parameter in the program.
3. Since the initial model usually contains many coordinate errors, use a simulated annealing protocol.
4. When the structure contains a nonstandard amino acid (and ligands), include a parameter file (CIF file) with the input files.
5. After finishing the model correction, run one cycle of crystallographic refinement.
6. After one cycle of crystallographic refinement calculation, correct the model using COOT. First, check the Ramachandran plot. If residues have inappropriate phi-psi angles, check the residues with the obtained electron density map and correct the

model. In general, the MR model contains several errors. Electron densities calculated from the error-containing model are frequently biased. As it is difficult to correct the model with a biased electron density, it is better to use a composite omit map for model correction.

7. Pick water with great care when the resolution is worse than 2.5 Å. Only clear (sigma level of more than three) and spherical difference densities should be selected as water molecules. Ambiguous electron densities should not be interpreted as water molecules. All the water molecules and their corresponding densities should be checked using COOT after auto water picking. Water molecules with poor density will not improve the model or the statistics of the crystallographic refinement.
8. Perform cycles of crystallographic refinement and model building until convergence.
9. Select the diffraction data for the crystallographic refinement while considering the quality of the electron density of the bound peptide, Rwork/Rfree, resolution, etc. In our examples, some crystals diffracted at nearly 2.0 Å resolution. However, these crystals showed a large twin fraction and poor Rwork/Rfree. Based on a comprehensive consideration of the crystallographic parameters and statistics, we selected a structure at 2.6 Å resolution with a twin fraction of 0.38. The crystal was cryoprotected by 30% (w/v) PEG4000, 0.1 M Tris–HCl pH 8.0, and 0.13 M sodium acetate for 30 seconds (*see* **Note 16**).

4 Notes

1. The anaerobic chamber used in this experiment was specially designed for crystallization experiments. It has three compartments and contains a stereomicroscope with an LCD monitor, a UV-vis spectrometer, and incubators (20 °C or 4 °C). This allows all experiments from crystallization to crystal freezing to be performed inside the chamber. As described in the Subheading 3, oxygen molecules absorbed on the surface of plasticwares are hard to remove. Therefore, plasticwares used in the crystallization experiments should be stored in the anaerobic chamber. The chamber maintains an anaerobic environment by removing the oxygen molecules using Pd catalysts. A gas mixture containing 96% N_2 and 4% H_2 is introduced for 1 min every 2 h to keep the anaerobic condition nearly constant.
2. Crystallization conditions of the SH2–EPIpYA complex were screened using only Crystal Screen (Hampton Research)

because the screening was carried out manually in the anaerobic chamber. However, screening is usually performed using robots such as MOSQUITO (SPT Labtech) and Crystal GRYPHON (Art Robbins Instruments). Fully automated crystallization systems are also available, such as the fully automated crystallization screening robot, PXS2, in use at the Structural Biology Research Center (SBRC) of the High Energy Accelerator Research Organization (KEK) [20] (*see also* **Note 6**).

3. The characteristics of polyethylene glycols (PEGs) differ from maker to maker. There are also significant variations among lot numbers within the same manufacturer. Therefore, we recommend recording the lot number and maker of the PEG used in the crystallization experiment for reproducibility. We used PEGs from Merck in the crystallization of the EPIpYA complexes. PEGs from other manufacturers did not work as well.
4. We did not use a fixed condition for the crystallization even after optimizing the crystallization conditions. In the case of the EPIpYA-C and -D complexes, we used relatively small concentration ranges of the precipitant (PEG4000) in every crystallization experiment (*see also* **Note 8**).
5. Most beamlines have their own preference of cryo-pin. Consult the beamline to be used to learn the recommended cryo-pins before freezing crystals. This information is also usually available on the beamline web page.
6. Although we used only a Crystal Screen kit (Hampton Research) for crystallization screening of the SH2–peptide complex, it has become common to carry out crystallization screening using several kits, such as Crystal Screen 2, PEG/Ion 1, PEG/Ion 2, Index, and MembFac from Hampton Research, Wizard Classic 1 and 2 and Wizard Cryo 1 and 2 from Rigaku, the Protein Complex and PEGs II Suites from Qiagen, and Stura Footprint from Molecular Dimensions. At the Structural Biology Research Center at KEK, crystallization screening is carried out using a fully automated robot, PXS, as mentioned above [20]. This machine can screen nearly 800 conditions in 30 min. Once setting the protein solution to the robot, PXS2 dispenses crystallization solutions to a dedicated crystallization plate, sets up protein and crystallization solutions on the crystallization plate, seals the well, and stores the plate in an incubator (20 °C or 4 °C). Moreover, each droplet is observed automatically with a microscope following a programmed schedule [20].
7. When crystallizing a protein–peptide complex, the protein concentration is a critical parameter. Therefore, the protein concentration should be measured correctly. Of the several methods used to measure the protein concentration, we

recommend measuring the concentration at 280 nm absorption with a molar extinction coefficient calculated theoretically. Methods using a reference protein such as bovine serum albumin sometimes show a significant deviation from the actual concentration, which may cause trouble in determining the molar ratio to the peptide.

8. Since crystallization requires a high concentration of proteins (typically 5–10 mg/mL), proteins are usually concentrated using a filter with a molecular weight cutoff. In typical protocols, the protein concentration is adjusted to a defined value for reproducibility. Therefore, the protein solution is diluted with a buffer when the proteins are overconcentrated. However, in our experience, a diluted protein solution sometimes gives only tiny crystals. Therefore, we did not dilute the concentrated tandem SH2 solution. Instead, we tested several precipitant concentrations (PEG4000) to obtain high-quality crystals (see Subheading 3.3, **step 6**). Another potential obstacle to a successful crystallization is the handling of the solution after the protein concentration. In our case, the concentrated protein solution was filtered with a membrane to remove dust particles. Following this procedure, however, no crystals were obtained from the filtered protein solution, probably because the filtration removed small aggregates required for crystal nucleation.
9. Observing droplets just after crystallization sometimes provides preliminary information about crystallization conditions. It has been demonstrated that buffer conditions that yield soluble aggregates of the protein are often a good starting point for optimization of crystallization condition. After mixing the protein and crystallization solutions, proteins often form precipitates. To recognize the conditions that lead to soluble aggregates, one must observe the droplets at a later time point (typically 1 day after the crystallization) and note which of the droplets that formed precipitates immediately after mixing have now become clear.
10. When crystallizing a protein–protein (or peptide or DNA) complex, mixing solutions of each component on a coverslip frequently works better than using a protein complex sample purified by size exclusion chromatography (SEC). Since SEC requires dilution of the protein complex, it can cause the protein complex to be partially dissociated. As a result, the protein solution contains both free and complex forms of the protein, hampering the crystallization. Therefore, to obtain better crystals, it is more effective to change the molar ratio of the components in the complex.

11. Crystals prepared under anaerobic conditions are sometimes cracked when placed in aerobic conditions. The difference in the partial pressure of oxygen may cause crystal damage. However, the real reason for this phenomenon is unknown. In such cases, the crystal should be frozen under anaerobic conditions.
12. When optimizing the cryoprotectant conditions of the EPIpYA complexes, we used snapshot patterns from the crystals to evaluate the crystal quality. However, the diffraction data collection method has been substantially improved since our optimizations. Most advanced beamlines for macromolecular crystallography (MX) can automatically collect diffraction data. Moreover, due to the availability of pixel array detectors, the speed of the diffraction data collection has accelerated substantially. Since diffraction data collection from one crystal takes only a few minutes, we strongly recommend collecting a full dataset of each crystal for cryoprotectant optimization. Most MX beamlines take only a few hours to collect full datasets from the crystals in one UniPuck cassette, which can contain 16 crystals. The full dataset provides more information than the snapshots.
13. Model bias is a critical problem in the structure determination with the MR method. Indeed, crystallographic refinement after the MR phasing takes a relatively long time due to the model bias. MR-SAD phasing is an effective method to avoid the model-bias problem. Moreover, because it combines experimental phases with the MR phases, MR-SAD frequently provides a better electron density map than the usual MR method. While Se-Met has frequently been used for SAD phasing, anomalous scattering from sulfur atoms can also be used in the MR-native SAD method. For example, the Photon Factory at KEK has a beamline BL-1A dedicated to the native SAD method. It is possible to collect diffraction data with a relatively long wavelength, e.g., 2.7 Å, to enhance anomalous signals from sulfur atoms. An X-ray with a wavelength of 1.9 Å also works fine for the MR-native SAD method.
14. Crystal quality is frequently different, even if crystals were harvested from the same droplet. In the case of the EPIpYA complex, thin crystals were of better quality than thick ones. While thick crystals diffracted better than thin crystals, the twin fractions of thick crystals were close to 0.5 (perfect twinning), which was significantly larger than those of thin crystals. Since the diffraction data from a crystal with a large twin fraction usually cannot give a good result in crystallographic refinement, we selected thin crystals for crystallographic analysis.
15. The twin fraction of a crystal may differ depending on its position in the crystal. Therefore, a series of data collections

with a small X-ray beam sometimes shows positional differences in the twin fraction. This type of crystal property may cause trouble in the diffraction data collection using a helical scan. Since the irradiation point of the helical scan changes during the data collection, the twin fraction may also vary during data collection. This variation of the twin fraction causes problems with the data processing and scaling. The twin fraction may vary within the area of the X-ray beam even when collecting data using the standard method, but in many cases, the data collected can be processed and scaled better than those collected by the helical scan mode because the twin fraction is averaged within the X-ray beam area. Therefore, we recommend a standard data collection method when collecting data from twinned crystals.

16. As shown in Table 1, we found that crystals cryoprotected by PEG4000 diffracted less than those cryoprotected by other cryoprotectants. However, crystals diffracting at approximately 2 Å resolution showed large twin fractions (nearly perfect twinning) and did not provide good results for crystallographic refinement.

References

1. Kwok T, Zabler D, Urman S et al (2007) Helicobacter exploits integrin for type IV secretion and kinase activation. Nature 449:862–866. https://doi.org/10.1038/06187
2. Backert S, Meyer TF (2006) Type IV secretion systems and their effectors in bacterial pathogenesis. Curr Opin Microbiol 9:207–217. https://doi.org/10.1016/j.mib.2006.02.008
3. Higashi H, Tsutsumi R, Fujita A et al (2002) Biological activity of the Helicobacter pylori virulence factor CagA is determined by variation in the tyrosine phosphorylation sites. Proc Natl Acad Sci 99:14428–14433. https://doi.org/10.1073/pnas.222375399
4. Higashi H, Tsutsumi R, Muto S et al (2002) SHP-2 Tyrosine Phosphatase as an Intracellular Target of Helicobacter pylori CagA Protein. Science 295:683–686. https://doi.org/10.1126/science.1067147
5. Hayashi T, Senda M, Suzuki N et al (2017) Differential Mechanisms for SHP2 Binding and Activation Are Exploited by Geographically Distinct Helicobacter pylori CagA Oncoproteins. Cell Rep 20:2876–2890. https://doi.org/10.1016/j.celrep.2017.08.080
6. Senda M, Senda T (2017) Anaerobic crystallization of proteins. Biophys Rev 10:183–189. https://doi.org/10.1007/s12551-017-0345-8
7. Senda M, Senda T (2016) Advanced methods in structural biology. Springer protocols handbooks, pp 139–151. https://doi.org/10.1007/978-4-431-56030-2_8
8. Kabsch W (2010) XDS. Acta Crystallogr D Biol Crystallogr 66:125–132. https://doi.org/10.1107/s0907444909047337
9. Winn MD, Ballard CC, Cowtan KD et al (2011) Overview of the CCP4 suite and current developments. Acta Crystallogr Sect D Biol Crystallogr 67:235–242. https://doi.org/10.1107/s0907444910045749
10. Liebschner D, Afonine PV, Baker ML et al (2019) Macromolecular structure determination using X-rays, neutrons and electrons: recent developments in Phenix. Acta Crystallogr Sect D 75:861–877. https://doi.org/10.1107/s2059798319011471
11. Adams PD, Afonine PV, Bunkóczi G et al (2010) PHENIX: a comprehensive Python-based system for macromolecular structure solution. Acta Cryst D 66:213–221. https://doi.org/10.1107/s0907444909052925
12. Vagin A, Teplyakov A (1997) MOLREP: an Automated Program for Molecular Replacement. J Appl Crystallogr 30:1022–1025. https://doi.org/10.1107/s0021889897006766

13. McCoy AJ, Grosse-Kunstleve RW, Adams PD et al (2007) Phaser crystallographic software. J Appl Crystallogr 40:658–674. https://doi.org/10.1107/s0021889807021206
14. Emsley P, Lohkamp B, Scott WG, Cowtan K (2010) Features and development of Coot. Acta Cryst D 66:486–501. https://doi.org/10.1107/s0907444910007493
15. Senda M, Kishigami S, Kimura S, Senda T (2007) Crystallization and preliminary X-ray analysis of the reduced Rieske-type [2Fe–2S] ferredoxin derived from Pseudomonas sp. strain KKS102. Acta Crystallogr Sect F Struct Biol Cryst Commun 63:311–314. https://doi.org/10.1107/s1744309107009992
16. Senda M, Muto S, Horikoshi M, Senda T (2008) Effect of leucine-to-methionine substitutions on the diffraction quality of histone chaperone SET/TAF-Iβ/INHAT crystals. Acta Crystallogr Sect F Struct Biol Cryst Commun 64:960–965. https://doi.org/10.1107/s1744309108028704
17. Senda M, Hayashi T, Hatakeyama M et al (2016) Use of multiple cryoprotectants to improve diffraction quality from protein crystals. Cryst Growth Des 16:1565–1571. https://doi.org/10.1021/acs.cgd.5b01692
18. Hamdane D, Lechauve C, Marden MC, Golinelli-Pimpaneau B (2009) Pseudo-merohedral twinning in monoclinic crystals of wild-type human brain neuroglobin. Acta Crystallogr Sect D Biol Crystallogr 65:388–392. https://doi.org/10.1107/s0907444909003382
19. Jumper J, Evans R, Pritzel A et al (2021) Highly accurate protein structure prediction with AlphaFold. Nature 596:583–589. https://doi.org/10.1038/s41586-021-03819-2
20. Kato R, Hiraki M, Yamada Y et al (2021) A fully automated crystallization apparatus for small protein quantities. Acta Crystallogr Sect F Struct Biol Commun 77:29–36. https://doi.org/10.1107/s2053230x20015514

Chapter 4

Revealing Allostery in *PTPN11* SH2 Domains from MD Simulations

Massimiliano Anselmi and Jochen S. Hub

Abstract

Src-homology 2 (SH2) domains are protein interaction domains that bind to specific peptide motifs containing phosphotyrosine. SHP2, a tyrosine phosphatase encoded by *PTPN11* gene, which has been emerged as positive or negative modulator in multiple signaling pathways, contains two SH2 domains, respectively, called N-SH2 and C-SH2. These domains play a relevant role in regulating SHP2 activity, either by recognizing its binding partners or by blocking its catalytic site. Considering the multiple functions that these domains carry out in SHP2, N-SH2 and C-SH2 represent an interesting case of study. In this chapter, we present a methodology that permits, by means of the principal component analysis (PCA), to study and to rationalize the structures adopted by the SH2 domains, in terms of the conformations of their binding sites. The structures can be distinguished, grouped, classified, and reported in a diagram. This approach permits to identify the accessible conformations of the SH2 domains in different binding conditions and to eventually reveal allosteric interactions. The method further reveals that the conformation dynamics of N-SH2 and C-SH2 strongly differ, which likely reflects their distinct functional roles.

Key words Molecular dynamics simulations, Principal component analysis, SH2 domains, Phosphatases, Protein dynamics, Protein function, Allosteric regulation

1 Introduction

A cell needs to continuously respond to external and internal stimuli to perform its processes. For this purpose, the cell response is often mediated by covalent modifications of existing proteins. Several posttranslational modifications are found in nature, including phosphorylation and dephosphorylation of several amino acid residues, such as serine, threonine, and tyrosine. Posttranslational modifications are recognized and transduced into signal cascades via protein interaction domains, which often recognize short peptide motifs of their target protein. The interaction domains usually contain a conserved binding site with high affinity for the modified residue, along with a more variable cleft that, engaging the flanking

Teresa Carlomagno and Maja Köhn (eds.), *SH2 Domains: Functional Modules and Evolving Tools in Biology*, Methods in Molecular Biology, vol. 2705, https://doi.org/10.1007/978-1-0716-3393-9_4,

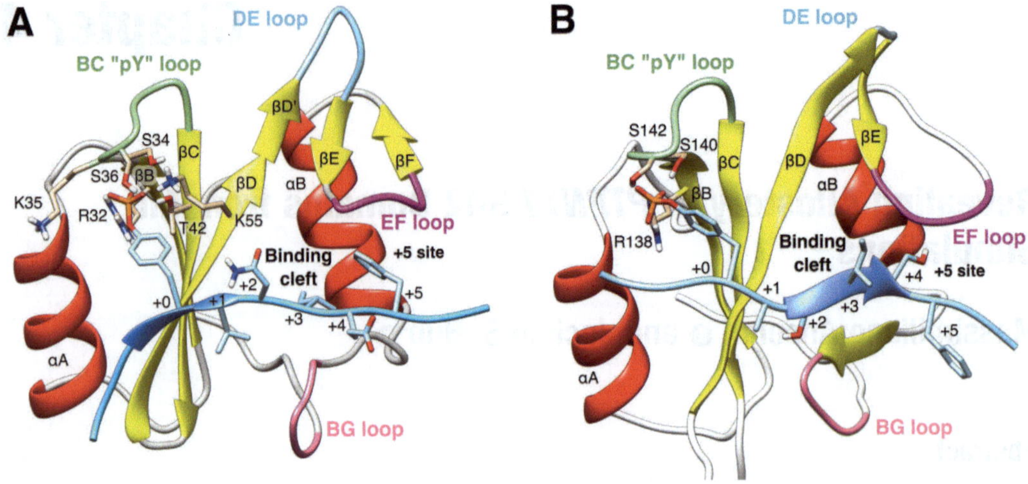

Fig. 1 Cartoon representations (**a**) of the N-SH2 domain in complex with the SPGEpYVNIEFGS (IRS-1 pY895) peptide (crystal structure, PDB ID 1AYB) [2] and (**b**) of the C-SH2 domain in complex with the LSApYASISFQK (IRS-1 pY1229) peptide, taken from an MD simulation. The peptide is shown in cyan, comprising the phosphotyrosine and residues from position +1 to +5 relative to the phosphotyrosine (see labels). Functionally important loops are highlighted in color: BC "pY" loop (green), DE loop (light blue), EF loop (magenta), and BG loop (deep pink). The phosphotyrosine binds the site delimited by the pY loop and the central β-sheet (βB, βC, βD strands). EF and BG loops delimit the "+5 site" where the peptide residue in position +5 is settled

residues, confers selectivity and allows to distinguish different peptide motifs containing the same posttranslational modification [1]. Protein tyrosine phosphorylation (PTP) is an important posttranslational modification that controls the cell signaling involved in a variety of biological processes, including cell growth, proliferation, differentiation, migration, survival, and apoptosis.

Src-homology 2 (SH2) domains are protein interaction domains of about 100 amino acids that bind to specific peptide motifs containing phosphotyrosine (pY). Atomic structures of a large number of SH2 domains have been determined. SH2 domains share a common fold consisting of a central antiparallel β-sheet flanked by two α-helices (Fig. 1). Phosphopeptides bind in extended conformation along a cleft that lies perpendicular to the central β-sheet. Conserved residues contribute to the hydrophobic core or are involved in pY recognition, while more variable residues contribute to specific recognition of the residues on the C-terminal side of the pY [3].

Among all SH2 domains, those belonging to SHP2, a tyrosine phosphatase encoded by *PTPN11* gene, represent a case of study with major physiological and medical impact. SHP2 is a regulator of signaling downstream of several cell surface receptors, functioning as positive or negative modulator in multiple signaling pathways. The structure of SHP2 includes two tandemly arranged SH2 domains, called N-SH2 (Fig. 1a) and C-SH2 (Fig. 1b), followed by the catalytic PTP domain, and a C-terminal tail. Under basal conditions, SHP2 is in an autoinhibited state where the N-SH2 domain occludes the catalytic site of the PTP domain. Association of SHP2 to its binding partners via the SH2 domains favors the release of the autoinhibitory N-SH2–PTP interactions, rendering the catalytic site accessible to substrates [4].

Thus, N-SH2 plays the twofold role (i) of blocking the catalytic site in the autoinhibited structure and (ii) of recognizing the binding partners before the activation takes place. For that reason, a coupling is required between the binding of a phosphopeptide to N-SH2 and the loss of affinity of N-SH2 for the PTP domain, which is likely realized via a series of conformational changes. On the other hand, C-SH2 recognizes one of the two phosphorylated motifs in BTAMs, namely, a kind of ubiquitous phosphorylated bisphosphoryl tyrosine-based activation motifs. BTAMs, which contain two phosphorylated motifs separated by a linker, activate SHP2 by binding to both N-SH2 and C-SH2. The second recognition site in C-SH2 confers a grade of conformational selection that favors, in presence of a linker of suitable length, the complete displacement of the N-SH2 domain from the PTP active site [4]. Hence, C-SH2 requires a pY recognition site for its function similar to N-SH2, but it may not require the allosteric mechanism that triggers the release of N-SH2 from the PTP domain upon the ligand binding.

In this chapter, we present a simple and effective methodology that permits, by means of the principal component analysis (PCA) [5], to study and to rationalize the structures adopted by the SH2 domains, in their apo and holo form. The conformational states are reported in an essential plane defined by two eigenvectors, respectively, describing the principal mode either of the affinity site or of the specificity site. The structures can be distinguished, grouped, and classified in terms of the conformations adopted by the affinity and specificity sites. This approach permits to report the conformational space explored by a SH2 domain in different binding conditions and to quickly compare conformational distributions. It is also possible to highlight the allosteric coupling between the affinity and specificity sites in a SH2 domain, or to compare the conformations adopted by two distinct SH2 domains, if the aminoacidic sequence is sufficiently conserved. Accordingly, a comparison between the N-SH2 domain and the C-SH2 domain of SHP2 suggests that the conformational rearrangement, put in place to accommodate a phosphopeptide, is more marked in N-SH2 than in C-SH2. We hypothesize that this difference is linked to the peculiar role of N-SH2 in SHP2 activation, namely, to recognize a binding partner and, at the same time, to control the phosphatase activity of SHP2.

2 Materials

1. Coordinates of the isolated N-SH2 and C-SH2 domains in solution, complexed with a set of high affinity phosphopeptides, with natural or synthetic sequence for the MD simulations. SLNpYIDLDLVK (IRS-1 pY1172) [6], IEEpYTEMMPAA (IRS-1 pY546) [7], QVEpYLDLDLD (Gab1 pY627) [8], SVLpYTAVQPNE (PDGFR-β pY1009) [7], AALNpYAQLMFP [9], RLNpYAQLWHR [9], SPGEpYVNIEFGS (IRS-1 pY895) [7], and VLpYMQPLNGRK [10] were complexed with the N-SH2 domain (Table 1), whereas HTEpYASIQTSP (SHPs-1 pY453) [11], FSEpYASVQVPRK (SHPs-1 pY496) [11], LSApYASISFQK (IRS-1 pY1229) [12], RVDpYVVVDQQK (Gab1 pY659) [13], and TMVpYGHLLQDS (CDCP1 pY743) [14] were complexed with the C-SH2 domain (Table 2). The initial atomic coordinates of N-SH2 domains were taken from crystallographic structures [2, 15]. The peptides bound to the N-SH2 domain were modeled by means of Molecular Operating Environment (MOE) [16] (Table 1). The structures of the C-SH2 domain bound to phosphopeptides were obtained by homology from the crystal structure of *PTPN11* tandem SH2 domains in complex with TXNIP peptide (PDB ID 5DF6) [17]. The peptides were docked to the C-SH2 binding cleft by CLUSPRO [18] (Table 2). N-SH2 domains comprised the residues from position 3 to 103. C-SH2 domains comprised the residues from position 109 to 217.

Table 1
Peptides bound to the N-SH2 domain

Simulated target sequence	Original sequence[a]	PDB ID	Target sequence source
Ac-SLNpYIDLDLVK-NH_2	GDK**QVE**pYLDLDL**D**	4QSY	IRS-1 (rat) 1169–1179
Ac-AALNpYAQLMFP-NH_2	**SVL**pY**TAVQP**NE	1AYA	Synthetic (Ref. [9])
Ac-SPGEpYVNIEFGS-NH_2	SPGEpYVNIEFGS	1AYB	IRS-1 (human) 892–903
Ac-VLpYMQPLNGRK-NH_2	SVLpY**TAVQPNE**	1AYA	Synthetic (Ref. [10])
Ac-IEEpYTEMMPAA-NH_2	**SVL**pYT**AVQ**P**NE**	1AYA	IRS-1 (human) 548–557
Ac-SVLpYTAVQPNE-NH_2	SVLpYTAVQPNE	1AYA	PDGFRB (human) 1006–1016
Ac-RLNpYAQLWHR-NH_2	RLNpYAQLWHR	3TL0	Synthetic (Ref. [9])
Ac-QVEpYLDLDLD-NH_2	GDKQVEpYLDLDLD	4QSY	GAB1 (human) 624–633

[a]The substituted residues are reported in bold. Missing residues are underlined

Table 2
Peptides bound to the C-SH2 domain

Simulated target sequence	PDB ID[a]	Target sequence source
Ac-HTEpYASIQTSP-NH_2	5DF6	SHPs-1 (human) 450–460
Ac-FSEpYASVQVPRK	5DF6	SHPs-1 (human) 493–504
Ac-LSApYASISFQK-NH_2	5DF6	IRS-1 (human) 1226–1236
Ac-RVDpYVVVDQQK-NH_2	5DF6	GAB1 (human) 656–666
Ac-TMVpYGHLLQDS-NH_2	5DF6	CDCP1 (human) 740–750

[a]Structure used as template for the homology modeling

2. GROMACS software package [19], using AMBER99SB force field [20], to perform MD simulations. Production simulations of 1 μs were performed. More details on the setup of MD simulations can be found in [21].

3 Methods

3.1 Theory

Identifying the motions of interest in a biomolecule, in order to uncover functional mechanisms, is a nontrivial task considering the complexity of the system. In molecular dynamics (MD) simulations, both the rapid local fluctuations and the slow collective motions occur simultaneously and in erratic manner, making it hard to distinguish functionally relevant motions from thermal noise. Covariance analysis, also called principal component analysis (PCA), is a useful tool to detect correlated motions in protein dynamics. The PCA assumes that the large-scale collective modes of motion dominate the functional dynamics, so that the majority of protein dynamics can be described by a low number of collective degrees of freedom. This approach has the advantage to identify the main modes of collective motions and to filter them from the noisier local fluctuations. Moreover, it is possible to easily describe complex structural rearrangement in the high-dimensional conformational space by means of a few collective variables [22].

After superposition to a common reference structure, a non-mass-weighted covariance matrix $\boldsymbol{C}$ of the atomic coordinates of N atoms is constructed:

$$C_{ij} = \langle (x_i - \langle x_i \rangle)(x_j - \langle x_j \rangle) \rangle$$

where $\langle \cdot \rangle$ denotes the average over all simulation frames. $\boldsymbol{C}$ is a symmetric $3N \times 3N$ matrix that can be diagonalized by an orthogonal coordinate transformation $\boldsymbol{R}$:

$$\boldsymbol{R}^T \boldsymbol{C} \boldsymbol{R} = \mathrm{diag}(\lambda_1, \lambda_2, \ldots, \lambda_{3N})$$

where $\lambda_1 \geq \lambda_2 \geq \ldots \geq \lambda_{3N}$

The eigenvalues λ_i correspond to the mean square fluctuation along the ith eigenvector and, therefore, represent the contribution of each principal component to the total fluctuations. The eigenvectors are sorted such that their eigenvalues are in decreasing order. Usually, six eigenvalues are exactly zero, since the corresponding eigenvectors describe the overall rotation and translation that are eliminated by the superposition. The columns of the orthogonal matrix $\boldsymbol{R}$ are the eigenvectors, also called principal or essential modes. Let $\boldsymbol{\xi}_i$ denote the ith eigenvector of $\boldsymbol{C}$ (the ith column of $\boldsymbol{R}$), and then the projection of the MD trajectory $\mathbf{x}(t)$ onto the ith principal mode gives the principal component p_i:

$$p_i(t) = \boldsymbol{\xi}_i \tilde{n}(\boldsymbol{x}(t) - \langle \boldsymbol{x} \rangle)$$

Finally, the trajectory can be filtered along one (or several) principal modes, thereby transforming the projections onto the eigenvectors back to Cartesian coordinates:

$$\boldsymbol{x}'_i(t) = p_i(t)\boldsymbol{\xi}_i + \langle \boldsymbol{x} \rangle$$

3.2 PCA of the Cumulative Trajectory

To characterize the conformations that the N-SH2 domain may adopt, apply principal component analysis (PCA) to the cumulative trajectory from all simulations of the N-SH2 domain bound to different phosphopeptides (Table 1) [21]. We assume that the simulation trajectories were previously converted so that the domain is whole in all frames. Since the phosphopeptides differ and the cumulative trajectory can include only a fixed number of ordered atom coordinates, consider only the part of the system shared among all simulations, namely, the N-SH2 domain. To this end, use the *trjconv* and *trjcat* modules of GROMACS to extract the N-SH2 domain coordinates from each trajectory and to concatenate the coordinates into a single cumulative trajectory:

```
$ gmx trjconv -f traj_1.xtc -n index_1.ndx -o
traj_N-SH2_only_1.xtc
$ gmx trjconv -f traj_2.xtc -n index_2.ndx -o
traj_N-SH2_only_2.xtc
...
$ gmx trjcat -f traj_N-SH2_only_*.xtc -o
cumulative_traj_N-SH2_only.xtc -settime
```

Then calculate and diagonalize the non-mass-weighted covariance matrix by the *covar* module:

```
$ gmx covar -f cumulative_traj_N-SH2_only.xtc -s
ref_structure_N-SH2.gro-n index.ndx
```

Build the covariance matrix using the backbone atoms from residues Trp[6] to Pro[101], excluding the flexible termini.

Superimpose the structures by a least squares fit onto the backbone considering only the residues with small root mean square fluctuation, representing the relatively rigid core of the domain. The core is defined by the sequence ranges Phe^{7}–Pro^{33}, Asp^{40}–Arg^{47}, Ala^{50}–Asn^{58}, Asp^{61}–Leu^{65}, Phe^{71}–Tyr^{81}, Leu^{88}–Glu^{90}, and Val^{95}–Pro^{101}. It is worth noting that the PCA eigenvectors do not necessarily represent the conformational transitions undergone by the N-SH2 domain during an individual simulation; instead, they describe the overall conformational space accessible to the N-SH2 domain, as collected from all simulations (*see* **Note 1**). Use the output file *eigenvec.trr* containing the eigenvectors for the following analysis.

Analyze the eigenvectors by the *anaeig* module:

```
$ gmx anaeig -f cumulative_traj_N-SH2_only.xtc -s
ref_structure_N-SH2.gro -v eigenvec.trr -first
1 -last 1 -proj proj.xvg -extr extreme.pdb nframes 30
```

In this example, only the first eigenvector is analyzed (options *-first* and *-last*), the trajectory of the projection onto the first eigenvector is saved in *proj.xvg*, while a trajectory of 30 frames joining the extreme configurations along the first eigenvector is saved in *extreme.pdb*.

3.3 Large-Scale Correlated Motion Revealed by Standard PCA

PCA suggests that the N-SH2 domain may adopt at least two main conformational states, hereafter called α and β [21]. In this example, these two states are resolved by the first PCA vector, which is representative of almost 40% of the overall domain fluctuations. The correlated structural rearrangements detected by the first PCA vector are visualized in Fig. 2a by means of the extreme projections onto the vector (*see* also Fig. 3a), and the related structural rearrangements are quantified in Fig. 2e–g (*see* **Note 2**), namely, the α-state shown in Fig. 2a (transparent) is characterized by (i) a closed pY loop (Lys^{35} Cα–Thr^{42} Cα distance ~9 Å); (ii) an increased distance between the ends of two β-strands, βC and βD, leading to breaking of three interstrand hydrogen bonds and spreading of the central β-sheet into a Y-shaped structure (Gly^{39} C–Asn^{58} N distance ~12 Å); and (iii) the closed +5 site with a narrow, less accessible cleft (Tyr^{66} Cα–Leu^{88} Cα distance ~7 Å; see also Fig. 3a, left). The β-state shown in Fig. 2a (opaque) is characterized by (i) an open pY loop (Lys^{35} Cα–Thr^{42} Cα distance ~11 Å), (ii) a closed central β-sheet with parallel β-strands (Gly^{39} C–Asn^{58} N distance ~4 Å), and (iii) an open +5 site with an accessible cleft (Tyr^{66} Cα–Leu^{88} Cα distance ~12 Å, *see* also Fig. 3a, right). The correlation analysis shows that the conformation adopted by the affinity binding site (pY loop) is strictly coupled to the spread of the central β-sheet (Fig. 2b, correlation $R = -0.82$) and strongly associated with the opening/closure of the specificity binding site (+5 site) (Fig. 2d, $R=0.64$) [21].

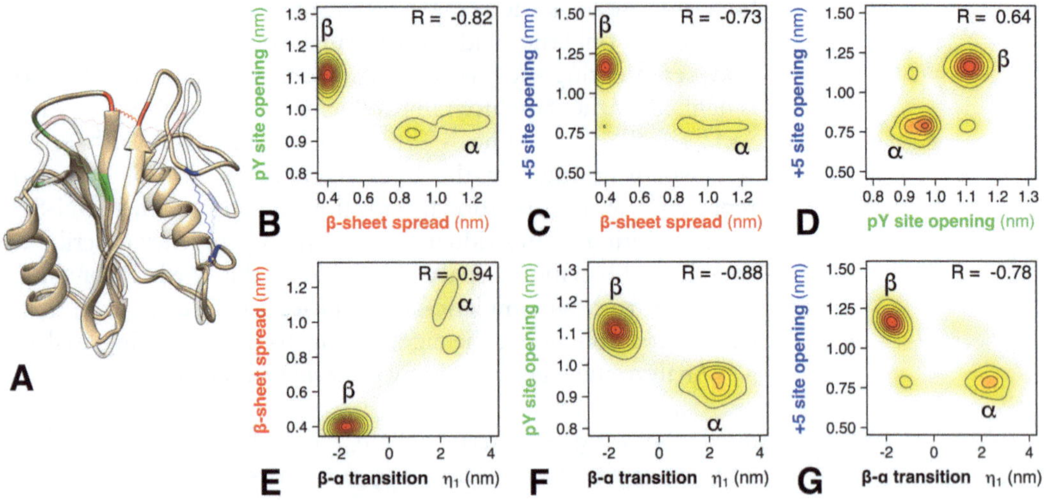

Fig. 2 Conformational states and correlations revealed by the first PCA vector. (**a**) Conformational transition from β (opaque) to α (transparent), representing the two main conformational states adopted by the N-SH2 domain, here visualized as the extreme projections onto the first PCA vector. The residues used to quantify the β-sheet spread, the pY loop opening, and the +5 site opening are highlighted in red, green, and blue, respectively. (**b**–**d**) Correlation between the β-sheet spread, pY loop opening, and +5 site opening, as taken from microsecond simulations of N-SH2 bound to eight different peptides. The distances were defined as described in the main text. (**e**–**g**) Correlation between the projection η_1 onto the first PCA vector and β-sheet spread, pY loop opening, and +5 site opening (*see* axis labels). Pearson correlation coefficients *R* are shown in each panel (**b**–**g**). (Reproduced from Scientific Reports 2020 with permission from Springer Nature [21])

3.4 Allosteric Coupling Highlighted by Decomposition of the PCA Vector into Structural Motifs

The covariance analysis on the cumulative trajectory suggests a coupling between the pY loop and the +5 site (*see* **Note 3**), revealing an allosteric interaction for controlling the state of the pY loop through binding of a specific amino acid sequences at the +5 site (Fig. 2d). To analyze the coupling during simulations with different peptides, split the PCA vector into two subvectors (comprising Ser^{34}–Phe^{41} and Gln^{57}–Glu^{97} backbone, respectively) that describe the motion either of the pY loop or of the +5 site.

Technically, these subvectors are generated by PCA considering only the respective subsets of atoms in a filtered trajectory which has been previously projected onto the full-length PCA vector $\boldsymbol{\xi}_1$. For this purpose, use the option *-filter* of the *anaeig* module, and then calculate and diagonalize the covariance matrix with the *covar* module for only the residues Ser^{34}–Phe^{41} or Gln^{57}–Glu^{97} (*see* **Note 4**):

```
$ gmx anaeig -f cumulative_traj_N-SH2_only.xtc -s
ref_structure_N-SH2.gro -v eigenvec.trr -first
1 -last 1 -filter cumulative_traj_filtered.xtc
```

```
$ gmx covar -f cumulative_traj_filtered.xtc -s re-
f_structure_N-SH2.gro -n index.ndx -o eigenvec_pY-
site.trr
$ gmx covar -f cumulative_traj_filtered.xtc -s re-
f_structure_N-SH2.gro -n index.ndx -o eigenvec_
+5site.trr
$ gmx anaeig -v eigenvec_pYsite.trr -f traj_N-
SH2_only_1.xtc -n index.ndx
-s ref_structure_N-SH2.gro -first 1 -last 1 -proj
proj_pYsite_1.xvg
$ gmx anaeig -v eigenvec_+5site.trr -f traj_N-
SH2_only_1.xtc -n index.ndx
-s ref_structure_N-SH2.gro -first 1 -last 1 -proj
proj_+5site_1.xvg
```

Finally, plot the projection onto the pY site subvector versus the projection onto the +5 site subvector using a combination of Unix commands, sed, awk, and paste:

```
$ paste <(sed '/^[#@&]/d' proj_pYsite_1.xvg | awk
'{print $2}') <(sed '/^[#@&]/d' proj_+5site_1.xvg |
awk '{print $2}') | column -t > 2dproj_pYsite_
+5site_1.xvg
```

In principle, projections on these two subvectors can characterize four conformational states given by the two possible states of the pY loop times two possible states of the +5 site (four shaded areas in each panel of Fig. 3b). The two additional states, relative to the α- and β-states characterized above, would be given by (i) a closed pY loop with an open +5 site (γ-state) and (ii) an open pY loop with a closed +5 site (δ-state) (*see* **Note 5**).

As reported in Fig. 3b, a correlation between the pY site and the +5 site is observed in all simulations, where most peptides strictly select for either the α- or for the β-state (Fig. 2b, blue/green areas), namely, when the pY site is closed, the +5 site is mostly closed (β-state); when the pY site is open, the +5 site is always open (α-state). Consequently, a conformational change at the pY site is accompanied by a concerted conformational change at the +5 site, spanning ~20 Å across the N-SH2 domain (*see* **Note 6**) [21].

The conformational equilibrium in the N-SH2 domain is determined by a balance of different energetic contributions: (i) the intermolecular interactions between the ligand and the N-SH2 domain and (ii) the intramolecular interactions between the residues of N-SH2. This can be rationalized for three typical binding modes using a diagram as reported in Fig. 4:

1. In absence of a ligand (Fig. 4a), the β-state represents the most populated conformational state, presumably favored by the

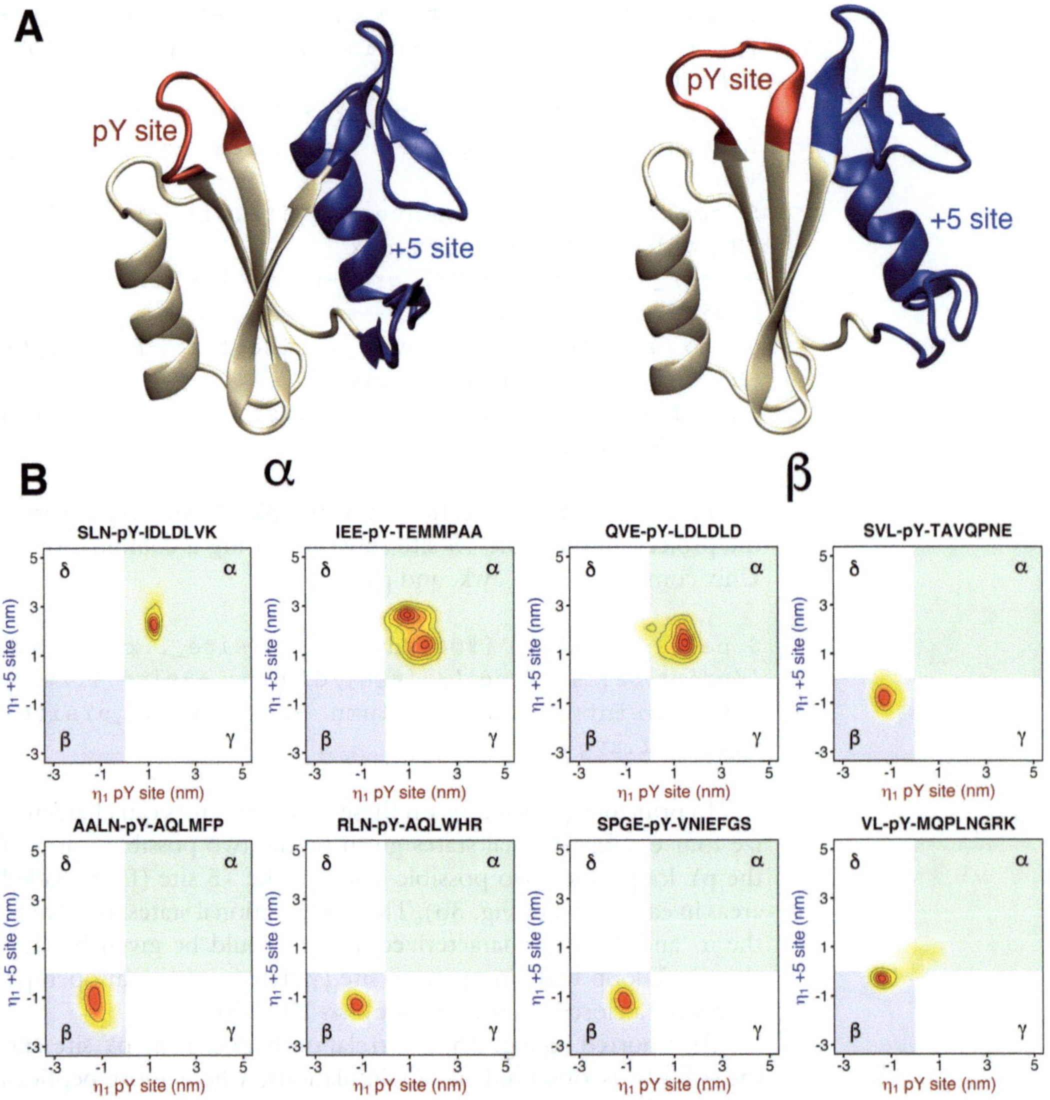

Fig. 3 (**a**) Cartoon representation of the two conformations, α (left) and β (right), adopted by the N-SH2 domain, and shown as the extreme projections onto the first PCA vector. From the first PCA vector, two subsectors were selected, describing the motions of the pY loop (Ser[34]–Phe[41], red cartoon) and of the +5 site (Gln[57]–Glu[97], blue cartoon), respectively. The residues used for defining the +5 site comprise the end of the βD strand, the DE loop, the EF loop, the αB helix, and the BG loop. (**b**) Projection of N-SH2 trajectories with different bound peptides (subplot titles) projected onto the PCA subvectors of the pY loop (abscissa) and of the +5 site (ordinate). The region corresponding to the α-state (pY loop closed, +5 site closed) is shaded in green, and the region of the β-state (pY loop open, +5 site open) is shaded in blue. The regions corresponding to the additional γ-state (pY loop closed, +5 site open) and δ-state (pY loop open, +5 site closed) are reported as white areas. (Reproduced from Scientific Reports 2020 with permission from Springer Nature [21])

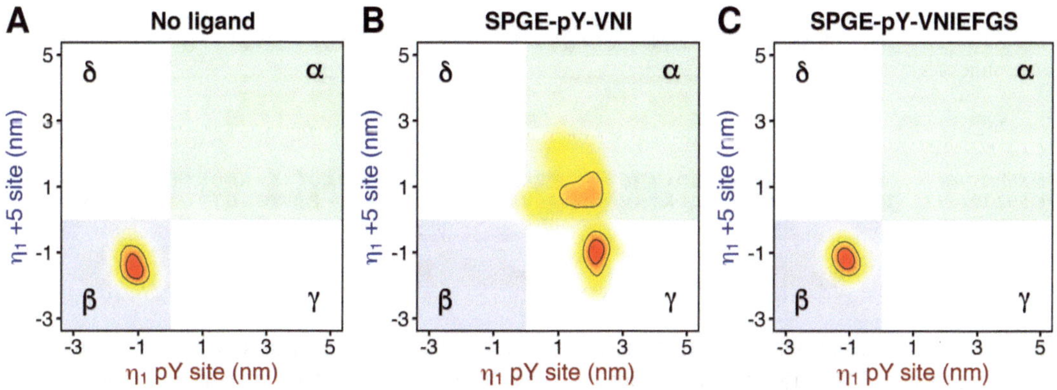

Fig. 4 Projection of the N-SH2 trajectories, either (**a**) unbound N-SH2 or (**b**–**c**) bound N-SH2 with peptides of different lengths (subplot titles), projected onto the PCA subvectors of the pY loop (abscissa) and of the +5 site (ordinate). The region corresponding to the α-state (pY loop closed, +5 site closed) is shaded in green, and the region of the β-state (pY loop open, +5 site open) is shaded in blue. The regions corresponding to the additional γ-state (pY loop closed, +5 site open) and δ-state (pY loop open, +5 site closed) are reported as white areas

formation of interstrand H-bonds between the closed central β-sheet.

2. In presence of a truncated ligand, such as SPGEpYVNI (Fig. 4b), which contains a phosphotyrosine but lacks the residues flanking the +5 site, the pY loop–phosphate interactions prevail over interstrand H-bonds of the central β-sheets, leading to closure of the pY loop and opening of the β-sheets. The α-state is one of the most populated conformations, along with the cognate γ-state. It is worth noting that, in absence of residues binding the specificity +5 site, the pY loop is stuck in closed conformation, while the conformation of the +5 site is formally decoupled, since it may be either closed (α-state) or open (γ-state). Therefore, the residues binding the specificity site play an essential role in determining the conformation of the N-SH2 domain. This is achieved by their electrostatic attraction or steric repulsion with the flexible loops lining the +5 site.
3. With a specific binding mode of the ligand SPGEpYVNIEFGS, the β-state results the most populated conformation (Fig. 4c). In this specific binding mode, a bulky phenylalanine at position +5 occupies the cleft of the +5 site, thereby keeping the +5 site open. This conformation stabilizes the central β-sheet, aiding the interstrand H-bonds of the β-sheet to prevail over the pY loop–phosphate H-bonds. Consequently, β-sheets close and pY loop opens.

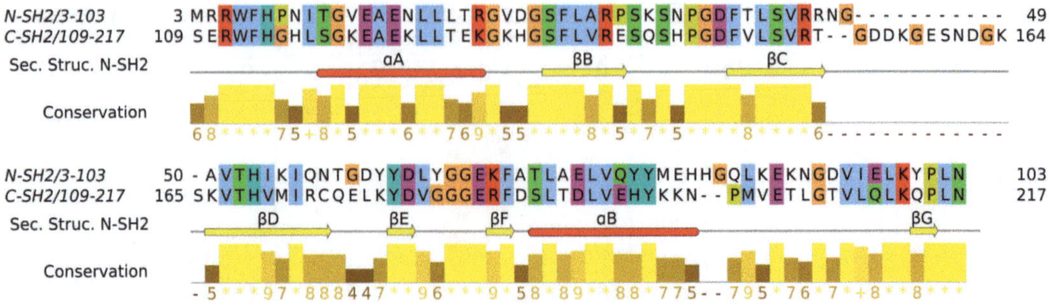

Fig. 5 Sequence alignment after structure superposition of the C-SH2 domain structure over the N-SH2 domain structure. The conserved amino acids are highlighted according to their properties and the following color scheme: hydrophobic, aliphatic (blue); basic, positively charged (red); acidic, negatively charged (magenta); polar (green); aromatic (cyan); glycines (orange); and prolines (yellow). Conservation is visualized on the alignment as a histogram giving the score for each column. The score measures the number of conserved physicochemical properties conserved for each column of the alignment. Score calculation is derived from the AMAS method of multiple sequence alignment analysis [23]. Fully conserved columns are indicated by "*" (score of 11, grouping all default amino acid properties). Columns with mutations, where all properties are conserved and are marked with a "+" (score of 10, indicating all properties are conserved)

3.5 Similarities and Differences in Dynamics Between C-SH2 and N-SH2

Can we use the eigenvectors obtained from the cumulative trajectory of the N-SH2 domain to analyze and to compare the conformations adopted by another SH2 domain such as C-SH2? This mostly depends on the structural homology and on the sequence conservation between the two SH2 domains. If possible, it represents a unique chance to disclose similarities and differences between the conformations adopted by the two SH2 domains of SHP2.

Figure 5 presents the sequence alignment after structural superposition of the C-SH2 domain over the N-SH2 domain. The two sequences are characterized by a high sequence conservation. The major difference is found in the length of the loop connecting the βC and βD strands, which is irrelevant for the PCA since it is part neither of the pY loop nor of the +5 site. The pY loop (Ser^{34}–Phe^{41} for N-SH2, Ser^{140}–Phe^{147} for C-SH2) is well conserved and of the same length in the two SH2 domains (8 a.a.), therefore the corresponding eigenvector may be used without modifications. On the other hand, the +5 site (Gln^{57}–Glu^{97} for N-SH2, Arg^{173}–Gln^{211} for C-SH2) is less conserved, as expected for a specificity site. The +5 site sequence in C-SH2 is shorter (39 a. a. vs. 41 a.a.) and misses two residues, corresponding to His^{85} and Gly^{86} in N-SH2. Therefore, it is not possible to bijectively map the residues of the +5 site in N-SH2 over the residues of the +5 site in C-SH2. However, all the residues of the +5 site in C-SH2 can be mapped over the corresponding residues in N-SH2, and therefore a comparison is possible only if the residues His^{85} and Gly^{86} are neglected in +5 site definition.

First to proceed with the PCA of C-SH2, create a reference structure of C-SH2, which overlaps to the reference structure of N-SH2 used previously. This task is performed by the *confrms* module that carries out the least squares fit using two corresponding groups of atoms and does not require that the two structures have the same number of atoms. Therefore, mapping the N-SH2 core, previously used, over C-SH2, obtain the following sequence ranges: Phe^{113}–Glu^{139}, Asp^{146}–Thr^{153}, Lys^{166}–Cys^{174}, Leu^{177}–Val^{181}, Phe^{187}–Tyr^{197}, Met^{202}–Glu^{204}, and Val^{209}–Pro^{215}:

```
$ gmx confrms -f1 ref_structure_N-SH2.gro -n1 in-
dex_N-SH2 -f2 any_structure_C-SH2.gro -n2 index_C-
SH2 -one -o ref_structure_C-SH2.gro
```

Then, extract from the filtered cumulative trajectory of N-SH2 the sub-eigenvector for the +5 site according the new definition, which excludes only the residues His^{85} and Gly^{86}:

```
$ gmx covar -f cumulative_traj_filtered.xtc -s re-
f_structure_N-SH2.gro -n index.ndx -o new_eigen-
vec_5site.trr
```

Finally, calculate the projections onto the subvectors for any trajectory of C-SH2 (*see* **Note 7**):

```
$ gmx anaeig -v eigenvec_pYsite.trr -s
ref_structure_C-SH2.gro -n index.ndx -f
any_traj_C-SH2.xtc -first 1 -last 1 -proj proj_pY-
site.xvg
$ gmx anaeig -v new_eigenvec_+5site.trr -s
ref_structure_C-SH2.gro -n index.ndx -f
any_traj_C-SH2.xtc -first 1 -last 1 -proj proj_
+5site.xvg
```

Again, plot the projection onto the pY site subvector versus the projection onto the +5 site subvector:

```
$ paste <(sed '/^[#@&]/d' proj_pYsite.xvg | awk
'{print $2}') <(sed '/^[#@&]/d' proj_+5site.xvg |
awk '{print $2}') | column -t > 2dproj_pYsite_
+5site.xvg
```

In absence of a ligand, the C-SH2 domain adopts a β-like conformation, with both the pY loop and the +5 site in open conformation (Fig. 6a). This is in line with that observed in the N-SH2 domain (Fig. 4a). However, in presence of a full-length phosphopeptide (Fig. 6b–f), the C-SH2 domain adopts a γ-like conformation, with the pY loop closed and the +5 site still open.

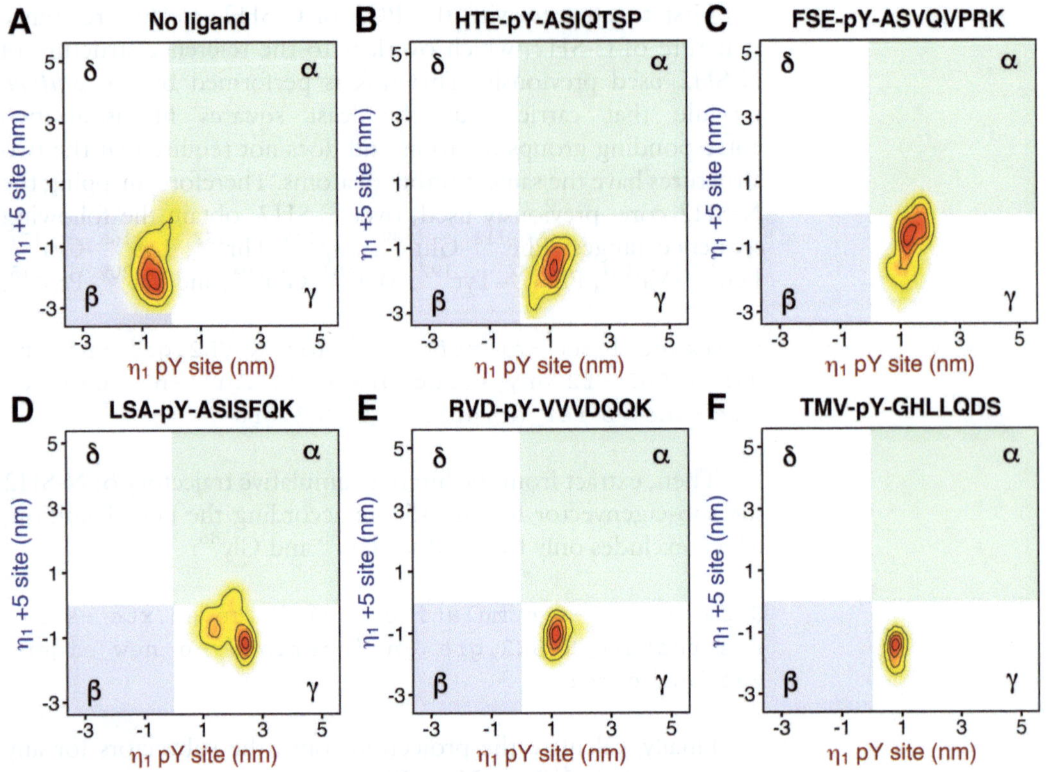

Fig. 6 Projection of C-SH2 trajectories, either (**a**) unbound C-SH2 or (**b**–**f**) bound C-SH2 with different peptides (subplot titles) projected onto the PCA subvectors of the pY loop (abscissa) and of the +5 site (ordinate). The region corresponding to the α-state (pY loop closed, +5 site closed) is shaded in green, and the region of the β-state (pY loop open, +5 site open) is shaded in blue. The regions corresponding to the additional γ-state (pY loop closed, +5 site open) and δ-state (pY loop open, +5 site closed) are reported as white areas

In case of the C-SH2 domain, a coupling between the affinity site and the specificity site is not observed, and the region of the domain lining the +5 site does not undergo any significant conformational change to accommodate the phosphopeptide. This behavior is in contrast to our findings for the N-SH2 domain, as discussed above (*see* **Note 8**).

4 Notes

1. More commonly, principal component analysis is performed on equilibrated trajectories of a biomolecule in a given state (*i.e.*, binding, redox, and folding). In that case, the principal modes of motion are representative of the largest structural rearrangements that the biomolecule undergoes during the simulations, and they provide information on the dynamics and the structural plasticity of the biomolecule in a particular condition.

Alternatively, the analysis can be performed on a cumulative trajectory encompassing several distinct states of the same biomolecule. Critically, the cumulative trajectory would contain only the part of the system shared among all original trajectories. This time, the principal modes of motion do not describe only the conformational transitions undergone by the biomolecule in a particular state or simulation; instead, they describe the overall conformational space accessible to the biomolecule, as collected from all simulations, namely, the principal modes of motion consider, besides the dynamics of the system in every state, even the conformational transitions that the biomolecule may undergo for switching from a state to another, even if these transitions were not observed in any individual trajectory. Hence, such analysis reveals conformational transitions between distinct states of a biomolecules, and it indicates how to describe such transitions by molecular structural features, such as interatomic distances or angles, which can be utilized in following analyses or simulations.

2. The trajectory joining the extreme configurations can be used to systematically search the interatomic distances that best linearly correlate with the projection onto the eigenvector.
3. This approach permits to overcome the problem of poor sampling that affect the individual trajectories and to detect an allosteric interaction between the affinity site and the specificity site of N-SH2.
4. It is important to perform the superposition using the same reference structure and the same reference group used in the previous covariance matrix calculation on the non-filtered cumulative trajectory. The covariance analyses performed on the filtered trajectory (i.e., projected onto the original full eigenvector) will generate every time only one eigenvector with a nonzero eigenvalue. Any first eigenvector will retain the normalized components of the original full eigenvector, corresponding to the group of atom coordinates used for the analysis.
5. The eigenvector describing a coupling between two sites can be always split into the subvectors representing each site. We can combine these subvectors to generate a symmetric configuration (where the two sites are correlated like in the original eigenvector) and an antisymmetric configuration (where the two sites are anticorrelated) and use the plane of the two subvectors to represent the correlated and anticorrelated motions. Such analysis is, to our knowledge, still rare in the literature. The advantage of this procedure, compared the use of a projection onto a single eigenvector, is that we can quickly

visualize the component representing the deviation from a perfect correlation between the two sites, which would be otherwise orthogonal to the original eigenvector.

6. Critically, it is unlikely that individual simulations explore all possible binding modes of a phosphopeptide within 1 μs of simulation time because the interconversion between binding modes would first involve a weakening of peptide–N-SH2 interactions, before the required side chain and backbone reorientations could occur. Therefore, each individual simulation likely showed only a partial view on the accessible binding modes of each peptide, and probably in every simulation the N-SH2 domain is mostly stuck in one of the possible conformations. Consequently, such equilibrium MD simulations, taken individually, are insufficient for deciding whether a specific peptide sequence selects for the α- or the β-state and for estimating an equilibrium constant between the two conformational states. However, the individual simulations taken together are representative of the N-SH2 domain in bound form to reveal a correlation between the affinity site and the specificity site.
7. Since the subvectors were obtained from the eigenvector of N-SH2, they are not strictly representative of the collective modes of motion in C-SH2. However, this procedure permits a direct comparison in order to test if the collective motions in N-SH2 are conserved in C-SH2, thereby revealing similarities and differences between the N-SH2 and C-SH2 dynamics.
8. This application confirms that PCA can be used to compare dynamics between homologous domains, not just SH2, evidencing similarities and differences.

References

1. Seet BT, Dikic I, Zhou M-M, Pawson T (2006) Reading protein modifications with interaction domains. Nat Rev Mol Cell Biol 7(7): 473–483. https://doi.org/10.1038/nrm1960
2. Lee CH, Kominos D, Jacques S et al (1994) Crystal structures of peptide complexes of the amino-terminal SH2 domain of the Syp tyrosine phosphatase. Structure 2(5):423–438. https://doi.org/10.1016/S0969-2126(00)00044-7
3. Waksman G, Kumaran S, Lubman O (2004) SH2 domains: role, structure and implications for molecular medicine. Expert Rev Mol Med 6(3):1–18. https://doi.org/10.1017/S1462399404007331
4. Poole AW, Jones ML (2005) A SHPing tale: perspectives on the regulation of SHP-1 and SHP-2 tyrosine phosphatases by the C-terminal tail. Cell Signal 17(11): 1323–1332. https://doi.org/10.1016/j.cellsig.2005.05.016
5. Amadei A, Linssen AB, Berendsen HJC (1993) Essential dynamics of proteins. Proteins 17(4): 412–425. https://doi.org/10.1002/prot.340170408
6. Sugimoto S, Wandless TJ, Shoelson SE et al (1994) Activation of the SH2-containing protein tyrosine phosphatase, SH-PTP2, by phosphotyrosine-containing peptides derived from insulin receptor substrate-1. J Biol Chem 269(18):13614–13622
7. Case RD, Piccione E, Wolf G et al (1994) SH-PTP2/Syp SH2 domain binding specificity is defined by direct interactions with platelet-derived growth factor beta-receptor, epidermal

growth factor receptor, and insulin receptor substrate-1-derived phosphopeptides. J Biol Chem 269(14):10467–10474

8. Koncz G, Tóth GK, Bökönyi G et al (2001) Co-clustering of Fcγ and B cell receptors induces dephosphorylation of the Grb2-associated binder 1 docking protein. Eur J Biochem 268(14):3898–3906. https://doi.org/10.1046/j.1432-1327.2001.02295.x
9. Imhof D, Wavreille A-S, May A et al (2006) Sequence specificity of SHP-1 and SHP-2 Src homology 2 domains: critical roles of residues beyond the pY+3 position. J Biol Chem 281(29):20271–20282. https://doi.org/10.1074/jbc.M601047200
10. Sweeney MC, Wavreille A-S, Park J et al (2005) Decoding protein–protein interactions through combinatorial chemistry: sequence specificity of SHP-1, SHP-2, and SHIP SH2 domains. Biochemistry 44(45):14932–14947. https://doi.org/10.1021/bi051408h
11. Fujioka Y, Matozaki T, Noguchi T et al (1996) A novel membrane glycoprotein, SHPS-1, that binds the SH2-domain-containing protein tyrosine phosphatase SHP-2 in response to mitogens and cell adhesion. Mol Cell Biol 16(12):6887–6899. https://doi.org/10.1128/MCB.16.12.6887
12. Pluskey S, Wandless TJ, Walsh CT, Shoelson SE (1995) Potent stimulation of SH-PTP2 phosphatase activity by simultaneous occupancy of both SH2 domains. J Biol Chem 270(7):2897–2900. https://doi.org/10.1074/jbc.270.7.2897
13. Cunnick JM, Mei L, Doupnik CA, Wu J (2001) Phosphotyrosines 627 and 659 of Gab1 constitute a bisphosphoryl tyrosine-based activation motif (BTAM) conferring binding and activation of SHP2. J Biol Chem 276(26):24380–24387. https://doi.org/10.1074/jbc.M010275200
14. Gandji LY, Proust R, Larue L, Gesbert F (2015) The tyrosine phosphatase SHP2 associates with CUB domain-containing protein-1 (CDCP1), regulating its expression at the cell surface in a phosphorylation-dependent manner. PLoS One 10(4):e0123472. https://doi.org/10.1371/journal.pone.0123472
15. Zhang Y, Zhang J, Yuan C et al (2011) Simultaneous binding of two peptidyl ligands by a Src homology 2 domain. Biochemistry 50(35):7637–7646. https://doi.org/10.1021/bi200439v
16. Molecular Operating Environment (MOE) Software (2014) Chemical Computing Group Inc.
17. Liu Y, Lau J, Li W et al (2016) Structural basis for the regulatory role of the PPxY motifs in the thioredoxin-interacting protein TXNIP. Biochem J 473(2):179–187. https://doi.org/10.1042/BJ20150830
18. Kozakov D, Hall DR, Xia B et al (2017) The ClusPro web server for protein-protein docking. Nat Protoc 12(2):255–278. https://doi.org/10.1038/nprot.2016.169
19. Van Der Spoel D, Lindahl E, Hess B et al (2005) GROMACS: fast, flexible, and free. J Comput Chem 26(16):1701–1718. https://doi.org/10.1002/jcc.20291
20. Hornak V, Abel R, Okur A et al (2006) Comparison of multiple Amber force fields and development of improved protein backbone parameters. Proteins 65(3):712–725. https://doi.org/10.1002/prot.21123
21. Anselmi M, Hub JS (2020) An allosteric interaction controls the activation mechanism of SHP2 tyrosine phosphatase. Sci Rep 10(1):18530. https://doi.org/10.1038/s41598-020-75409-7
22. Hayward S, de Groot BL (2008) Normal modes and essential dynamics. In: Kukol A (ed) Molecular modeling of proteins. Humana Press, Totowa, pp 89–106. https://doi.org/10.1007/978-1-59745-177-2_5
23. Livingstone CD, Barton GJ (1993) Protein sequence alignments: a strategy for the hierarchical analysis of residue conservation. Bioinformatics 9(6):745–756. https://doi.org/10.1093/bioinformatics/9.6.745

Chapter 5

Structure Determination of SH2–Phosphopeptide Complexes by X-Ray Crystallography: The Example of p120RasGAP

Amy L. Stiegler and Titus J. Boggon

Abstract

The p120RasGAP protein contains two Src homology 2 (SH2) domains, each with phosphotyrosine-binding activity. We describe the crystallization of the isolated and purified p120RasGAP SH2 domains with phosphopeptides derived from a binding partner protein, p190RhoGAP. Purified recombinant SH2 domain protein is mixed with synthetic phosphopeptide at a stoichiometric ratio to form the complex in vitro. Crystallization is then achieved by the hanging drop vapor diffusion method over specific reservoir solutions that yield single macromolecular co-crystals containing SH2 domain protein and phosphopeptide. This protocol yields suitable crystals for X-ray diffraction studies, and our recent X-ray crystallography studies of the two SH2 domains of p120RasGAP demonstrate that the N-terminal SH2 domain binds phosphopeptide in a canonical interaction. In contrast, the C-terminal SH2 domain binds phosphopeptide via a unique atypical binding mode. The crystallographic studies for p120RasGAP illustrate that although the three-dimensional structure of SH2 domains and the molecular details of their binding to phosphotyrosine peptides are well defined, careful structural analysis can continue to yield new molecular-level insights.

Key words SH2 domain, Phosphopeptide, Phosphotyrosine, Crystallization, X-ray crystallography

1 Introduction

The Src homology 2 (SH2) domain is one of the best studied and well-characterized modular protein signaling domains [1]. The SH2 domain was first identified as a conserved non-catalytic region in Src and other nonreceptor tyrosine kinases [2] and is well represented in the human genome, with approximately 120 SH2 domains in 111 human proteins [3]. The central function of the SH2 domain is to bind phosphotyrosine residues in target proteins, thereby regulating both recruitment and enzymatic activity in tyrosine kinase signaling cascades.

The three-dimensional structure of the SH2 domain is well described through a combination of many structural

Teresa Carlomagno and Maja Köhn (eds.), *SH2 Domains: Functional Modules and Evolving Tools in Biology*, Methods in Molecular Biology, vol. 2705, https://doi.org/10.1007/978-1-0716-3393-9_5,

determinations by both X-ray crystallography and NMR spectroscopy [4]. The SH2 domain is approximately 100 residues in length and adopts a conserved overall fold consisting of a central beta sheet flanked by two alpha helices [4–6]. The phosphotyrosine peptide binds in a two-pronged manner to two binding pockets on either side of the central beta sheet: the phosphotyrosine site and the specificity site [7, 8]. In canonical SH2 domains, the positively charged phosphotyrosine site contains a highly conserved arginine residue in a FLVR sequence (Phe-Leu-Val-Arg) that directly binds the phosphotyrosine and contributes a majority of the binding free energy [9]. The specificity pocket binds the peptide residue at the +3 position to the phosphotyorosine (defined as the 0 position), and the characteristics of the residues lining this pocket establish the binding specificity.

Co-crystal structures of isolated SH2 domains with phosphopeptide helped establish the two-pronged binding mode and specificity determinants [4–6]. Typically, studies utilize synthetic phosphopeptides derived from target protein sequences of approximately 4–10 residues in length, beginning at residues −3 to −1 from the 0 position phosphotyrosine and extending to the +3 or as long as the +5 position. Dissociation constant (K_d) values range from 0.1 to 10 μM [10]. Thus, the achievable in vitro concentrations of recombinantly expressed and purified SH2 domain protein (0.1 mM or higher) and phosphopeptide (~1 mM or higher) are amenable to complex formation and co-crystallization. Additionally, recombinant isolated SH2 domain proteins can often be easily overexpressed in bacterial systems and purified to high quality necessary for crystallization trials.

One of the earliest identified SH2 domain-containing proteins is p120RasGAP (*RASA1*). p120RasGAP was the first GTPase-activating protein (GAP) identified for the Ras small GTPases and the only one to contain Src homology domains, namely, two SH2 domains and one SH3 domain [11–13] (Fig. 1a). Additionally, p120RasGAP contains pleckstrin homology (PH) and C2 domains which may aid membrane binding and a C-terminal catalytic GAP domain (Fig. 1a). p120RasGAP enhances the intrinsic GTP hydrolysis activity of Ras GTPases by two to five orders of magnitude and promotes the GDP-bound form of Ras to turn off signaling [11, 14]. This is achieved by p120RasGAP binding directly to Ras-GTP and contributing a so-called "arginine finger" residue to the active site to stabilize the reaction intermediate and promote catalysis. Thus, p120RasGAP is an essential factor in the Ras GTPase cycle and a key regulator of Ras signaling.

Phosphotyrosine-containing binding partners for p120RasGAP SH2 domains have been identified. They include the Rho GTPase-activating protein p190RhoGAP [15, 16]; the adaptor proteins p62Dok, p65Dok [17, 18] and SH2 domain-containing adapter protein B (SHB) [19]; and receptor tyrosine kinases

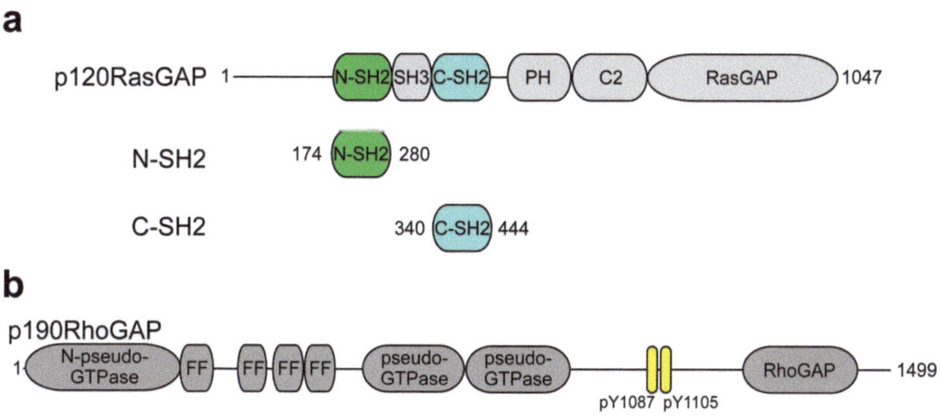

Fig. 1 Domain architecture of p120RasGAP and p190RhoGAP. (**a**) p120RasGAP (UniProt ID: P20936) contains a predicted unstructured N-terminus, N-terminal Src homology 2 (N-SH2) domain, Src homology 3 (SH3) domain, C-terminal Src homology 2 (C-SH2) domain, pleckstrin homology (PH) domain, protein kinase C conserved 2 (C2) domain, and Ras GTPase-activating protein (GAP) domain. The isolated N-SH2 (residues 174–280) and C-SH2 (340–444) domains are used for co-crystallization experiments with p190RhoGAP phosphopeptides. (**b**) p190RhoGAP (UniProt ID: Q9NRY4) contains N-pseudoGTPase, FF, pseudoGTPase, and RhoGAP domains

including the platelet-derived growth factor (PDGF) receptor [20, 21] and the ephrin type B (EphB) receptors [22]. Several binding partners contain multiple sites of tyrosine phosphorylation, and simultaneous engagement of the two SH2 domains to the same protein target has been postulated [23].

p190RhoGAP is a dual tyrosine-phosphorylated GAP protein for the Rho subfamily of GTPases and helps regulate the actin cytoskeleton. Like p120RasGAP, p190RhoGAP is a multidomain protein that, in addition to its catalytic GAP domain, contains other modular protein domains which may play a role in its recruitment and/or activation [24, 25] (Fig. 1b). Phosphorylation of tyrosine residues in the unstructured "middle domain" (Fig. 1b) by kinases such as Src and Abl has been shown to promote interaction with the SH2 domains of p120RasGAP [16, 26–28]. As p120RasGAP is believed to be membrane associated, formation of this complex leads to recruitment of p190RhoGAP to the membrane where it is in closer proximity to Rho GTPases and can regulate Rho signaling [29]. Therefore, understanding the molecular details of the interaction between p120RasGAP SH2 domains and p190RhoGAP phosphotyrosines has helped establish the structural requirements for this important signaling complex.

The three-dimensional structures of the individual p120RasGAP SH2 domains have been determined, both alone and in complex with phosphotyrosine peptides derived from the phosphotyrosine sites of p190RhoGAP [30, 31] (Fig. 2a–d). These studies revealed that both domains adopt the canonical SH2 domain fold,

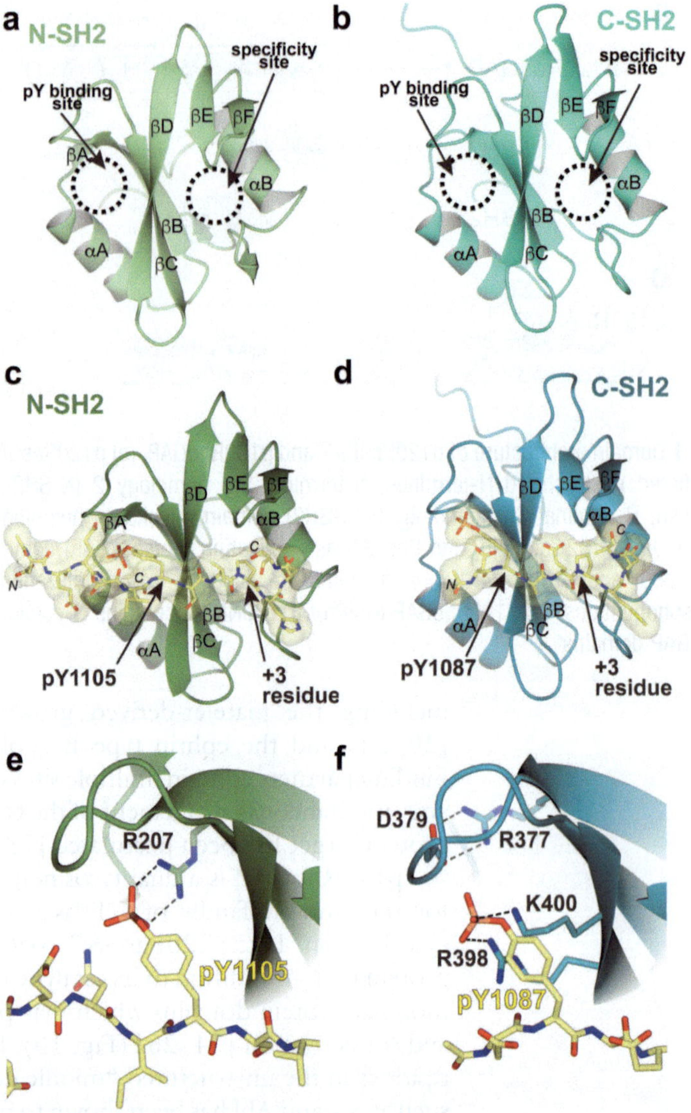

Fig. 2 Ribbon diagram of crystal structures of p120RasGAP SH2 domains. (**a**) N-SH2 apo crystal structure (PDB ID: 6PXB [31]). (**b**) Crystal structure of apo C-SH2 domain (PDB ID: 6WAX [30]). (**c**) Co-crystal structure of N-SH2 bound to p190RhoGAP pY1105-containing phosphopeptide (PDB ID: 6PXC [31]). (**d**) Co-crystal structure of C-SH2 bound to p190RhoGAP pY1087-containing phosphopeptide (PDB ID: 6WAY [30]). In all panels, secondary structure elements are labeled. In (**b**) and (**d**), the phosphopeptides are drawn as all atom cylinders and transparent surfaces in yellow, and the amino- ("N") and carboxy termini ("C") of the peptides are labeled. The positions of the phosphotyrosine and +3 residues are labeled and shown with arrows. (**e**) The FLVR arginine Arg-207 of N-SH2 bound directly to pY-1105. (**f**) FLVR-unique binding of pY-1087 to C-SH2. The FLVR arginine R377 is bound to D379; the phosphotyrosine is bound by R398 and K400 in the phosphotyrosine-binding site. All illustrations are rendered in CCP4mg [34]

comprised of a central beta sheet flanked by two alpha helices (Fig. 2a, b). The phosphotyrosine and specificity sites are located on either side of the long central beta strand βD (using canonical SH2 domain secondary structure nomenclature) and together create the two-pronged binding mode for phosphotyrosine-containing peptides. In detail, the crystal structures reveal a canonical binding mode for the N-SH2 in complex with a pTyr-1105 peptide in which the FLVR arginine, Arg-188, directly binds the phosphotyrosine, while the specificity pocket binds the preferred +3 Pro-1108 (Fig. 2c, e). In contrast, the structure of the C-SH2 in complex with a pTyr-1087 peptide determined by X-ray crystallography revealed atypical binding interactions, in which the FLVR arginine, Arg-377, does not directly bind the phosphotyrosine but rather is engaged with an intramolecular interaction with Asp-379 (Fig. 2d, f). Nonetheless, the position of the phosphotyrosine-binding pocket is conserved and is instead lined with alternative residues including Arg-398 and Lys-400 to directly bind the phosphotyrosine (Fig. 2f). Similar to N-SH2, the specificity pocket of C-SH2 reveals the structural basis for specificity for Pro-1090 at +3. Taken together, these structural studies reflect the importance of experimentally observing molecular-level interactions between SH2 domains and their target peptides. In this chapter, we detail the procedure for co-crystallization of these protein–peptide complexes by hanging drop vapor diffusion. The general procedures for co-crystallization can be applied to other SH2–phosphopeptide pairs.

2 Materials

2.1 SH2 Domain Proteins

1. Purified recombinant p120RasGAP N-SH2 domain and C-SH2 domain proteins.
2. Protein storage buffer: 20 mM Tris–HCl pH 8.0, 150 mM NaCl.
3. Centrifugal filters with nominal molecular weight limit (NMWL) 3 kilodalton. We use Amicon Ultra-4 Centrifugal Filters (Sigma Millipore).
4. Refrigerated centrifuge that fits 15 mL conical tube equivalent.
5. Spectrophotometer to measure protein concentration. We use NanoDrop (Thermofisher).
6. 18% SDS-PAGE gel, Tris–glycine running buffer.
7. Coomassie blue staining solution: 1 g Coomassie brilliant blue, 40% ethanol, 10% acetic acid.
8. Refrigerated microcentrifuge.
9. Microcentrifuge tubes, 1.5–1.7 mL capacity.

2.2 Phosphopeptide

1. Lyophilized phosphopeptides, commercially synthesized and HPLC purified to greater than 98%. The peptides are modified at the amino and carboxy termini with acetyl and amide groups, respectively, to neutralize charge and improve stability. In this example, we use pTyr-1105, EEENI(pY)SVPHDST (p190RhoGAP residues 1100–1112, UniProt Q9NRY4) and pTyr-1087, DpYAEPMD (p190RhoGAP residues 1086–1092, UniProt Q9NRY4).
2. Peptide reconstitution buffer: 10 mM Tris–HCl pH 7.4.

2.3 Crystallization

1. Crystallization plate with 18 mm well diameter. We use VDXm™ crystallization plate with sealant.
2. Plastic coverslips, 18 mm square or circle.
3. Tweezers.
4. 70% ethanol and lint-free labwipe or lens paper.
5. Reservoir stock solutions for apo N-SH2 crystals.
   (a) 50% (w/v) polyethylene glycol (PEG) 10,000.
   (b) 1 M ammonium acetate.
   (c) 1 M Tris–HCl pH 8.0, pH adjusted with HCl.
6. Reservoir stock solutions for N-SH2–pY-1105 co-crystals.
   (a) 1 M Bis–Tris pH 6.5, pH adjusted with HCl.
   (b) 100% Polyethylene glycol monomethyl ether 550 (PEG MME 550).
7. Reservoir stock solutions for apo C-SH2 crystals.
   (a) 4 M ammonium sulfate.
   (b) 1 M Bis–Tris pH 6.5, pH adjusted with HCl.
8. Reservoir stock solutions for C-SH2–pY-1087 co-crystals.
   (a) 1.6 M sodium citrate tribasic dihydrate.
   (b) 1 M Tris–HCl pH 8.5, pH adjusted with HCl.
9. Stereo light microscope for viewing crystallization drops.

3 Methods

3.1 Prepare Purified SH2 Domain Protein for Crystallization

1. Purify recombinant SH2 domain protein using optimal protocols for your SH2 domain; for example, purification procedures for the SH2 domains of p120RasGAP are established [30, 31]. *See* **Note 1**.
2. Buffer exchange approximately 2 mg protein into protein storage buffer in the centrifugal filter device. *See* **Notes 2** and **3**.

Table 1
p120RasGAP SH2 domain parameters

	Molecular weight (Da)	Extinction coefficient (M^{-1} cm^{-1}) at 280 nm	Residue #s	Length (# amino acids)
N-SH2	12,251	15,930	174–280	109
C-SH2	12,453	17,420	340–444	107

3. Using a pipet, transfer 2 mg of purified protein to the filter device. Bring final volume up to 4 mL with protein storage buffer.
4. Centrifuge in centrifugal filter device at 4 °C 7500 × *g* maximum speed in 10 min intervals until the protein sample volume is less than 0.5 mL.
5. Remove the filtrate from the bottom tube and set aside.
6. Add up to 4 mL protein storage buffer to the protein solution in the sample reservoir. Resuspend solution carefully.
7. Repeat **steps 2–4** two additional times, for a total of three buffer exchanges.
8. Centrifuge in centrifugal filter device until volume reaches ~100 μL.
9. Using a micropipettor, transfer volume from filter device to fresh microcentrifuge tube.
10. Determine protein concentration: measure absorbance at 280 nm (A_{280}) on the spectrophotometer. Zero the instrument using protein storage buffer.
11. From A_{280} values, calculate the concentration using the extinction coefficient and the molecular weight values from Table 1. 2 mg of protein in 100 μL should be approximately 20 mg/mL or 1.6 mM.
12. Aliquot protein to 20–50 μL in microcentrifuge tubes. Flash freeze in liquid nitrogen and store at −80 °C. *See* **Note 4**.
13. Assess protein purity by SDS-PAGE chromatography and Coomassie staining. Purity should be >98%.

3.2 Preparation of Phosphopeptides

1. Reconstitute lyophilized phosphopeptides to maximum solubility in peptide reconstitution buffer in a microcentrifuge for final concentrations as shown in Table 2. For pY-1087 peptide, we add 500 μL of peptide reconstitution buffer to 5 mg lyophilized powder, to obtain 10 mg/mL (9.7 mM). For pY-1105 phosphopeptide, we add 1000 μL of peptide reconstitution buffer to 5 mg lyophilized powder, to obtain 5 mg/mL (3.0 mM). *See* **Notes 5**, **6**, and 7.

Table 2
p190RhoGAP phosphopeptide parameters

Peptide	Residue #s	Sequence	Length (# amino acids)	Molecular weight (g/mol)	Maximum solubility (mg/mL)	Final stock concentration (mM)
pY-1087	1086–1093	D(pY)AEPMDA	8	1032	10	9.7
pY-1105	1100–1112	EEENI (pY)SVPHDST	13	1641	5	3.0

2. Centrifuge solution for 15 min at 4 °C at 20,000 RCF to pellet any remaining particulates or aggregates. Aliquot supernatant to microcentrifuge tubes in 50 μL. Flash freeze in liquid nitrogen. Store at −80 °C.

3.3 Formation of Complex Between SH2 Domain and Phosphopeptide

1. Calculate the volume of protein and phosphopeptide required to achieve a 1:1 molar ratio in solution. For example, 85 μL of 1.6 mM SH2 domain protein plus 15 μL of 9.7 mM phosphopeptide equals 1.4 mM each component in 100 μL. *See* **Notes 8** and **9**.
2. Mix the components on ice. Incubate for 1 h.
3. Centrifuge the sample to pellet particulates and/or aggregates prior to setting crystallization drops.

3.4 Crystallization Using Hanging Drop Vapor Diffusion Method

1. Calculate a series of reservoir solutions according to the crystallization conditions specified for each sample type in Table 3 (example in Table 4). 500 μL reservoir solution is needed per single well of the 18 mm diameter crystallization plate. *See* **Notes 10**, **11**, and **12**.
2. Mix reservoir components thoroughly in separate tubes.
3. Pipet 500 μL of reservoir solution to a single well of a pre-greased 24-well crystallization plate (e.g., VDX plate).
4. Prepare a single coverslip by cleaning with 70% ethanol and drying with lint-free labwipe or lens paper. *See* **Note 13**.
5. Pipet 1 μL of protein/peptide mixture onto the center of a coverslip.
6. Pipet 1 μL of reservoir solution from the well and add to the protein/peptide drop on the coverslip. *See* **Notes 14**, **15**, and **16**.
7. Handle the coverslip by its edges using tweezers or your fingers and carefully invert, so that the drop is hanging.
8. Center coverslip over greased reservoir, place down gently, and then press along the edge of the coverslip to seal.

Table 3
Crystallization conditions

		Precipitant			Buffer (0.1 M)	
SH2 domain	Phosphopeptide	Precipitant	Concentration range[a]	Additive	Reagent	pH[b]
N-SH2	–	PEG 10,000	18–23% (w/v)	0.2 M ammonium acetate	Tris–HCl	8.0
C-SH2	–	Ammonium sulfate	1.9–2.4 M	–	Bis–Tris	6.5
N-SH2	pY-1105	Sodium malonate	1.4–1.9 M	2% PEG MME 550	Bis–Tris	6.5
C-SH2	pY-1087	Sodium citrate tribasic	0.8–1.3 M	–	Tris–HCl	8.5

[a]Vary this reagent stepwise. For example, 18–19–20–21–22–23% PEG 10,000, or 1.9–2.0–2.1–2.2–2.3–2.4-M ammonium sulfate
[b]Tris–HCl and Bis–Tris buffers are pH adjusted with HCl

Table 4
Example optimization screen for crystallization of N-SH2 with pY-1105 peptide

1	2	3	4	5	6
1.4 M sodium malonate 2% PEG MME 550 0.1 M Bis–Tris pH 6.5	1.5 M sodium malonate 2% PEG MME 550 0.1 M Bis–Tris pH 6.5	1.6 M sodium malonate 2% PEG MME 550 0.1 M Bis–Tris pH 6.5	1.7 M sodium malonate 2% PEG MME 550 0.1 M Bis–Tris pH 6.5	1.8 M sodium malonate 2% PEG MME 550 0.1 M Bis–Tris pH 6.5	1.9 M sodium malonate 2% PEG MME 550 0.1 M Bis–Tris pH 6.5

9. Repeat **steps 3–8** with additional reservoir positions.
10. Incubate crystallization plate at ambient (20 °C) temperature for several days. Monitor crystal growth by viewing under stereo light microscope daily. *See* **Note 17**.
11. These crystals are suitable for X-ray diffraction data collection.

4 Notes

1. p120RasGAP SH2 domain expression constructs include substitutions at codons for native cysteine residues to serine, to prevent disulfide-mediated dimerization or heteromerization of purified proteins in solution which can introduce heterogeneity and interfere with crystallization. These substitution sites are C236S and C261S in N-SH2 and C372S and C402S in C-SH2.

2. Amicon-4 centrifuge filters with nominal molecular weight limit (NMWL) of 3 kDa are appropriate for SH2 domain protein which has a molecular weight of approximately 10–12 kDa.

3. Alternative to using a centrifugal filter device, buffer exchange may be performed by size exclusion chromatography or dialysis.

4. Avoid repeated freeze–thaw cycles of protein and peptide.

5. Phosphopeptides have variable solubility in aqueous buffers, and solubility may be limited. Solubility testing in various reconstitution solutions may be offered by the commercial peptide vendor. Phosphopeptide reconstitution may be achieved in aqueous solutions including water or peptide reconstitution buffer. Alternatively, the peptide may be initially dissolved in a small amount of DMSO, ammonium hydroxide, or acetic acid, followed by dilution in water or peptide reconstitution buffer.

6. Commercially synthesized peptides may be provided in aliquots of lyophilized powder in premeasured mass (e.g., 5 mg per vial) to simplify the reconstitution process.

7. Phosphopeptide concentration can be verified by measuring A_{280} on the spectrophotometer blanked with peptide reconstitution buffer. Use the extinction coefficient difference for phosphotyrosine versus tyrosine as reported in [32]. Alternatively, phosphopeptide concentration can be quantified by amino acid analysis [33].

8. The protein:peptide molar ratio can be varied to account for errors in protein or phosphopeptide concentration determination and/or systematic errors in pipetting. For example, protein:peptide ratios of 1:1.5 or 1:0.8 may be trialed.

9. A phosphopeptide titration into SH2 domain may be monitored by native PAGE electrophoresis, isothermal titration calorimetry, etc., to experimentally establish the condition under which a stoichiometric complex is formed.

10. All solutions are prepared with deionized or doubly distilled water and sterile filtered through a 0.22 μm membrane. Store stock solutions at 4 °C.

11. Stock crystallization solutions can be made from powder or can be purchased as premade formulations directly from multiple suppliers.

12. Crystallization conditions are identified by screening using commercially available crystallization kits. After initial hits are identified, the conditions are further optimized by testing the effects of altering the reservoir solution by changing precipitant concentration, buffer pH values, etc. Further optimization

can include changing protein concentration and adjusting the SH2:phosphopeptide molar ratio.

13. Plastic or siliconized glass coverslips can be used.
14. Be careful not to introduce bubbles into crystallization drop when pipetting protein/peptide or reservoir solutions.
15. Mixing of drop is not necessary.
16. Total drop volume can vary from 1 to 10 μL.
17. Crystals should appear within the first 4 days but may take longer to grow to full size, up to 2 weeks.

Acknowledgments

We acknowledge Rachel Jaber Chehayeb and Jessica Wang. Kimberly Vish is acknowledged for helpful comments. This research was supported by 1R01NS117609 and 1R01GM138411 to T.J.B.

References

1. Liu BA, Jablonowski K, Raina M et al (2006) The human and mouse complement of SH2 domain proteins-establishing the boundaries of phosphotyrosine signaling. Mol Cell 22(6): 851–868. https://doi.org/10.1016/j.molcel.2006.06.001
2. Sadowski I, Stone JC, Pawson T (1986) A noncatalytic domain conserved among cytoplasmic protein-tyrosine kinases modifies the kinase function and transforming activity of Fujinami sarcoma virus P130gag-fps. Mol Cell Biol 6(12):4396–4408. https://doi.org/10.1128/mcb.6.12.4396
3. Liu BA, Shah E, Jablonowski K et al (2011) The SH2 domain-containing proteins in 21 species establish the provenance and scope of phosphotyrosine signaling in eukaryotes. Sci Signal 4(202):ra83. https://doi.org/10.1126/scisignal.2002105
4. Waksman G, Kumaran S, Lubman O (2004) SH2 domains: role, structure and implications for molecular medicine. Expert Rev Mol Med 6(3):1–18. https://doi.org/10.1017/S1462399404007331
5. Waksman G, Kuriyan J (2004) Structure and specificity of the SH2 domain. Cell 116(2 Suppl):S45–S48, 43 p following S48. https://doi.org/10.1016/s0092-8674(04)00043-1
6. Liu BA, Engelmann BW, Nash PD (2012) The language of SH2 domain interactions defines phosphotyrosine-mediated signal transduction. FEBS Lett 586(17):2597–2605. https://doi.org/10.1016/j.febslet.2012.04.054
7. Eck MJ, Shoelson SE, Harrison SC (1993) Recognition of a high-affinity phosphotyrosyl peptide by the Src homology-2 domain of p56lck. Nature 362(6415):87–91. https://doi.org/10.1038/362087a0
8. Waksman G, Kominos D, Robertson SC et al (1992) Crystal structure of the phosphotyrosine recognition domain SH2 of v-src complexed with tyrosine-phosphorylated peptides. Nature 358(6388):646–653. https://doi.org/10.1038/358646a0
9. Bradshaw JM, Mitaxov V, Waksman G (1999) Investigation of phosphotyrosine recognition by the SH2 domain of the Src kinase. J Mol Biol 293(4):971–985. https://doi.org/10.1006/jmbi.1999.3190
10. Kaneko T, Joshi R, Feller SM, Li SS (2012) Phosphotyrosine recognition domains: the typical, the atypical and the versatile. Cell Commun Signal 10(1):32. https://doi.org/10.1186/1478-811X-10-32
11. Trahey M, McCormick F (1987) A cytoplasmic protein stimulates normal N-ras p21 GTPase, but does not affect oncogenic mutants. Science 238(4826):542–545. https://doi.org/10.1126/science.2821624
12. Vogel US, Dixon RA, Schaber MD et al (1988) Cloning of bovine GAP and its interaction with oncogenic ras p21. Nature 335(6185):90–93. https://doi.org/10.1038/335090a0

13. Bos JL, Rehmann H, Wittinghofer A (2007) GEFs and GAPs: critical elements in the control of small G proteins. Cell 129(5):865–877. https://doi.org/10.1016/j.cell.2007.05.018
14. Gideon P, John J, Frech M et al (1992) Mutational and kinetic analyses of the GTPase-activating protein (GAP)-p21 interaction: the C-terminal domain of GAP is not sufficient for full activity. Mol Cell Biol 12(5):2050–2056. https://doi.org/10.1128/mcb.12.5.2050
15. Ellis C, Moran M, McCormick F, Pawson T (1990) Phosphorylation of GAP and GAP-associated proteins by transforming and mitogenic tyrosine kinases. Nature 343(6256): 377–381. https://doi.org/10.1038/343377a0
16. Moran MF, Polakis P, McCormick F et al (1991) Protein-tyrosine kinases regulate the phosphorylation, protein interactions, subcellular distribution, and activity of p21ras GTPase-activating protein. Mol Cell Biol 11(4):1804–1812
17. Moran MF, Koch CA, Anderson D et al (1990) Src homology region 2 domains direct protein-protein interactions in signal transduction. Proc Natl Acad Sci U S A 87(21):8622–8626. https://doi.org/10.1073/pnas.87.21.8622
18. Di Cristofano A, Carpino N, Dunant N et al (1998) Molecular cloning and characterization of p56dok-2 defines a new family of RasGAP-binding proteins. J Biol Chem 273(9): 4827–4830. https://doi.org/10.1074/jbc.273.9.4827
19. Wagner MJ, Hsiung MS, Gish GD et al (2020) The Shb scaffold binds the Nck adaptor protein, p120 RasGAP, and Chimaerins and thereby facilitates heterotypic cell segregation by the receptor EphB2. J Biol Chem 295(12): 3932–3944. https://doi.org/10.1074/jbc.RA119.009276
20. Anderson D, Koch CA, Grey L et al (1990) Binding of SH2 domains of phospholipase C gamma 1, GAP, and Src to activated growth factor receptors. Science 250(4983):979–982
21. Kaplan DR, Morrison DK, Wong G et al (1990) PDGF beta-receptor stimulates tyrosine phosphorylation of GAP and association of GAP with a signaling complex. Cell 61(1): 125–133. https://doi.org/10.1016/0092-8674(90)90220-9
22. Holland SJ, Gale NW, Gish GD et al (1997) Juxtamembrane tyrosine residues couple the Eph family receptor EphB2/Nuk to specific SH2 domain proteins in neuronal cells. EMBO J 16(13):3877–3888. https://doi.org/10.1093/emboj/16.13.3877
23. Hu KQ, Settleman J (1997) Tandem SH2 binding sites mediate the RasGAP-RhoGAP interaction: a conformational mechanism for SH3 domain regulation. EMBO J 16(3): 473–483. https://doi.org/10.1093/emboj/16.3.473
24. Heraud C, Pinault M, Lagree V, Moreau V (2019) p190RhoGAPs, the ARHGAP35- and ARHGAP5-encoded proteins, in health and disease. Cell 8(4):351. https://doi.org/10.3390/cells8040351
25. Stiegler AL, Boggon TJ (2017) p190RhoGAP proteins contain pseudoGTPase domains. Nat Commun 8(1):506. https://doi.org/10.1038/s41467-017-00483-x
26. Koch CA, Moran MF, Anderson D et al (1992) Multiple SH2-mediated interactions in v-src-transformed cells. Mol Cell Biol 12(3): 1366–1374. https://doi.org/10.1128/mcb.12.3.1366-1374.1992
27. Roof RW, Haskell MD, Dukes BD et al (1998) Phosphotyrosine (p-Tyr)-dependent and -independent mechanisms of p190 RhoGAP-p120 RasGAP interaction: Tyr 1105 of p190, a substrate for c-Src, is the sole p-Tyr mediator of complex formation. Mol Cell Biol 18(12): 7052–7063
28. Hernandez SE, Settleman J, Koleske AJ (2004) Adhesion-dependent regulation of p190RhoGAP in the developing brain by the Abl-related gene tyrosine kinase. Curr Biol 14(8):691–696. https://doi.org/10.1016/j.cub.2004.03.062
29. Bradley WD, Hernandez SE, Settleman J, Koleske AJ (2006) Integrin signaling through Arg activates p190RhoGAP by promoting its binding to p120RasGAP and recruitment to the membrane. Mol Biol Cell 17(11): 4827–4836. https://doi.org/10.1091/mbc.E06-02-0132
30. Jaber Chehayeb R, Wang J, Stiegler AL, Boggon TJ (2020) The GTPase-activating protein p120RasGAP has an evolutionarily conserved "FLVR-unique" SH2 domain. J Biol Chem 295(31):10511–10521. https://doi.org/10.1074/jbc.RA120.013976
31. Jaber Chehayeb R, Stiegler AL, Boggon TJ (2020) Correction: crystal structures of p120RasGAP N-terminal SH2 domain in its apo form and in complex with a p190RhoGAP phosphotyrosine peptide. PLoS One 15(2): e0229627. https://doi.org/10.1371/journal.pone.0229627
32. Zhang ZY, Maclean D, Thieme-Sefler AM et al (1993) A continuous spectrophotometric and fluorimetric assay for protein tyrosine phosphatase using phosphotyrosine-containing

peptides. Anal Biochem 211(1):7–15. https://doi.org/10.1006/abio.1993.1224

33. Ozols J (1990) Amino acid analysis. Methods Enzymol 182:587–601. https://doi.org/10.1016/0076-6879(90)82046-5

34. McNicholas S, Potterton E, Wilson KS, Noble ME (2011) Presenting your structures: the CCP4mg molecular-graphics software. Acta Crystallogr D Biol Crystallogr 67(Pt 4):386–394. https://doi.org/10.1107/S0907444911007281

Part II

Biophysics and Bioinformatics of SH2 Domain–Phosphopeptide Binding

Chapter 6

Fluorescence Anisotropy and Polarization in the Characterization of Biomolecular Association Processes and Their Application to Study SH2 Domain Binding Affinity

Sara Bobone, Claudia Storti, Paolo Calligari, and Lorenzo Stella

Abstract

Fluorescence anisotropy (or polarization) is a powerful technique to study biomolecular association processes, by following the rotational motions of one of the two partners in the interaction, labeled with a fluorophore. It can be used to determine dissociation constants in solution, down to nM values, and unlabeled ligands can be characterized, too, by using competition experiments. In this chapter, we introduce the basic principles of the technique, compare it with other experimental approaches, and discuss the experimental details with specific examples regarding SH2 domain/phosphopeptide association processes. The experimental protocols to be used in binding experiments and displacement studies are described, as well as the caveats to be considered in performing accurate measurements.

Key words Fluorescence anisotropy, Fluorescence polarization, Association processes, SH2-domains, Phosphopeptides

1 Introduction

1.1 Why Measure SH2 Domain-Binding Affinity?

Src homology 2 (SH2) domains are structural units with sequences of ~100 aa, which selectively recognize polypeptides containing a phosphotyrosine (pTyr) along with specific contiguous residues. In tyrosine phosphorylation, which plays a critical role in eukaryotic cell regulation [1], tyrosine kinases act as "writers". SH2 domains are capable of recognizing or "reading" this information, which can be subsequently "erased" by protein tyrosine phosphatases (PTPs) [2]. Considering their function, determining the binding affinity and selectivity of SH2 domains is essential for understanding their biological behavior and for the rational design of new drugs targeting them [3].

Teresa Carlomagno and Maja Köhn (eds.), *SH2 Domains: Functional Modules and Evolving Tools in Biology*, Methods in Molecular Biology, vol. 2705, https://doi.org/10.1007/978-1-0716-3393-9_6,

The affinities of SH2 domains for their binding partners (i.e., the dissociation constants) are usually in the order of 0.1–10 μM [2, 4–7]. Generally, phosphorylation of the pTyr contributes approximately 30–50% of the overall standard binding free energy [3, 8], ensuring that significant binding takes place only to phosphorylated sequences. At the same time, the residues surrounding pTyr (usually from −3 to +5 with respect to its position) account for the selectivity of binding [3, 9]. A notable exception to moderate affinities is provided by the SH2 domains of the SHP2 phosphatase. Dissociation constants in the order of 1–10 nM have been reported for the binding of the N-terminal SH2 domain to insulin receptor substrate 1 (IRS-1, pTyr 1172) [3], GRB2-associated binding protein 1 (Gab1, pTyr 627) [10], Gab2, pTyr 614 [11], and to an artificial peptide sequence [3], and for binding of the C-terminal SH2 domain to programmed cell death protein 1 (PD-1, pTyr 225) [12].

1.2 Experimental Approaches to Measure SH2 Domain-Binding Affinity

Several different approaches can be used to quantitatively determine the binding affinity of SH2 domains for phosphopeptides and phosphorylated proteins. Surface plasmon resonance (SPR) [13–17] is based on the immobilization of one of the two interaction partners on a gold surface and on the measurement of the change in refractive index caused by adsorption of the ligand molecules. Isothermal titration calorimetry (ITC) [14, 18] measures the heat exchanged by the association process. Quantitative microarrays [19] use surface-immobilized SH2 domains and measure the fluorescence signal of bound labeled phosphopeptides. Microscale thermophoresis (MST) [20, 21] measures the movement of particles in a microscopic temperature gradient and how the association process affects this process. Fluorescence spectroscopy [11, 22] follows changes in the fluorescence intensity of a labeled phosphopeptide caused by binding. Fluorescence anisotropy and polarization [3] are the topic of this chapter and will be discussed in detail in the following sections.

These methods can measure binding down to low nM or even sub-nM affinities, but each of them has some limitations. SPR and quantitative microarrays require immobilization to a surface of one of the two interacting components, a procedure that can affect the properties of a macromolecule or introduce artifacts in the measurements caused by a high density of binding sites [14]. ITC is performed in solution, requires no labeling, and directly determines the enthalpic and entropic contributions to the free energy of binding. However, it requires relatively large sample quantities [18]. MST, fluorescence spectroscopy, and anisotropy experiments are all performed in solution and are based on the measurement of a fluorescence signal. These approaches have a high sensitivity, do not require the separation of free and bound ligand molecules, and are easily adapted to high-throughput methods that utilize volumes of

10–100 μL [23]. In general, the intrinsic fluorescence signal of the protein or ligand can be used for these methods. However, this is not possible in the case of phosphopeptide/SH2 domain complex formation, due to the low brightness of intrinsic fluorophores and the overlap between the emission spectra of Trp and Tyr protein residues with the pTyr emission of the phosphopeptide ligands. Therefore, when studying SH2 domains, all these approaches use phosphopeptides labeled with extrinsic fluorophores. In principle, derivatization with a fluorophore could affect the binding properties of the peptide ligand [3]. However, the binding of unlabeled peptides can be characterized, too, using displacement experiments (*see* below) [3].

The concentration of labeled ligand, which is titrated with the SH2 domain, determines the range of dissociation constants that can be quantified accurately (*see* below). In the case of MST, it has been reported that the minimum concentration of labeled molecule that can be used is 5 nM [24]. With fluorescence anisotropy, we could perform experiments down to 0.1 nM using a carboxyfluorescein label [3]. Therefore, this latter technique allows the measurement of affinities over the entire range encountered in the case of SH2 domains (*see* above). Measurements of fluorescence intensities in principle can be performed even at lower concentrations since they do not require the use of polarizers that strongly reduce the sensitivity of a fluorometer. However, in some systems, the fluorescence spectrum does not change appreciably upon binding, and this is more likely with extrinsic probes emitting in the visible with high brightness [25].

1.3 Fluorescence Anisotropy and Polarization: A Primer

Fluoresce anisotropy and polarization detect the motions of fluorescent molecules by measuring the polarization of their emitted light. Here, we will briefly introduce these techniques. For further details, the interested reader is referred to the reviews and to the fascinating textbook by Jameson [26–28].

Natural light is unpolarized, i.e., its electric field oscillates in all possible directions perpendicular to the direction of propagation of the beam. In fluorescence anisotropy or polarization experiments, the exciting light is linearly polarized, i.e., its electric field oscillates in a well-defined direction, and it is obtained by filtering unpolarized light through an optical element called a polarizer (Fig. 1). Fluorescence emission is then measured through a second polarizer, oriented parallel or perpendicular to the excitation polarizer. The two measured intensities are termed $I_{\parallel}$ and $I_{\perp}$, respectively. Fluorescence polarization (P) and anisotropy (r) are two distinct dimensionless measures of the difference between these two values and are defined as

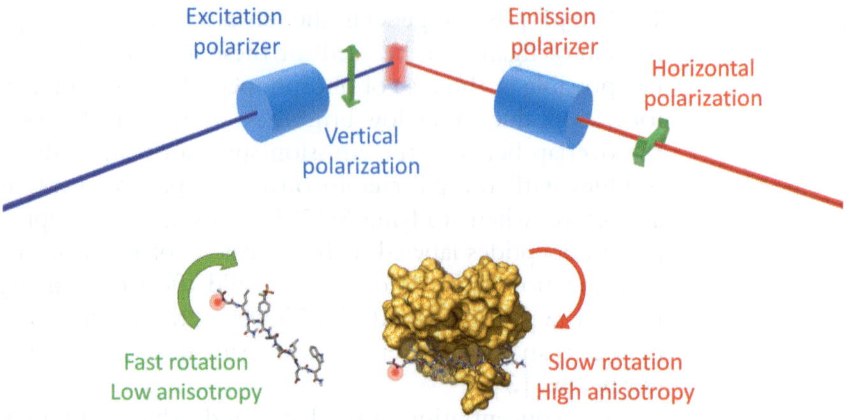

Fig. 1 Principles of anisotropy measurements. An excitation polarizer sends vertically polarized light on the sample (blue beam). Using an emission polarizer, the intensity of emitted light (red beam) polarized parallel or perpendicular to the direction of polarization of the exciting light is measured. Rotations of the molecules in solution decrease the degree of polarization of the emitted light. Therefore, a free peptide (which rotates fast with respect to the lifetime of the fluorophore that is labeling it) has a lower fluorescence polarization (and anisotropy) than a peptide bound to an SH2 domain

$$P = \frac{I_{\parallel} - I_{\perp}}{I_{\parallel} + I_{\perp}} \tag{1}$$

and

$$r = \frac{I_{\parallel} - I_{\perp}}{I_{\parallel} + 2I_{\perp}}. \tag{2}$$

Both parameters have a value of 1 when the emitted light is completely polarized parallel to the direction of polarization of the exciting light and 0 when the fluorescence is unpolarized ($I_{\parallel} = I_{\perp}$). However, when the light is completely polarized in the perpendicular direction, P is -1, while r is -0.5. Based on their definition (Eqs. 1 and 2), r can be derived from the value of P and vice versa:

$$r = \frac{2P}{3 - P}. \tag{3}$$

Therefore, the two quantities can be used interchangeably, but each of them is preferred, by tradition, by specific scientific communities. The definition of P is more intuitive, but only r has the important property of additivity (the anisotropy of a sample containing two fluorescent molecules is the weighted average of their anisotropies) [26]. For this reason, as discussed below, the use of anisotropy is advised for quantitative binding studies.

It must be noted that fluorimeter sensitivity often depends significantly on the polarization of the light to be detected. Let us indicate with I_{VV} and I_{VH} the fluorescence intensities measured with the excitation polarizer oriented vertically with respect to the plane of the instrument (for a cuvette fluorimeter) and the emission polarizer oriented vertically or horizontally (respectively). Due to the differential instrument sensitivity for vertically and horizontally polarized light, these measured values generally do not correspond to the ideal quantities $I_{\|}$ and $I_{\perp}$. For instance, for equal values of $I_{\|}$ and $I_{\perp}$ emitted light, the instrument would measure different values of I_{VV} and I_{VH} (resulting in nonzero anisotropy). This phenomenon must be accounted and corrected for. This is accomplished by measuring the so-called G factor, defined as

$$G = \frac{I_{HV}}{I_{HH}} \tag{4}$$

In Eq. (4), I_{HH} and I_{HV} are the fluorescence intensities measured with the excitation polarizer oriented horizontally and the emission polarizer oriented horizontally or vertically (respectively). In an ideal instrument, these values are expected to be identical, while in reality the G factor is often very different from 1. The G factor does not depend on the sample but only on the instrumental settings (slit width, excitation and emission wavelengths, and filters). The corrected anisotropy value is then obtained as

$$r = \frac{I_{VV} - GI_{VH}}{I_{VV} + 2GI_{VH}} \tag{5}$$

The possibility of detecting molecular motions by using fluorescence anisotropy or polarization is due to the phenomenon of photoselection. When polarized light is used for excitation, the probability that each specific molecule in the sample absorbs a photon depends on its orientation or, more specifically, on the direction of its transition dipole (a quantum property linked to the change in the electronic cloud during the transition from the ground to the excited state). Molecules that, due to their random orientation, happen to have a transition dipole along the direction of polarization of the incident light are excited preferentially (the probability of absorption is proportional to the cosine squared of the angle formed by the transition dipole and the electric field). As a consequence of photoselection, only a subpopulation of molecules absorbs light in the originally isotropic sample, resulting in a predominant orientation of the transition dipoles of the excited molecules. Since only molecules that have reached the excited state can then emit light, an "oriented" sample has been generated.

Excited molecules emit fluorescence, on average, after a period corresponding to the lifetime of the excited state, which is typically some nanoseconds long. At emission, they will have lost their original orientation to a degree that depends on the extent of motions experienced during this time. In a similar way to what happens in absorption, the emitted light is preferentially polarized in the direction of the transition dipole (again with a cosine squared distribution). Therefore, the extent of fluorescence polarized parallel to the direction of polarization of the exciting light provides an indication of the amplitude of molecular motions during the excited state lifetime. For instance, if the only relevant motion is rotational diffusion of the fluorescent macromolecule, which can be approximated to a rigid sphere, the fluorescence anisotropy corresponds to

$$r = \frac{r_o}{1 + \frac{\varphi}{\tau}}. \tag{6}$$

where r_o is the anisotropy in the absence of motions [28], τ is the excited state lifetime, and φ is the rotational correlation time (the characteristic time of rotational diffusion [29]), given by the Stokes–Einstein relation

$$\varphi = \frac{\eta V}{k_B T}. \tag{7}$$

η is the viscosity of the solution, V the (hydrodynamic) volume of the molecule, k_B Boltzmann's constant, and T temperature.

Performing an anisotropy measurement can be compared to shooting a photograph, with a specified shutter speed. If the subject of the photo moves much faster than the time in which the shutter stays open, the image will be blurry. On the other hand, if the subject moves slowly on that timescale, the image will be sharp, just like if there were no motions at all. Similarly, if the molecule rotates rapidly during the lifetime, the anisotropy will be close to zero. If it rotates slowly, the anisotropy value will correspond to r_o.

A fluorescently labeled phosphopeptide free in solution is much smaller than a peptide/SH2 domain complex, and therefore it rotates much faster. As a consequence, the bound peptide usually has a significantly higher anisotropy (or polarization) value than the free peptide, and this property can be used to follow the association process in a homogeneous assay (without the need of separating the sample components) (Fig. 1).

Considering that anisotropy is additive, in the presence of both free and bound peptides, the total anisotropy is given by

$$r = f_{\text{free}} r_{\text{free}} + f_{\text{bound}} r_{\text{bound}} \tag{8}$$

where f_{free} and f_{bound} are the fractions of fluorescence emitted by the free and bound species (and their sum is 1). Similarly, r_{free} and r_{bound} are the anisotropy values corresponding to the two peptide states. From the anisotropy values, the fraction of fluorescence emitted by the bound peptide can be calculated:

$$f_{\text{bound}} = \frac{r - r_{\text{free}}}{r_{\text{free}} - r_{\text{bound}}} \tag{9}$$

If the quantum yield of the fluorophore in the free and bound conditions is similar, f_{bound} coincides with the molar fraction of bound peptide. This latter condition is not always valid, since the different environments of the emitting dye in the two states can lead to different probabilities of non-radiative decay. This possibility can be verified (and corrected for) by observing variations in the fluorescence intensity during the binding titration [27]. Another point worth highlighting is that the additivity condition, on which Eqs. (8 and 9) are based, is not valid for polarization. Therefore, in principle, this latter parameter should not be used directly to determine quantitatively the fraction of free and bound species. However, P can always be converted to r, by using Eq. (3) [27].

An important point to remember is that scattered light is completely polarized in the parallel direction. Therefore, any spurious contribution of scattered excitation light to the measured emission signal results in an artifactual increase in the anisotropy and polarization values. Therefore, it should be avoided or, at least, minimized and subtracted.

1.4 Basic Concepts in Anisotropy Measurements of SH2/Peptide Binding

1.4.1 Binding Experiments

For anisotropy studies of SH2 binding, at least one of the phosphopeptides of interest should be labeled with a fluorophore (the others can be characterized in displacement studies, *see* below). In direct binding experiments, a fluorescently labeled phosphopeptide is titrated with increasing concentrations of the SH2 domain (or of the protein containing it) while measuring the fluorescence anisotropy, which increases due to binding.

With a few exceptions [30], each SH2 domain binds a single peptide (1:1 stoichiometry). Therefore, the fraction of bound peptide x_{bound} is expected to vary as follows [22]:

$$x_{\text{bound}} = \frac{[P]_{\text{T}} + [L]_{\text{T}} + K_{\text{d}} - \sqrt{\left([P]_{\text{T}} + [L]_{\text{T}} + K_{\text{d}}\right)^2 - 4[P]_{\text{T}}[L]_{\text{T}}}}{2[L]_{\text{T}}} \quad (10)$$

Here, $[P]_{\text{T}}$ and $[L]_{\text{T}}$ are the total protein and ligand concentrations, and K_{d} is the dissociation constant. The anisotropy is therefore expected to change according to the following Eq. (11):

$$r = r_{\text{free}} + \left(r_{\text{bound}} - r_{\text{free}}\right)\frac{[P]_{\text{T}} + [L]_{\text{T}} + K_{\text{d}} - \sqrt{\left([P]_{\text{T}} + [L]_{\text{T}} + K_{\text{d}}\right)^2 - 4[P]_{\text{T}}[L]_{\text{T}}}}{2[L]_{\text{T}}} \quad (11)$$

When $[L]_{\text{T}} \gg K_{\text{d}}$, the above equation is reduced to stoichiometric binding, i.e., all the protein molecules added to the sample bind a phosphopeptide. In this case, the experiment cannot provide any information on the K_{d}. Since the value of the dissociation constant is not known before measuring, it is advisable to start by using a peptide concentration close to the lowest value allowed by the instrumental sensitivity. Once a tentative value of the dissociation constant has been determined, the experiment can be repeated with a higher peptide concentration, to increase the fluorescence signal and to reduce the measurement noise. In any case, values of $[L]_T$ higher than $2K_d$ should be avoided, because the error in the determined value of the dissociation constant would be unacceptable [31].

When $[L]_T \ll K_d$, then the protein concentration can be varied up to 10 K_d while remaining always much higher than $[L]_T$. In this case, the above equation can be simplified by assuming $[P]_T \cong [P]$ and obtaining

$$r = r_{\text{free}} + (r_{\text{bound}} - r_{\text{free}}) \frac{[P]_T/K_d}{1 + [P]_T/K_d} \tag{12}$$

1.4.2 Displacement Studies

Binding of unlabeled peptides can be determined by performing displacement experiments. A fluorescently labeled phosphopeptide, interacting with the SH2 domain, is displaced by titration with an unlabeled peptide. Knowing the K_d and concentration of the fluorescent peptide and the concentration of the domain, the dissociation constant of the unlabeled peptide (K_d^u) can be quantitatively determined. Very often, several approximations are used for this purpose, which however are not always justified [32]. Fortunately, the displacement curves can be analyzed without any approximations [3, 33].

The displacement curve is fit with a phenomenological Hill equation [34] to determine the EC_{50}, i.e., the concentration of unlabeled peptide that displaces half of the bound fluorescent peptide:

$$r = r_{\text{start}} - (r_{\text{start}} - r_{\text{free}}) \frac{\left\{[I]_T/EC_{50}\right\}^n}{1 + \left\{[I]_T/EC_{50}\right\}^n} \tag{13}$$

Using this value, the dissociation constant of the unlabeled peptide is determined:

$$K_d^u = \frac{EC_{50} - [P]_T + \frac{[PL]_0}{2}\left(1 + \frac{K_d}{[L]_T - \frac{[PL]_0}{2}}\right)}{\left(\frac{2[P]_T}{[PL]_0} - 1\right)\frac{[L]_T - \frac{[PL]_0}{2}}{K_d} - 1} \tag{14}$$

Here, $[PL]_0$ is the concentration of bound fluorescent peptide in the absence of unlabeled peptide, i.e., at the beginning of the titration experiment, and it can be expressed as

$$[PL]_0 = \frac{[P]_T + [L]_T + K_d - \sqrt{([P]_T + [L]_T + K_d)^2 - 4[P]_T[L]_T}}{2} \tag{15}$$

An alternative approach is to fit the displacement curve directly with an equation where the anisotropy is expressed as a function of $[I]_T$, with $[L]_T$, $[P]_T$, K_d, r_{free}, and r_{bound} as fixed parameters and K_d^u as a fitting parameter [35].

Regarding the conditions to be used in the experiment, the concentration of fluorescent peptide is typically the same employed in the direct binding study. For the protein concentration, a compromise between two requirements is needed [36]. On one hand, it is desirable to have a significant fraction of the labeled peptide bound to the protein, to maximize the dynamic range in the anisotropy signal, which decreases during the displacement experiment. On the other hand, the protein concentration should be comparable to or lower than the dissociation constant of the unlabeled peptide (K_d^u), to allow a quantitative and reliable determination of its binding affinity (i.e., to avoid stoichiometric binding). In our experience, $[P]_T \sim K_d$ is a good compromise (where K_d is the dissociation constant for the labeled peptide). The dissociation constant of the fluorescent peptide is a limiting factor for the lowest K_d^u values that can be determined with this approach [36]. Therefore, the higher the affinity of the labeled peptide for the SH2 domain, the better.

2 Materials

2.1 Phosphopeptides

1. A phosphopeptide labeled with a fluorescence probe attached to the N- or C-terminus, or to a side chain of the peptide, not involved in the binding process (*see* **Notes 1–3**), for the direct binding assays.
2. Unlabeled phosphopeptides, whose affinity is to be determined by displacement assays.
3. Phosphopeptide stock solutions in methanol, ethanol, or dimethyl sulfoxide, depending on the solubility properties of the peptides. Do not use aqueous solutions, to maximize the long-term chemical stability of the peptides. Prepare stock solutions of 1–5 mM, so that the final percentage of organic solvent in the measurement sample (where the stock is diluted in measurement buffer) does not exceed 1%, and store them at −20 °C. The exact concentration of the stock solutions should be determined by absorbance measurements (*see* **Note 4**).

2.2 Buffer

1. Measurement buffer: a buffer with physiological pH and ionic strength (such as 10 mM HEPES or phosphate buffer, pH 7.4, containing 140 mM NaCl) to ensure protein stability (*see* **Note 5**). For experiments at nM concentrations of the labeled peptide, it is essential that the measurement buffer does not show any significant spurious fluorescence; any emission signal should not exceed 30% of the intensity of the water Raman peak and, in any case, it should not be more than 10% of the fluorophore signal under the final measurement conditions (*see* below).

2.3 Cuvette

All protocols are written for a fluorometer with a single cell holder (*see* **Note 6**):

1. 10 mm × 4 mm cell, with a compartment for a magnetic stirrer at the bottom (e.g., Hellma cuvette 109004F-10-40) for sample volumes of 1.1 mL or larger. This cuvette gives the advantage of allowing automatic stirring during titrations.
2. 5 mm × 5 mm or 3 mm × 3 mm cuvettes (e.g., Hellma 111-057-40), or even the so-called ultramicro cells, can be alternatively used for sample volumes smaller than 1.1 mL. In this case, gentle manual mixing after every protein addition is recommended.

3 Methods

1. Determine the exact concentration of the protein stock solution by absorbance, using the extinction coefficient determined based on the protein sequence [37]. Due to possible aggregation and precipitation phenomena, it is advisable to collect a spectrum of the protein stock before every experimental session (*see* **Note 7**).
2. Thoroughly clean the cuvette by treating it with Hellmanex, according to the manufacturer's instructions, or with chromic acid (for 5–10 min). Then, profusely rinse the cell with deionized water and ethanol and dry carefully. In some cases, it is necessary to avoid protein adsorption to the cuvette walls, by increasing the hydrophilicity of the quartz surfaces (*see* **Note 8**).

3.1 Direct Binding Assay

1. Choose the peptide concentration to be comparable or lower than the dissociation constant to be determined. If the order of magnitude of the K_d, is not known, perform a first tentative binding curve with a 1 nM peptide concentration and determine a tentative value for the dissociation constant. For this purpose, start with a protein concentration of 10 pM, and then increase it by a factor of ten with each successive addition, until a plateau in the anisotropy is observed (*see* **Note 9**). Then, repeat the experiment using a more appropriate peptide concentration and more closely spaced protein concentrations (*see* **step 10**). For high K_d values, the peptide concentration can be increased up to $K_d/10$, to have a higher and more stable fluorescence signal. For high-affinity interactions, the peptide concentration can be decreased down to 0.1 nM, which is typically the lowest value that provides a reasonable fluorescence signal.

2. Select the excitation wavelength to coincide with the wavelength of maximum absorption of the S_0–S_1 transition of the fluorophore (i.e., the longest wavelength absorption peak). In this way, both the fluorescence intensity and the value of r_o are maximized (however, *see* **Note 10**). For emission, select the maximum of the fluorescence spectrum.
3. Set the slits of the excitation and emission monochromators to obtain a suitable signal while still avoiding an overlap of the excitation and emission bandpass. When a single-photon counting instrument is used, the linear range of detection rates usually extends up to ~10^6 counts per second. Therefore, set the monochromator slits so that the detection rate for the highest intensity signal (typically $I_{||}$) is below this value and the integrated signal is at least 10^4 counts for the lowest intensity signal (typically $I_{\perp}$), since this is the minimum number of counts that provides a suitable signal to noise ratio (10^2, according to Poisson statistics). When a fluorimeter with analog detection is used, the instrumental settings are selected so that the signal falls in the linear dynamic range of the instrument (which depends on the specific setup) and possibly close to the upper limit.
4. If necessary, modify the integration time. An integration time of 1 s is used by default, but if the signal rate is too low to reach 10^4 counts in 1 s, increase this time.
5. If possible, use a cutoff filter in the emission channel, to block the scattered exciting light, while transmitting the fluorescence photons.
6. Determine the G factor. It is recommended to check for the G value before every session of measurements, even if the experimental conditions have not been changed.
7. Keep the temperature of the sample constant during the acquisition of the whole binding curve.
8. Acquire each point of the binding titration (*see* below) several times (~15 times). Average the replicates and determine the standard deviation. Typically, a standard deviation of about 0.001 should be obtained.
9. Measure first the anisotropy of the free peptide in the measurement buffer. Successively, titrate the peptide with increasing protein concentrations and record the value of r for each point of the titration curve. When a magnetic stirrer is inserted in the cuvette, the anisotropy must be recorded after switching the stirring off. Check that, after each addition, the value obtained for r remains stable.

10. Acquire the peptide anisotropy for increasing protein concentrations logarithmically spaced by a factor of 2, ideally ranging roughly from $K_d/100$ to 100 K_d [38] or at least from $K_d/10$ to 10 K_d.
11. Acquire at least three independent binding curves, possibly using three distinct protein batches.

3.2 Data Analysis for Direct Binding Assays

1. For subtraction of background signal, repeat the titration in the absence of the labeled peptide, with the same protein concentrations used in the binding experiment.
2. Calculate the anisotropy by subtracting the blank counts for each position of the polarizers, according to the following equation:

$$r = \frac{[I(\text{peptide})_{VV} - I(\text{background})_{VV}] - G[\text{I}(\text{peptide})_{VH} - I(\text{background})_{VH}]}{[I(\text{peptide})_{VV} - I(\text{background})_{VV}] + 2G[\text{I}(\text{peptide})_{VH} - I(\text{backgrond})_{VH}]} \tag{16}$$

3. Calculate averages and standard deviations from the three datasets (*see* Subheading 3.1, **step 11**).
4. Fit the resulting value with Eq. (11) (or (12), when appropriate), to determine the value of K_d. A typical example of a binding curve is reported in Fig. 2.

3.3 Displacement Assay

The same concentration of labeled peptide used in the binding assay and therefore the same instrumental settings are employed:

1. After measuring the anisotropy of the free fluorescent peptide, add a protein concentration equivalent to the K_d for the labeled peptide (*see* **Note 11**).
2. Titrate the peptide–protein complex with increasing amounts of unlabeled peptide (the displacing peptide). The concentration of displacing peptide to be added can be initially estimated by assuming that the fluorescent peptide and the unlabeled peptide have comparable K_d values. Therefore, the titration can be performed with displacing peptide concentrations ranging from $K_d/100$ to 100 K_d. If the affinity of the unlabeled peptide turns out to be significantly different from that of the fluorescent peptide, the titration can be repeated with an improved choice of the concentrations. In any case, the maximum concentration of displacing peptide should cause the anisotropy to decrease to the value obtained for the free labeled peptide (*see* **Note 12**) (Fig. 2).
3. Fit data with Eq. (13), to determine EC_{50}; then use this value to obtain K_d^u, via Eqs. (14 and 15).

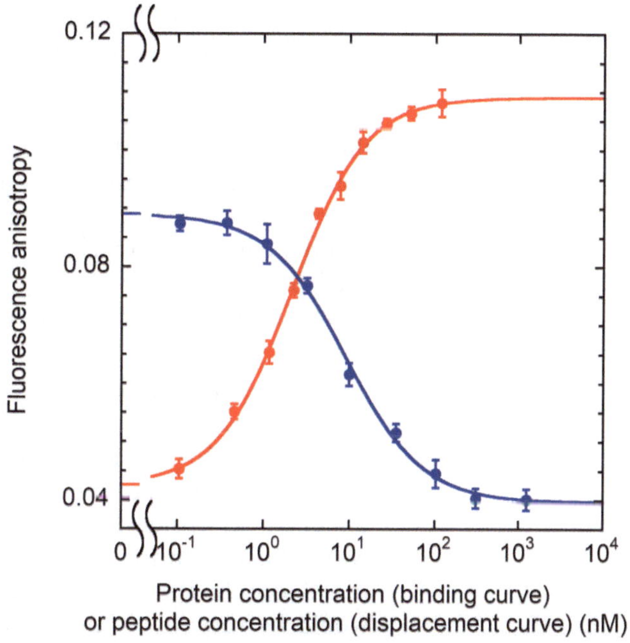

Fig. 2 Anisotropy curves recorded in a binding (red) and displacement (blue) experiment. The binding curve was obtained by titrating a carboxyfluorescein-labeled peptide (CF-P9W5, 0.1 nM) with the N-terminal SH2 domain of the SHP2 phosphatase. The displacement curve was obtained by titrating the fluorescent analog (CF-P9W5, 0.1 nM, SH2 domain 3.3 nM) with an unlabeled peptide (P8W5) [3]

4 Notes

1. An optimal fluorophore would have a fluorescence spectrum that does not overlap with the emission of aromatic amino acids (and therefore is in the visible region), a high brightness (the product of extinction coefficient and quantum yield), a high limiting anisotropy (r_o), a lifetime that optimizes the differentiation of the motions of the free and bound peptide, and a quantum yield close to 1 (so that this value is not expected to change significantly in the free and bound states). Regarding the fluorophore's lifetime, it should be considered that the rotational correlation times of an SH2 domain and of an octa-phosphopeptide are roughly 5 and 0.5 ns, respectively, based on molecular weights and approximating the molecules to a sphere [39]. The lifetime of the fluorophore should not be much shorter or longer than these two values; otherwise, the anisotropy would be maximal or 0 for both the free and bound peptide. The maximum difference in anisotropy between these two species is obtained with a lifetime falling between the two rotational correlation times. The value of r_o depends on the

angle between the absorption and emission transition dipoles, and it is maximal (0.4) when they are parallel to each other. This parameter is a property of the fluorescing molecule, and therefore it can be found in the literature for most common fluorophores. r_0 also depends on the excitation wavelength. To maximize it, the lowest energy absorption band should be used for excitation, so that the state excited by photon absorption and the emitting state usually coincide [28]. Some commonly used fluorophores that satisfy many of the conditions listed above are molecules of the fluorescein and rhodamine families [25, 31, 40].

2. The fluorescing peptide to be used for characterizing unlabeled peptides should have the highest possible affinity for the SH2 domain (*see* above), and therefore usually, it is desirable that the fluorophore does not interfere with binding. This can be obtained by separating the probe from the binding sequence with a linker. On the other hand, the linker should not allow the fluorophore to rotate freely; otherwise, the anisotropy would be very low both in the free and in the bound state. We have obtained a good compromise by separating the binding octapeptide sequence from the fluorophore by the introduction of a glycine residue at the N-terminus, labeled with carboxyfluorescein [3].
3. There are instances of SH2 domains comprised in protein tyrosine phosphatases (SHP1 and SHP2). In these cases, measuring phosphopeptide binding to the whole protein (rather than to the isolated domain) can be complicated by dephosphorylation of the pTyr by the phosphatase itself, during the assay. To solve this issue, a non-hydrolyzable pTyr mimic can be used [3].
4. Due to hygroscopicity of peptide powders, the concentration determined by weight is usually significantly overestimated. Therefore, the exact concentration of the stock solutions should be determined by absorbance measurements. In the case of labeled peptides, the absorbance of the probe can be used for this purpose. For unlabeled peptides, the extinction coefficient in the near UV can be estimated by using the published values for Phe, Tyr, Trp, and pTyr [37, 41]. Since all these molar extinction coefficients have been determined in aqueous solutions, the measurement should be performed after appropriate dilution of the stock solution in the measurement buffer, to obtain an absorbance value in the 0.1–1.5 range. To the best of our knowledge, no extinction coefficient has been previously reported for pTyr mimic phosphonodifluoromethyl phenylalanine (F2Pmp), which is used to replace pTyr in some phosphopeptide mimetics [3]. Therefore, to allow the quantification of the concentrations of peptides containing this amino

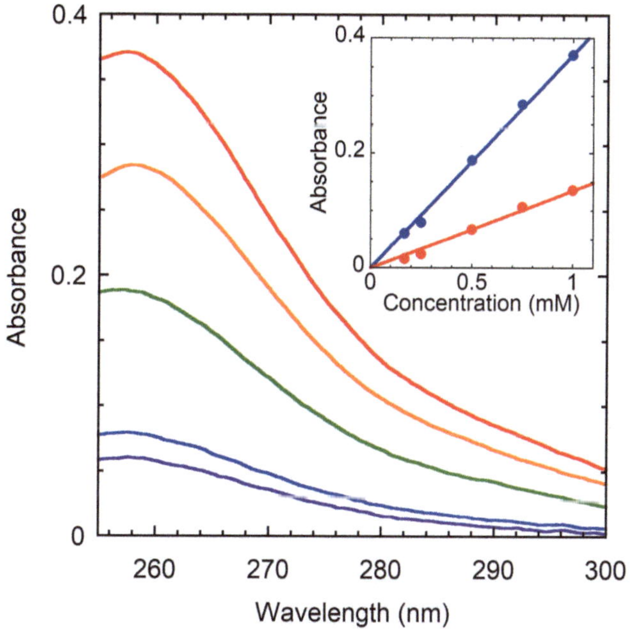

Fig. 3 Absorption spectra of F2Pmp. F2Pmp was at concentrations of 0.17, 0.25, 0.50, 0.75, and 1.0 mM in phosphate-buffered saline (pH 7.7). The inset shows a linear fit to the absorbances at 258 nm (the absorption maximum, in blue) and at 280 nm (where Trp has its absorption maximum, in red), resulting in molar extinction coefficients of 370 M^{-1} cm^{-1} and 140 M^{-1} cm^{-1}, respectively

acid, we determined the extinction coefficient ourselves. Fmoc-F2Pmp-OH (Iris Biotech) was bonded on 2-chlorotrityl chloride resin, and two cycles of deprotection (5 and 15 min, respectively) were performed to remove the Fmoc protecting group, using piperidine 20% in dimethylformamide (DMF). After washes with DMF and dichloromethane (DCM), the amino acid was detached using 1% trifluoroacetic acid (TFA) in DCM. The amino acid, obtained as TFA salt, was dried overnight in a dryer. A stock solution of the amino acid (5.6×10^{-3} M) in phosphate-buffered saline (pH 7.7) was prepared by weighing the amino acid powder. Solutions for absorbance measurements were obtained by dilution. Absorbance spectra were recorded with a UV-2501PC spectrophotometer (Shimadzu). Linear fitting of the data resulted in molar extinction coefficients of 370 M^{-1} cm^{-1} at 258 nm (the absorption maximum) and 140 M^{-1} cm^{-1} at 280 nm (Fig. 3).

5. The use of 5–10% glycerol is often reported in the purification protocols of SH2 domains or of the proteins that contain them, to increase protein stability [42, 43]. However, this compound should not be present in the measurement buffer, because it significantly increases the viscosity of the medium and consequently decreases the rotational correlation time of the labeled

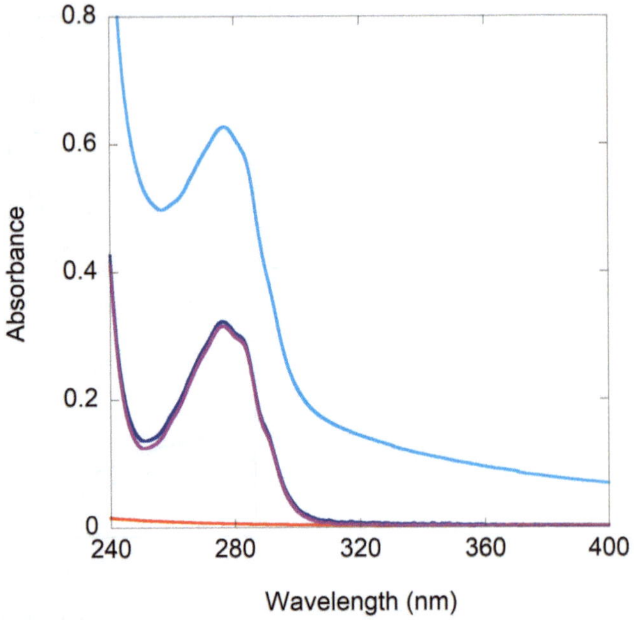

Fig. 4 Spectra of the N-SH2 domain of SHP2. SH2 domain samples either show aggregation (light blue) or are mostly in the monomeric state (dark blue). The red line represents the estimated light scattering contribution to the dark blue spectrum, which has been subtracted in the purple spectrum

peptide. If possible, glycerol should be avoided directly in the purification protocol. Alternatively, it can be removed before the anisotropy experiments by dialysis or by using a buffer exchange column.

6. The protocols can be adapted for the use of a plate reader fluorometer. A 96-well plate will typically be employed, with a volume of 200 μL for each well. In this case, every point of the titration is measured in a separate well, and the samples are prepared by serial dilutions [31, 44].

7. The presence of protein aggregates is detectable by the presence of artifactual absorption (caused by light scattering) above 310 nm (where the protein does not absorb) [45]. An example is reported in Fig. 4. If aggregates are present, they must be removed by centrifugation. When free Cys residues are present in the SH2 domain or in the protein, to avoid the formation of intermolecular disulfides, a small concentration (1 μM) of a reducing agent, such as TCEP (tris(2-carboxyethyl) phosphine) or DTT (dithiothreitol), can be added to the protein solution. If a minor contribution of apparent absorption, caused by aggregates that scatter the incident light, remains even after centrifugation, it must be subtracted before determining the protein absorbance. This can be achieved by fitting the absorbance at wavelengths above 310 nm with the

following phenomenological equation [45] and then determining the apparent absorbance value at the near UV wavelength used for concentration determination, usually 280 nm (Fig. 4):

$$A = a + \left(\frac{b}{\lambda}\right)^n \tag{17}$$

8. When a specific SH2 domain is assayed for the first time, it is advisable to check the stability of the anisotropy value of the protein/peptide complex, by measuring it every 5 min for at least 30 min (i.e., the approximate time needed to perform a whole titration). In some cases, particularly in the displacement studies where a nM protein concentration is used, the anisotropy signal of the complex might decrease over time. This problem can be caused by protein adsorption to the cuvette walls, which can be significant at the low protein concentrations used for high-affinity complexes. The presence of protein adsorption can be verified by measuring the intensity of intrinsic protein fluorescence as a function of time. Since the fluorescence light is collected at the center of the cuvette, emission by protein molecules adsorbed on the cuvette walls will not be detected. Therefore, protein adsorption will result in a progressive decrease in the signal. If a significant decrement is observed, protein adsorption can be minimized by treating the cuvette with hydrophilic polymers. After incubation with 5% (in weight) aqueous solutions of poly(vinyl alcohol) or poly (ethylene glycol), for at least 30 min, the cuvette must be rinsed with ultrapure water only and used for the experiments.
9. To completely characterize the binding curve, the protein concentration should be increased up to values of at least 10 K_d, because under these conditions the peptide will be almost completely bound. This is not always possible if the affinity of the interaction is low (and therefore the K_d is high) and if the domain or protein cannot be produced and purified at high enough concentrations or if they aggregate above a concentration threshold (causing a high intensity of scattered light to reach the detector and thus an artifactual increase in the anisotropy values). In these cases, the value of the anisotropy corresponding to the bound state (r_{bound}) can be left as a fitting parameter. Even in this way, the attainment of a significant degree of binding is necessary to determine r_{bound} reliably. Alternatively, r_{bound} can be derived from a binding curve corresponding to a phosphopeptide with a higher affinity interaction (labeled with the same fluorophore), for which complete binding has been attained. This is possible because r_{bound} is determined essentially by the fluorophore's properties and by the protein/domain hydrodynamic volume, which are the same for both peptides.

10. For fluorophores with a relatively small Stoke's shift, such as fluorescein of rhodamine, it might be convenient to select an excitation wavelength shorter than the value corresponding to the absorption maximum, to minimize the contribution of scattered excitation light to the fluorescence signal. Similarly, the wavelength of maximum emission is typically selected for the emission, but longer values can be used to increase the separation between excitation and emission (and consequently reduce the contribution of scattered light to the collected signal).
11. If the unlabeled peptide has a significantly higher affinity for the SH2 domain than the fluorescent peptide, then experiments performed in the presence of a protein concentration equal to K_d might yield excessive errors in the value of K_d^u, because an almost stoichiometric binding takes place. In these cases, the protein concentration can be decreased down to $K_d/2$. Lower values cannot be used, because the fraction of bound labeled peptide would be too small to provide a measurable signal change upon displacement.
12. Additions of unlabeled peptide should be performed from solutions obtained by appropriately diluting the stock solution in buffer. In the case of unlabeled peptides of low affinity, high concentrations can be reached by performing additions directly from the stock solution in organic solvent. However, the final concentration of solvent in the sample should always be lower than 1%.

Acknowledgments

This work was supported by AIRC Foundation for Cancer Research in Italy (grant IG24940, to L. S.), Regione Lazio (grant A0375-11322020-36719, to L. S.), and Italian Ministry of University and Research (grant PRIN 2020833Y75_005, to L. S.; grant MUR-PNRR M4C2I1.3 PE6 project PE00000019 Heal Italia, to L. S.).

References

1. Gopalasingam P, Quill L, Jeeves M, Overduin M (2015) SH2 domain structures and interactions. In: Kurochkina N (ed) SH domains. Springer, Cham, pp 159–185. https://doi.org/10.1007/978-3-319-20098-9_8
2. Kaneko T, Joshi R, Feller SM, Li SS (2012) Phosphotyrosine recognition domains: the typical, the atypical and the versatile. Cell Commun Signal 10:1–20. https://doi.org/10.1186/1478-811X-10-32
3. Bobone S, Pannone L, Biondi B et al (2021) Targeting oncogenic Src homology 2 domain-containing phosphatase 2 (SHP2) by inhibiting its protein–protein interactions. J Med Chem 62:15973–15990
4. Bradshaw JM, Waksman G (2002) Molecular recognition by SH2 domains. Adv Protein Chem 61:161–210. https://doi.org/10.1016/s0065-3233(02)61005-8
5. Machida K, Mayer BJ (2005) The SH2 domain: versatile signaling module and

pharmaceutical target. Biochim Biophys Acta (BBA) Proteins Proteomics 1747:1–25. https://doi.org/10.1016/j.bbapap.2004.10.005

6. Kuriyan J, Cowburn D (1997) Modular peptide recognition domains in eukaryotic signaling. Annu Rev Biophys Biomol Struct 26:259–288. https://doi.org/10.1146/annurev.biophys.26.1.259
7. Wagner MJ, Stacey MM, Liu BA, Pawson T (2013) Molecular mechanisms of SH2-and PTB-domain-containing proteins in receptor tyrosine kinase signaling. Cold Spring Harb Perspect Biol 5:a008987. https://doi.org/10.1101/cshperspect.a008987
8. Waksman G, Kumaran S, Lubman O (2004) SH2 domains: role, structure and implications for molecular medicine. Expert Rev Mol Med 6:1–18. https://doi.org/10.1017/S1462399404007331
9. Anselmi M, Calligari P, Hub JS et al (2020) Structural determinants of phosphopeptide binding to the N-terminal Src homology 2 domain of the SHP2 phosphatase. J Chem Inf Model 60:3157–3171
10. Koncz G, Tóth GK, Bökönyi G et al (2001) Co-clustering of Fcγ and B cell receptors induces dephosphorylation of the Grb2-associated binder 1 docking protein. Eur J Biochem 268:3898–3906. https://doi.org/10.1046/j.1432-1327.2001.02295.x
11. Bonetti D, Troilo F, Toto A et al (2018) Mechanism of folding and binding of the N-terminal SH2 domain from SHP2. J Phys Chem B 122:11108–11114. https://doi.org/10.1021/acs.jpcb.8b05651
12. Marasco M, Berteotti A, Weyershaeuser J et al (2020) Molecular mechanism of SHP2 activation by PD-1 stimulation. Sci Adv 6:eaay4458. https://doi.org/10.1126/sciadv.aay4458
13. Ladbury JE, Lemmon MA, Zhou M et al (1995) Measurement of the binding of tyrosyl phosphopeptides to SH2 domains: a reappraisal. Proc Natl Acad Sci U S A 92:3199–3203. https://doi.org/10.1073/pnas.92.8.3199
14. Panayotou G, Ladbury J (2000) Analysis of SH2 domain-phosphopeptide interactions by isothermal titration calorimetry and surface plasmon resonance. In: Reith AD (ed) Protein kinase protocols. Methods in molecular biology, vol 124. Humana Press, pp 295–311. https://doi.org/10.1385/1-59259-059-4:295
15. de Mol NJ, Dekker FJ, Broutin I et al (2005) Surface plasmon resonance thermodynamic and kinetic analysis as a strategic tool in drug design. Distinct ways for phosphopeptides to plug into Src-and Grb2 SH2 domains. J Med Chem 48:753–763. https://doi.org/10.1021/jm049359e
16. Gunzburg MJ, Ambaye ND, Hertzog JT et al (2010) Use of SPR to study the interaction of G7-18NATE peptide with the Grb7-SH2 domain. Int J Pep Res Ther 16:177–184. https://doi.org/10.1007/s10989-010-9222-z
17. Watson GM, Lucas WAH, Gunzburg MJ, Wilce JA (2017) Insight into the selectivity of the G7-18NATE inhibitor peptide for the Grb7-SH2 domain target. Front Mol Biosci 4:64. https://doi.org/10.3389/fmolb.2017.00064
18. McKercher MA, Wuttke DS (2017) Calorimetric measurement of SH2 domain ligand affinities. In: Machida K, Liu B (eds) SH2 domains. Methods in molecular biology, vol 1555. Humana Press, New York, pp 291–305. https://doi.org/10.1007/978-1-4939-6762-9_16
19. Jones RB, Gordus A, Krall JA, MacBeath G (2006) A quantitative protein interaction network for the ErbB receptors using protein microarrays. Nature 439:168–174. https://doi.org/10.1038/nature04177
20. Majkut P, Claußnitzer I, Merk H et al (2013) Completion of proteomic data sets by Kd measurement using cell-free synthesis of site-specifically labeled proteins. PLoS One 8:e82352. https://doi.org/10.1371/journal.pone.0082352
21. Kuropka B, Witte A, Sticht J et al (2015) Analysis of phosphorylation-dependent protein interactions of adhesion and degranulation promoting adaptor protein (ADAP) reveals novel interaction partners required for chemokine-directed T cell migration. Mol Cell Proteomics 14:2961–2972. https://doi.org/10.1074/mcp.M115.048249
22. Van de Weert M, Stella L (2011) Fluorescence quenching and ligand binding: a critical discussion of a popular methodology. J Mol Struct 998:144–150. https://doi.org/10.1016/j.molstruc.2011.05.023
23. Huang X, Aulabaugh A (2016) Application of fluorescence polarization in HTS assays. In: Janzen W (ed) High throughput screening. Methods in molecular biology, vol 1439, pp 115–130. https://doi.org/10.1007/978-1-4939-3673-1_7
24. Rainard JM, Pandarakalam GC, McElroy SP (2018) Using microscale thermophoresis to characterize hits from high-throughput screening: a European lead factory perspective. SLAS

Discov 23:225–241. https://doi.org/10.1177/2472555217744728

25. Jameson DM, MoczG (2005) Fluorescence polarization/anisotropy approaches to study protein-ligand interactions. In: Nienhaus U (ed) Protein-ligand interactions. Methods in molecular biology, vol 305. Humana Press, pp 301–322. https://doi.org/10.1385/1-59259-912-5:301
26. Jameson DM (2014) Introduction to fluorescence. Taylor & Francis, UK
27. James NG, Jameson DM (2014) Steady-state fluorescence polarization/anisotropy for the study of protein interactions. In: Engelborghs Y, Visser AJWG (eds) Fluorescence spectroscopy and microscopy. Methods in molecular biology (methods and protocols), vol 1076. Humana Press, Totowa, pp 29–42. https://doi.org/10.1007/978-1-62703-649-8_2
28. Jameson DM, Ross JA (2010) Fluorescence polarization/anisotropy in diagnostics and imaging. Chem Rev 110:2685–2708. https://doi.org/10.1021/cr900267p
29. Chalk SJ (2019) Online version of the IUPAC compendium of chemical terminology, 2nd edn. https://doi.org/10.1351/goldbook
30. Zhang Y, Zhang J, Yuan C et al (2011) Simultaneous binding of two peptidyl ligands by a Src homology 2 domain. Biochemistry 50: 7637–7646. https://doi.org/10.1021/bi200439v
31. Moerke NJ (2009) Fluorescence polarization (FP) assays for monitoring peptide-protein or nucleic acid-protein binding. Curr Protoc Chem Biol 1:1–15. https://doi.org/10.1002/9780470559277.ch090102
32. Goldstein A, Barrett RW (1987) Ligand dissociation constants from competition binding assays: errors associated with ligand depletion. Mol Pharmacol 31:603–609
33. Nikolovska-Coleska Z, Wang R, Fang X et al (2004) Development and optimization of a binding assay for the XIAP BIR3 domain using fluorescence polarization. Anal Biochem 332:261–273. https://doi.org/10.1016/j.ab.2004.05.055
34. Barlow R, Blake JF (1989) Hill coefficients and the logistic equation. Trends Pharmacol Sci 10: 440–441. https://doi.org/10.1016/S0165-6147(89)80006-9
35. Wang ZX (1995) An exact mathematical expression for describing competitive binding of two different ligands to a protein molecule. FEBS Lett 360:111–114. https://doi.org/10.1016/0014-5793(95)00062-e
36. Huang X (2003) Fluorescence polarization competition assay: the range of resolvable inhibitor potency is limited by the affinity of the fluorescent ligand. J Biomol Screen 8:34–38. https://doi.org/10.1177/1087057102239666
37. Pace CN, Vajdos F, Fee L et al (1995) How to measure and predict the molar absorption coefficient of a protein. Protein Sci 4:2411–2423. https://doi.org/10.1002/pro.5560041120
38. Rossi AM, Taylor CW (2011) Analysis of protein-ligand interactions by fluorescence polarization. Nat Protoc 6:365–387. https://doi.org/10.1038/nprot.2011.305
39. Cantor CR, Schimmel PR (1980) Biophysical chemistry: Part II: techniques for the study of biological structure and function. Macmillan, p 461
40. Zhang H, Wu Q, Berezin MY (2015) Fluorescence anisotropy (polarization): from drug screening to precision medicine. Expert Opin Drug Discov 10:1145–1161. https://doi.org/10.1517/17460441.2015.1075001
41. Bradshaw JM, Mitaxov V, Waksman G (1999) Investigation of phosphotyrosine recognition by the SH2 domain of the Src kinase. J Mol Biol 293:971–985. https://doi.org/10.1006/jmbi.1999.3190
42. Liu BA, Ogiue-Ikeda M, Machida K (2017) Expression and production of SH2 domain proteins. In: Machida K, Liu B (eds) SH2 domains. Methods in molecular biology, vol 1555. Humana Press, New York, pp 117–162. https://doi.org/10.1007/978-1-4939-6762-9_8
43. Qiu W, Wang X, Romanov V et al (2014) Structural insights into Noonan/LEOPARD syndrome-related mutants of protein-tyrosine phosphatase SHP2 (PTPN11). BMC Struct Biol 14:10. https://doi.org/10.1186/1472-6807-14-10
44. Yin L, Stern LJ (2014) Measurement of peptide binding to MHC class II molecules by fluorescence polarization. Curr Protoc Immunol 106:5.10.1–5.10.12. https://doi.org/10.1002/0471142735.im0510s106
45. Castanho MA, Santos NC, Loura LM (1997) Separating the turbidity spectra of vesicles from the absorption spectra of membrane probes and other chromophores. Eur Biophys J 26: 253–259. https://doi.org/10.1007/s002490050078

Chapter 7

Computational Evaluation of Peptide–Protein Binding Affinities: Application of Potential of Mean Force Calculations to SH2 Domains

Paolo Calligari, Lorenzo Stella, and Gianfranco Bocchinfuso

Abstract

Many biological functions are mediated by protein–protein interactions (PPIs), often involving specific structural modules, such as SH2 domains. Inhibition of PPIs is a pharmaceutical strategy of growing importance. However, a major challenge in the design of PPI inhibitors is the large interface involved in these interactions, which, in many cases, makes inhibition by small organic molecules ineffective. Peptides, which cover a wide range of dimensions and can be opportunely designed to mimic protein sequences at PPI interfaces, represent a valuable alternative to small molecules. Computational techniques able to predict the binding affinity of peptides for the target domain or protein represent a crucial stage in the workflow for the design of peptide-based drugs. This chapter describes a protocol to obtain the potential of mean force (PMF) for peptide–SH2 domain binding, starting from umbrella sampling (US) molecular dynamics (MD) simulations. The PMF profiles can be effectively used to predict the relative standard binding free energies of different peptide sequences.

Key words Molecular dynamics simulations, Umbrella sampling, Potential of mean force, Relative standard binding free energy, Peptide design, Protein–protein interactions

1 Introduction

1.1 Biological Relevance of SH2 Domain Binding Affinity and Selectivity

Protein–protein interactions (PPIs) mediate many functions in cells, including cell proliferation, apoptosis, and angiogenesis, mainly in the context of cascades transducing signals from the extracellular environment to the nucleus. Therefore, it is not surprising that misregulated PPIs cause several pathologies, including tumors. Proteomics studies estimate that 650,000 PPIs are present in the human interactome, compared to only 20,000 protein-coding genes [1, 2]. These interactions represent a huge number of potential targets in oncology and many other therapeutic areas. In spite of these possibilities, only a small fraction of the components in the cellular signaling pathways misregulated in the

Teresa Carlomagno and Maja Köhn (eds.), *SH2 Domains: Functional Modules and Evolving Tools in Biology*, Methods in Molecular Biology, vol. 2705, https://doi.org/10.1007/978-1-0716-3393-9_7,

disease has been targeted by drugs [3]. A couple of decades ago, PPIs were commonly considered "undruggable." This conclusion was based, in part, on the evidence that protein–protein interacting surfaces have areas in the range of 1000–2000 Å^2. These values are quite larger than those of typical ligand binding or catalytic sites that can be inhibited by small organic molecules, whose surfaces are in the range of 300–500 Å^2 [4]. For their intrinsic similarity to proteins, peptides are probably the most promising alternative to small molecules for the inhibition of PPIs. Even if their pharmacokinetics is not optimal, the skepticism on the usage of peptides as drugs is steadily decreasing, as more and more success stories reach the clinic [5], and nowadays, peptides represent one of the most investigated classes of molecules for inhibiting PPIs [6, 7].

The interactions of multi-domain proteins are often mediated by different classes of specific modules, including Src homology 2 (SH2) domains [8, 9], which selectively recognize polypeptides containing a phosphotyrosine (pY) residue [10]. For their role in many signaling pathways, they have received attention as potential pharmaceutical targets [11]. The affinity of SH2 domains for their partners comes from the pY pocket, which is quite invariant in this class of domains, and from the binding clefts interacting with residues immediately preceding and following pY, which account for the domain selectivity. The relative weights of these two contributions vary in different SH2 domains.

Computational tools are a key component of the drug-design workflow, mainly in the early stages of the process [12]. In this protocol, we discuss how the relative affinities for the SH2 domain of different pY-containing sequences can be predicted by molecular dynamics (MD) simulations. This kind of study can be used to investigate the determinants of selective binding by SH2 domains and to design artificial peptide sequences with high affinity and selectivity for a particular domain. As an example, we will use the N-terminal SH2 domain (N-SH2) of the SH2-domain-containing protein tyrosine phosphatase 2 (SHP2). This domain is a particularly interesting test case, for its high pharmacological relevance. Mutations in the gene coding for SHP2 are causative of developmental disorders [13] and childhood hematologic malignancies [14], and many of these mutations affect residues in the N-SH2 domain [15–18]. Furthermore, SHP2 is a hot target in cancer research, as it is required for the survival of RTK-driven cancer cells [19]; it plays a role in the intrinsic and acquired resistance to targeted cancer drugs [20], as a mediator of immune checkpoint pathways [21, 22] and in the induction of gastric carcinoma by *Helicobacter pylori* [23, 24]. The inhibition of the interaction of SHP2 with cognate proteins represents a promising pharmacological strategy to downregulate its function [25].

1.2 Statistical Thermodynamics of Equilibrium Association Processes

Many biological processes are mediated by non-covalent association between biomolecules. Therefore, a big effort has been made to develop computational protocols to predict the affinity in such complexes, and many tools have been proposed for this purpose over the years. Several reviews are available, which compare different approaches [26–29], and the readers interested in an overview of these methods are referred to these papers. Here, only a brief introduction is given, with a particular focus on the evaluation of standard binding free energy by potential of mean force (PMF) from umbrella sampling (US) simulations, on which this protocol is focused.

In solution, the reaction describing an equilibrium of association between a host molecule (H) and a guest (G), to form a complex (H:G), can be written as

$$\mathrm{H} + \mathrm{G} \rightleftharpoons \mathrm{H:G} \tag{1}$$

To estimate the standard free energy of binding ($\Delta G°$) of this equilibrium, the system is simulated by considering one molecule of solute (H, G, or H:G) in a simulation volume V and a large number N molecules of solvent (with $N >> 1$). In this case, $\Delta G°$ can be expressed as a function of the configuration integrals Z of H, G, and H:G according to [30]

$$\Delta G° = -kT \ln\left(\frac{V}{V_0}\frac{Z_{\mathrm{H:G},N} Z_{0,N}}{Z_{\mathrm{H},N} Z_{\mathrm{G},N}}\right) \tag{2}$$

where V is the volume of the simulation and V_0 is the volume corresponding to the 1 M standard concentration (1660°A^3).

$$Z_{i,N} = \int e^{-\frac{U(r_i)}{kT}} d\mathbf{r}_i \tag{3}$$

represents the configuration integrals of a system containing the solutes i (i.e., H, G, or H:G) solvated in a box of volume V containing N molecules of solvent and

$$Z_{0,N} = \int e^{-\frac{U(r_0)}{kT}} d\mathbf{r}_i \tag{4}$$

refers to a system containing only the N solvent molecules. In Eq. (3), $U(\boldsymbol{r}_i)$ is the potential energy function depending on the coordinates of both the solute (i) and the solvent ($\boldsymbol{r}_i$); $U(\boldsymbol{r}_o)$ in Eq. (4) is the potential energy function depending on the coordinate of the solvent only ($\boldsymbol{r}_o$); the integrals run on all the possible configurations of the system.

The terms $Z_{\mathrm{H},\ N}$ and $Z_{\mathrm{G},\ N}$ in the denominator of Eq. (2) represent two different boxes of volume V, containing one molecule of H and G, respectively, and N molecules of solvent. If V is sufficiently large, we can transfer the guest G from its own solvent box to the solvent box containing H, without affecting the product of the two configurational integrals. In this way, we will obtain two new systems, for which the configuration integrals are $Z_{H,\,G,\,N}$, referred to the box in which H and G are both present, but they do not interact, and $Z_{0,\,N}$, in which only the N solvent molecules are present. After these changes, $Z_{0,\,N}$ elides in Eq. (2), which becomes

$$\Delta G^{\circ} = -kT\ln\left(\frac{V}{V_0}\frac{Z_{H:G,N}}{Z_{H,G,N}}\right) \tag{5}$$

In Eq. (5), both $Z_{H\,:\,G,\,N}$ and $Z_{H,\,G,\,N}$ refer to systems containing the H, G, and N solvent molecules (in a volume V), but in $Z_{H\,:\,G,\,N}$ H and G form a complex, while in $Z_{H,\,G,\,N}$ they are not interacting.

1.3 Evaluation of Binding Affinity by Potential of Mean Force/Umbrella Sampling Simulations

1.3.1 The Potential of Mean Force

Equation (5) serves as a basis for the development of several computational strategies aimed at the evaluation of binding affinities, by determining the term $\frac{Z_{H:G,N}}{Z_{H,G,N}}$.

In many cases, a reaction coordinate (ξ) can be defined, which provides a clear distinction between the starting and final states, in our case the isolated H and G molecules and the H:G complex, respectively. ξ can be a mono or multidimensional coordinate, and it can be defined in different ways, including simple geometrical parameters, such as a distance [31].

The probability distribution of the system along this coordinate $P(\xi)$ can be calculated by integrating out all degrees of freedom except ξ:

$$P(\xi) = \frac{\int \delta[\xi(\boldsymbol{r}) - \xi]e^{-\frac{U(\mathbf{r})}{kT}}d\boldsymbol{r}}{\int e^{-\frac{U(\mathbf{r})}{kT}}d\boldsymbol{r}} \tag{6}$$

where δ represents the Dirac delta function and $\xi(\boldsymbol{r})$ the ξ value corresponding to the set of coordinates $\mathbf{r}$ of the system; the integral runs on all possible sets of $\mathbf{r}$ coordinates. In these terms, $P(\xi)\ d\xi$ can be considered as the probability of finding the system in a small interval $d\xi$ around ξ. Starting from $P(\xi)$, one can define the potential of mean force (PMF) with respect to a generic reference state, opportunely defined (ξ^*):

$$W(\xi) - W(\xi^*) = -kT\ln\left[\frac{P(\xi)}{P(\xi^*)}\right]. \tag{7}$$

Roughly speaking, PMF, first introduced by Kirkwood [32], is the potential acting on the system along the selected reaction coordinate, having averaged out all the other degrees of freedom of the system. PMF as defined in Eq. (7) can be seen also as the reversible work needed to move H and G from the distance ξ^* to ξ.

Equation (6) can be integrated in the ξ space corresponding to H:G (bound) or in that corresponding to H and G isolated (unbound); the ratio between these two quantities corresponds to the $\frac{Z_{\mathrm{H:G},N}}{Z_{\mathrm{H,G},N}}$ ratio in Eq. (5). In other words

$$\frac{Z_{\mathrm{H:G},N}}{Z_{\mathrm{H,G},N}} = \frac{\int_{\mathrm{bound}} P(\xi)d\xi}{\int_{\mathrm{unbound}} P(\xi)d\xi} \tag{8}$$

1.3.2 Umbrella Sampling Simulations

In MD simulations, a trajectory is built, in which the time evolution of the system is followed to explore the conformational and configurational space. If the system is ergodic, $P(\xi)$ could be estimated directly, by looking at the normalized frequency $f(\xi)$ of finding the system in the vicinity of a given value of ξ. Unfortunately, very often, plain MD simulations are not long enough for the ergodic hypothesis to be reasonably valid, and some kind of bias has to be introduced to improve the sampling along selected coordinates. Among the computational techniques employed to ensure an enhanced sampling along a selected reaction coordinate ξ, the US method is one of the most commonly used. US was first introduced by Torrie e Valleau [33] for Monte Carlo simulations but is now frequently applied also to MD simulations.

The basic idea of the US approach is to drive the system along all the ξ values by small discrete steps, each one sampled by means of independent simulations. In each simulation, the system is "forced" to sample a small interval of ξ values; to fully sample the ξ coordinate, it is important to ensure partial overlap between adjacent intervals. Thus, one can divide the range of ξ values into Λ intervals ("umbrella windows") and simulate the system in each interval separately, with a ξ-dependent bias potential. In each simulation, the system's Hamiltonian will read as

$$H_\lambda = H_0 + V_\lambda \tag{9}$$

where H_λ is the Hamiltonian of the λ-window, H_0 is the unperturbed contribution to H_λ, and V_λ is the biasing potential, which usually has the form of a harmonic potential

$$V_\lambda = \frac{1}{2} K_{\mathrm{restraint}} (\xi - \xi_\lambda)^2. \tag{10}$$

The Hamiltonian in Eq. (9) produces a perturbed distribution along the ξ coordinate. Simulations carried out in each "umbrella window," with different biasing potentials, give rise to biased probability densities ($P_\lambda^b(\xi)$). These biased probabilities are related to their unbiased counterparts ($P_\lambda^u(\xi)$) by [34, 35]:

$$P_{\lambda}^{u}(\xi) = P_{\lambda}^{b}(\xi) / \left[\frac{e^{-V_{\lambda}/kT}}{\langle e^{-V_{\lambda}/kT} \rangle} \right] \tag{11}$$

where $\langle \ldots \rangle$ indicates an average over the configuration space [35]. In this context, the global probability density $P^{u}(\xi)$ and the related PMF $W(\xi)$ result from the Λ different unbiased probability densities $P_{\lambda}^{u}(\xi)$. Different methods have been proposed to calculate the global $P^{u}(\xi)$ from a set of biased simulations [35, 36]. Probably the most used method to obtain the unbiased PMF, starting from US simulations, is the Weighted Histogram Analysis Method (WHAM) [34].

The application of WHAM analysis requires attention, beyond the normal problems usually faced in classical MD simulations (for a non-exhaustive discussion, *see* **Note 1**). In particular, the application of WHAM to "umbrella windows" assumes an equal quality of sampling in each window. When this assumption is not satisfied, artifacts in PMF evaluation may appear; in these cases, longer simulation times are required.

Performing different independent evaluations of the same PMF can give an estimation of the statistical errors associated with the obtained data (*see* below). When this is too expensive to obtain, bootstrapping analysis can help [37]. Interestingly, Hub and coworkers suggest that to estimate uncertainties in the presence of long (possibly unknown) autocorrelations, many short umbrella simulations are preferable with respect to a few long simulations [37].

Finally, when the US–PMF is performed along a single reaction coordinate, the choice of the coordinate itself may influence the PMF profile. Different from other techniques for $\Delta G°$ evaluation, US–PMF can be used also to evaluate kinetic parameters, including the residence time of the guest in the binding site. In these cases, the coordinate influences the results significantly and has to be selected with care [38–40]. This problem is less relevant for the calculation of thermodynamic parameters, which do not depend strongly on the selected coordinates. However, the height of the potential energy barrier along the path can still affect the convergence of the calculated PMF, and a good choice of the coordinate remains desirable [39]. The usage of a single reaction coordinate could be considered an oversimplification. This is probably true to evaluate absolute binding free energies. For example, Roux and Woo [41] decomposed the unbinding process into several stages, each one evaluated independently. Of interest, in their paper, Roux and Woo used as an example the binding between an SH2 domain and a pY-containing peptide. Notwithstanding the recent attempts to include these procedures in automated protocols [42–44], they remain quite hard to manage for inexperienced researchers. However, if relative binding energies between different molecules are needed, simpler approaches can be used [27]. In paragraph 2.5, we will discuss this aspect, showing that the calculation of PMF along a single coordinate, namely, the relative distance of peptide and protein, was shown to efficiently capture the relative affinities of different peptides for an SH2 domain [25]. The simplicity of the discussed protocol makes its application feasible also for researchers without well-established expertise in this kind of calculation and demands lower CPU resources with respect to other available approaches.

1.4 PMF Evaluation from US Simulations to Investigate Peptide Binding to SH2 Domains

In the previous sections, we have summarized the general theoretical principles of the US–PMF technique. Here, we want to discuss aspects more directly related to the application of US–PMF to the binding of peptides to SH2 domains.

In the case of peptide–SH2 domain complexes, many crystallographic structures are available, so that the binding site and binding pose are usually known. In addition, these complexes typically have dissociation constants in the order of 10^{-6} M or lower (*see* Bobone et al., in this same volume and Ref. [25]). These values are amenable to the prediction by computational methods, while association processes with lower affinities can be more challenging [29].

In addition to the US–PMF methods discussed here, a valuable alternative approach for the analysis of binding processes is provided by free energy perturbation (FEP) methods. Here below, we briefly compare the two techniques, for the specific case of peptide–SH2 domain complexes. Electrostatic interactions play a major role in the binding between SH2 domains and cognate sequences containing pY residues (pY has a −2 charge at physiological pH and it interacts with a positively charged pocket in the SH2 domain). Charged residues can be present also in the rest of the sequence interacting with the SH2 domain, or they could be introduced during the design process [25, 45]. Notwithstanding the recent progress in FEP methods [46], their application to binding processes where electrostatic interactions play a major role remains challenging. US–PMF works better than FEP also in the case of large and flexible ligands, as is the case for the peptides considered here. Finally, US–PMF is not ideal for buried binding sites, for which the FEP approach is generally considered to be more indicated. However, the binding sites of SH2 domains can generally be considered superficial, even if in some cases they can be partially occluded by loops (*see* below). Overall, US–PMF is a good approach to study the binding between peptides and SH2 domains.

2 Materials

1. GROMACS simulation package [47]. Basic knowledge about GROMACS is required to set up the MD simulations and analyze the trajectories. The WHAM equations were solved using the implementation available in the GROMACS package [37].

2. UCSF Chimera software package [48]. The UCSF Chimera software package has been used to obtain the structural figures. For the analysis of the trajectories, the tools available in both the GROMACS and Chimera software have been used.
3. The simulations can be performed on any recent Linux workstation equipped with a graphic card capable of 3D acceleration, and the GROMACS software package can be easily compiled by following the instruction in the GROMACS user manuals; however, the usage of a cluster dedicated to parallel computations is definitively preferable.
4. Initial structures. These have been obtained from Protein Data Bank (PDB; www.pdb.org) and, when needed, modified using Chimera [48].

3 Methods

In the following, we will present the protocol for the calculation, by US simulations, of the PMF for the binding of a peptide sequence to an SH2 domain. The N-SH2 domain of SHP2 complexed with an octapeptide fragment from IRS1 (centered on pY 1172) will be used as an example.

3.1 System Preparation

The first step is to prepare the initial structure of the SH2 domain complexed with the peptide ligand. Here, we give the example of the SHP2 N-SH2 domain in complex with the octapeptide from IRS1-pY 1172, with sequence LN-pY-IDLDL (Fig. 1). Different

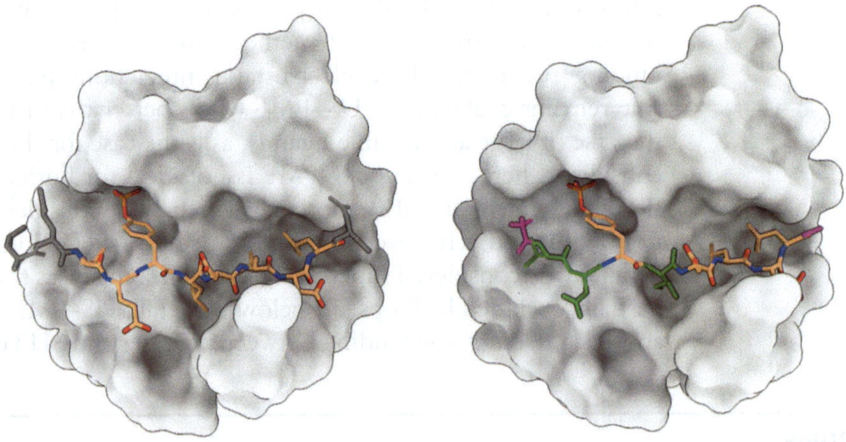

Fig. 1 Preparation of the initial structure. The crystallographic structure of N-SH2 domain from SHP2 in complex with GAB1 peptide (PDB code: 4QSY) is shown on the left. The same structure with amino acid substitutions in the complexed peptide IRS1-pY1172 is shown on the right side. Some peptide residues are highlighted in both structures: residues not directly interacting with the SH2 domain, mutated residues, and terminal caps are shown in dark gray, green, and pink, respectively

structures are available for peptide–SH2 domain complexes [45]. In each simulative investigation of a specific system, start by analyzing the available structures to identify the one closer to the complex of interest:

1. Start from an experimentally obtained structure (NMR, X-ray) of the SH2 domain in a complex with a peptide. In our example, we start from the crystallographic structure (PDB code 4QSY) of the protein complexed with the peptide sequence from the SHP2 partner GAB1 (sequence GDKQVE-pY-LDLDLD). The sequence is longer than the one we are interested in, and it presents different amino acids at positions −1, −2, and +1 (here and hereafter, the peptide residues are counted with respect to the pY). The next two steps below are necessary to remove these differences.
2. Remove from the initial structure the peptide residues exceeding the range of residues known to directly interact with the SH2 domain [45]. In our example, we remove residues beyond −2 and +5 from the experimental structure.
3. *Amino acid substitutions.* Where needed, apply amino acid substitutions on the peptide sequence presents in the available structure. In our example, we change residues of GAB1 peptide at positions −2, −1, and +1 to obtain IRS1-pY1172 sequence:

Position −2: Val → Leu
Position −1: Glu → Asn
Position +1: Leu → Ile

Side-chain configurations for substituted residues can be chosen with different methods (*see* **Note 2**). The most straightforward way to perform this task is to take the most probable configuration from the backbone-dependent rotamer library [49]. For this purpose, we used the UCSF Chimera software package, which was employed also to add ACE and NHE terminals. These additions were needed to allow comparison of the binding free energy obtained from PMF calculations with experimental ones obtained by fluorescence anisotropy assay (discussed in Bobone et al., this volume and in Ref. [25]).

3.2 System Equilibration

The following steps are necessary to set up the whole system before the production simulations. GROMACS commands are explicitly provided only when nonstandard options are used [50]. In the following, interactions for protein and phosphopeptide atoms are modeled by a variant of AMBER99SB force field with parameters for phosphorylated residues [51]. Water molecules are described by the TIP3P model [52]. Long-range electrostatic interactions are calculated using the particle mesh Ewald method [53], and the cutoff distance for the nonbonded interaction is set equal to 12.0 Å. The LINCS constraint is applied to all the hydrogen atoms and a 2 fs time-step is used [54]:

1. *Creation of topology files.* Convert initial PDB file into coordinate and topology files for GROMACS, with *pdb2gmx.* This tool allows the interactive selection of the force field, and the corresponding parameters are inserted in the topology file (a user-defined force field, like the one used here, should be placed in the working directory with a ".ff" suffix). Here, when simulating SH2 domains, particular attention should be paid to the protonation state of the highly conserved histidines that may interact with the phosphopeptide. It is strongly recommended to estimate local pKa for histidines taking into account their molecular environment in the peptide–protein bound state. For this task, PropKa [55, 56] in its version 3.1 (available in www.github.com/schrodinger/propka-3.1) can be used for peptide–protein complexes containing pY residues. As an alternative, the H++ algorithm [57] is implemented in a webserver (newbiophysics.cs.vt.edu/H++/), where it is possible to estimate the pKa simply by providing the structure of the pY-containing complex in PDB format. For specific issues of SH2 domains, *see* **Note 3**:

```
gmx pdb2gmx -f complex.pdb -ignh -his -o conf.gro
```

2. *Simulation box setup.* Insert the initial structure in a periodic box. The orientation of the complex and the dimension of the periodic box must be set with care. The ligand–protein complex is usually placed in the box by making the ligand unbinding direction (*see* below) correspond to one of the system axes. In the next step, the ligand will be pulled apart from the protein along the chosen direction, i.e., the chosen axis, to explore different configurations along the reaction coordinate. For this reason, in order to avoid artifacts due to periodic boundary conditions, the size of the simulation box along the chosen axis has to be significantly larger than in the other two dimensions: for SH2 domains, whose structure is almost spherically shaped, a ratio of 2:1 should be considered as a good choice. For the example given here, we place the complex at one side of the orthorhombic box with the complex center of mass at position (1.8 nm, 3.0 nm, 3.5 nm) (Fig. 2). Since the reaction coordinate is aligned along the X-axis, create a periodic box with dimensions 15.0 nm × 7.0 nm × 7.0 nm:

```
gmx editconf -f conf.gro -o conf_newbox.gro
-center 1.8 3.0 3.5 -box 15.0 7.0 7.0
```

3. *Solvation.* Add solvent molecules to the system. Ions have to be added to neutralize the whole system's total charge;

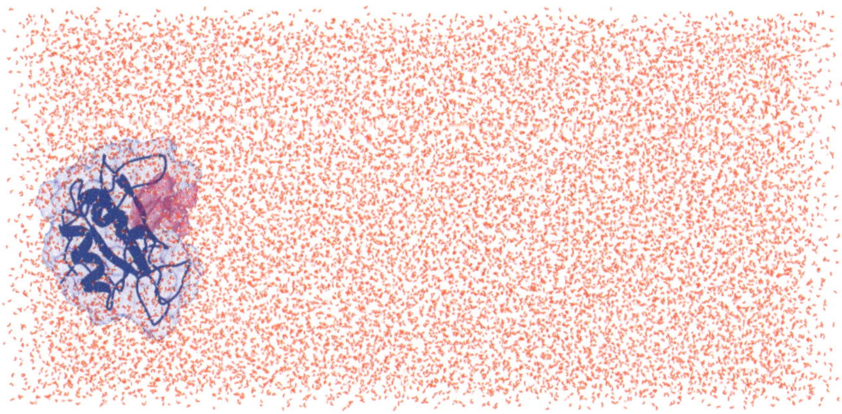

Fig. 2 Initial system setup. Initial position of the peptide–protein complex with respect to the longest axis in the simulation box (X-axis). The protein and the peptide are in blue and pink, respectively

sometimes, further ions are added to assure a specific ionic strength. In the given example, we add 24,000 TIP3P pre-equilibrated water molecules and 6 Na^+ counterions to neutralize the total system charge.

4. *Energy minimization*. Minimize the energy potential of the system. In the example given here, we apply a "steepest descent" algorithm for 50,000 cycles or until a maximum force equal to 1000 kJ/(mol·nm) is reached, by keeping protein positions fixed and leaving water and ions to adjust freely.
5. *System heating*. Progressively heat the system, increasing temperature from 100 to 300 K.
6. *Thermal equilibration*. Equilibrate the system for 100 ps in the NVT ensemble at 300 K, using velocity rescaling with a stochastic term, with a relaxation time of 1 ps [58].
7. *Pressure equilibration*. Equilibrate the system for 500 ps in the NPT ensemble at constant pressure (10^5 Pa), using the Parrinello–Rahman barostat, with a relaxation time of 5 ps.

3.3 Sampling of the Initial Configurations for US Simulations

Before starting with the US simulations, two preliminary steps have to be performed: (i) the choice of a binding/unbinding pathway that will define the reaction coordinate and (ii) the preparation of a set of configurations of the system with the ligand and the protein at different distances along the unbinding pathway, i.e., at different values of the reaction coordinate. In the following, we will illustrate how to perform these two stages by simulating the progressive separation of the protein and peptide center of masses (COMs).

In US simulations, the choice of the binding/unbinding pathway along which calculate PMF influences the convergence of the system, and pathways with lower energetic barriers are desirable.

This is particularly relevant if the binding pocket is not completely exposed to the solvent (*see* **Note 4**). In the given example, we define the collective variable for PMF calculation as the distance along a vector that is approximately perpendicular to the cavity enclosed by the EF and BG loops in the N-SH2 domain (*see* **Notes 4** and **5**):

1. *Pull dynamics*. Pull apart the COMs of protein and phosphopeptide by applying a harmonic potential between them and by moving outward the equilibrium distance at a constant rate. This simulation must be long enough to reach configurations in which the phosphopeptide is far from the SH2 domain and does not interact with it. On the other hand, as stated before, the distance between the two COMs should not be affected by periodic boundary conditions. To avoid this problem, the maximal distance reached during the pulling simulation must be less than or equal to half of the X-dimension (the largest dimension) of the simulation box built in the previous step. The pull dynamics could also serve as an a posteriori check of the chosen ratio between X and the other dimensions of the simulation box: if the pulling force reaches a plateau around zero at distances smaller than the half of the X-dimension, then the choice is correct, and one can go to the US steps.

   To prepare the pulling direction, align the chosen vector along the X-axis, using, the "Axis/Planes/Centroids" tool ("Structural Analysis" menu) of the UCSF Chimera package. First, create the two centroids that define the vector; then, create an axis between the two centroids; finally, align the axis along the system's X-axis.

   In the example given here, to define the reaction coordinate in GROMACS, we first define the SH2 domain and the peptide as "Chain_A" and "Chain_B," respectively, in the index.ndx file. Then, we set the coordinate geometry type as "distance." Then, we apply a harmonic potential with a force constant $K = 1000$ kJ/(mol·nm^2) and with a constant rate of 0.0025 nm/ps. The length of each simulation is about 2.5 ns. During this step, a positional restraint (1000 kJ/(mol·nm^2) is applied to all heavy atoms in the SH2 domain except for atoms in loops around the binding region (residues 30–45, 52–75, 80–100, in this example).

   The pull-dynamics options in the input *mdp* file for GROMACS would read as.

```
pull = yes
pull_ngroups = 2
pull_ncoords = 1
```

```
pull_group1_name = Chain_A
pull_group2_name = Chain_B
pull_coord1_type = umbrella
pull coord1_geometry = distance
pull_coord1_groups = 1 2
pull_coord1_dim = Y N N
pull_coord1_rate = 0.0025
pull_coord1_k = 1000
pull_coord1_start = yes
```

2. *Definition of the US initial configurations.* The chosen unbinding pathway can be used now to generate the set of configurations that will serve as starting points for the US simulation "umbrella windows." Save system's configurations from the pull-dynamics trajectory at every 1 Å in the COM distance space. For distances larger than 25 Å, configurations can be saved every 2 Å since, at these distances, peptide–protein interactions are unlikely to give significant contributions to the binding. In our example, we save configurations with COM distances in the range 10–35 Å. We obtain ~20 replicas of the system along the reaction coordinate.

3.4 US Simulations

The final stage is the US simulation itself. For the general principles, please see the introduction. In the following, we will present the specific issues related to the applications to SH2 domains:

1. *Equilibration of predetermined initial configurations.* Equilibrate the system in each of the previously obtained configurations for 1 ns. Here, the global roto-translational fit of the protein domain, which is customarily used in plain molecular dynamics simulations, cannot be used, because it would alter the intermolecular interactions between the SH2 domain and the peptide. For this reason, global rotations of the SH2 domain during the simulations are avoided by applying a strong positional restraint (1000 kJ/(mol·nm ^{2})) to all alpha carbons atoms except for those in loops flanking the binding region (as in the pull simulation).
2. *Production run.* Start the production run (150 ns in the discussed example) with the same restraints as in the previous step. During this stage, a harmonic potential (K = 1000 kJ/mol·nm^2) is applied on the distance between the two COMs. The US options in the *mdp* input file for GROMACS read as

```
pull = yes
pull_ngroups = 2
pull_ncoords = 1
pull group1 name = Chain_A
pull_group2_name = Chain_B
```

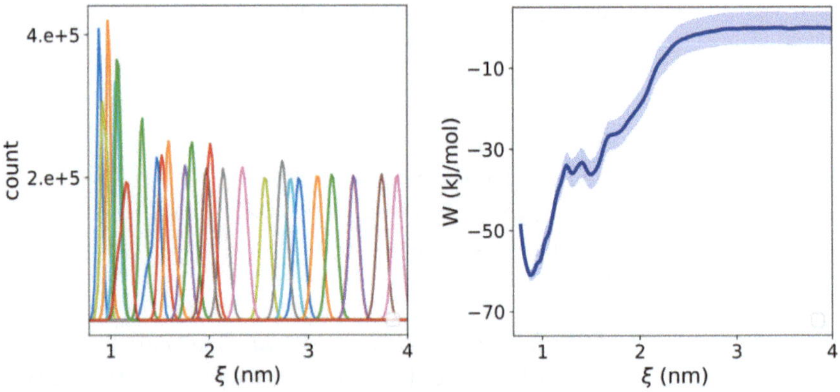

Fig. 3 US simulation and PMF calculation. Left panel: histograms of values of the reaction coordinate explored in each "umbrella window"; 200 bins have been considered for the whole range. Right panel: final PMF obtained by applying the WHAM analysis to data from "umbrella windows." Errors are reported as blue shadow and they have been estimated by the bootstrap method

```
pull_coord1_type = umbrella
pull_coord1_geometry = distance
pull_coord1_groups = 1 2
pull_coord1_dim = Y N N
pull_coord1_rate = 0.0
pull_coord1_k = 1000
pull_coord1_start = yes
```

These lines are very similar to those used in the previous step of pull dynamics, except for the `pull_coord1_rate` which is now set to zero to restrain the two groups at the initial position.

3. *Calculation of the potential of mean force (PMF)*. Evaluate the PMF by applying WHAM [34] to the histograms of the probability distribution $P(\xi)$ of the system along the reaction coordinate from each "umbrella window" (*see* Fig. 3 and **Notes 6** and 7). The confidence of each point can be evaluated by estimating the statistical errors (*see* **Note 8**). Good overlap between the distributions of adjacent windows is essential for the correct prediction of the PMF profile. However, a compromise is necessary between this requirement and the overall computational cost of the simulation. As a general rule, overlaps of at least 5% can be considered sufficient, but higher overlaps are desirable for lowering the statistical errors. For overlaps falling below this critical value, further windows should be added to improve the quality of the PMF (*see* **Note** 7). In our example, we calculate the PMF from the "umbrella windows," with default settings (50 bins and tolerance of 10^{-6} kJ mol^{-1}), by using the gmx wham GROMACS tool [37]. Statistical errors

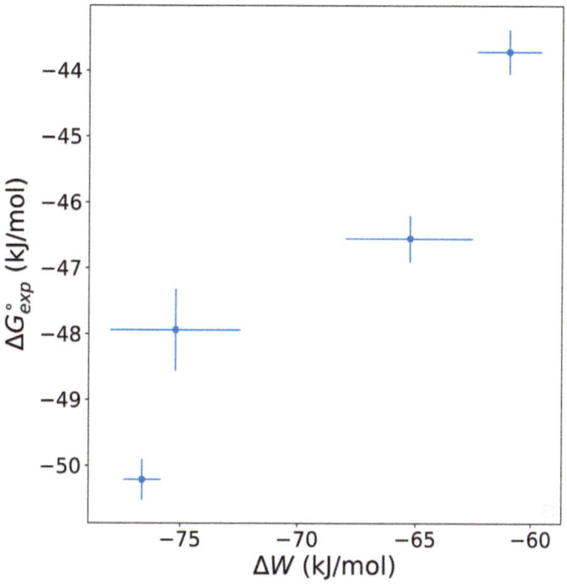

Fig. 4 Comparison between computational and experimental binding free-energy values. Data taken from Ref. [25]. $\Delta G^{\circ}_{\text{exp}}$ is calculated from the experimental dissociation constants and ΔW is obtained from US–PMF simulations

are determined with bootstrap analysis by considering the histograms as independent data points:

```
gmx wham -it tpr-files.dat -if pullf-files.dat -o -hist -unit kJ
-bsres profile_av-std.xvg -nBootstrap 100 -bs-method hist
```

Figure 3 shows the final PMF profile for the N-SH2 domain in complex with the peptide of sequence LN-pY-IDLDL together with an estimation of the statistical error for each point.

3.5 Standard Binding Free-Energy Estimation and Comparison with Experimental Data

Use PMF to estimate the relative binding free energy. In Fig. 4, we report a comparison between the experimentally obtained free-energy $\Delta G^{\circ}_{\text{exp}}$ (determined from the experimental dissociation constant) and the $\Delta W = W_{\text{min}} - W_{\text{unbound}}$, where W_{min} and W_{unbound} represent the PMF value corresponding to the global minimum and to the plateau of the curve in Fig. 3 (right panel), for different phosphopeptide sequences, in association to the N-SH2 domain of SHP2. The data are taken from [25] and they show that the PMF value captures the standard free energy differences between the investigated systems. This approximate approach works well when similar systems are compared (as in this case), for which the form of the PMF profile remains quite similar so that the depth of the well in the PMF curve is representative of the whole PMF profile [25, 59, 60].

4 Notes

1. Systematic errors can arise from computational techniques based on MD simulations. First, simulations usually employ a classical force field: different assumptions in this approach can bias the obtained results (e.g., fixed charges, harmonic potentials, and absence of terms for polarizability). Second, the limited dimension of the simulated systems could introduce problems deriving from unreal periodicities. Finally, the MD protocol can introduce persistent longtime correlations or integration errors [61]. Assessing the effects of these potential issues is not simple, but they can be limited by using well-established simulation protocols. Comparison of simulative and experimental results, at least for a subset of the systems being investigated, is the best validation of the computational methods, whenever it is possible [62].
2. Side-chain conformational rearrangement upon amino acid substitution can be estimated by using rotamer libraries as a reference, as in the present protocol, and selecting the most probable conformation compatible with the local steric constraints. Several rotamer libraries are available in the literature [63, 64]. When steric clashes are likely to happen during the first steps of simulation, a preliminary local energy minimization could also be performed [65]. An alternative approach is based on comparative models, which select the most probable conformations on the basis of analogous local conformations found in the PDB database [66].
3. The application of simulative techniques based on classical force fields can be difficult if titratable residues play a role in the investigated phenomena. SH2 usually presents a conserved histidine in the binding pocket of the pY residue (the βD4 residue) [67], and, in principle, its protonation state could change between the bound and unbound states. This could be a problem if absolute standard free energies have to be calculated, but it is a minor issue if the goal is to predict the relative affinities of different sequences, as in the protocol described here. In this case, it can be assumed that any issue in the protonation state of histidine residue in βD4 plays, *ceteris paribus*, the same role for all the sequences. Of note, in the example discussed here, the N-SH2 domain of SHP2 does not contain a histidine in the βD4 position.
4. The binding pockets of SH2 domains can be considered as borderline between embedded and superficial. On the one hand, the binding site involves a wide region on the SH2 domain surface. On the other hand, loops of the domain can act as regulatory gates, shielding part of the bound peptide

from the solvent. For the N-SH2 domain of SHP2, two loops, EF (residues 66–69) and BG (residues 84–96), partially occlude the hydrophobic region of the binding pocket that should allocate the C-terminal region of the peptide partner. In these cases, a search for the best unbinding pathway needs to be performed, taking into account the main interactions that guide the binding process. For the N-SH2 domain of SHP2, the binding of peptides is essentially dominated by two contributions: the pY, located in its binding pocket, which is highly conserved in SH2 domains, and the interaction of the peptide C-terminal region with the region flanked by the EF and BG loops.

These observations suggest three different possible unbinding directions defined by the following:

(i) The vector joining the phosphate and the alpha carbon in pY (shown as a green arrow in Fig. 5).

(ii) The vector defined by the initial position of the two COMs (blue arrow in Fig. 5).

(iii) A vector approximately in the perpendicular direction to the surface of the cavity flanked by the EF and BG loops (orange arrow in Fig. 5).

Figure 5 shows that pulling along the three directions requires rather different unbinding forces. The latter, plotted against the distance between peptide and protein COMs, shows not only the maximal force needed to start detaching the peptide from the N-SH2 domain along the different pathways but also the presence of local minima that indicate transient interactions between the two counterparts in the pulling

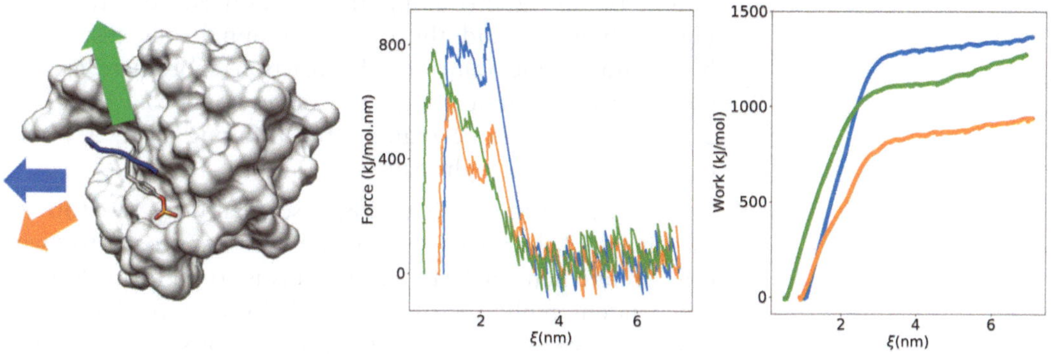

Fig. 5 Search for the optimal unbinding pathway. Left panel: tested unbinding directions shown as arrows with the SH2–peptide structure as reference: the vector joining the phosphate and the alpha carbon in pY (green); the vector defined by the initial position of the two COMs (blue); a vector approximately in the perpendicular direction to the surface of the cavity flanked by the EF and BG loops (orange). Middle panel: forces exerted for peptide unbinding in pulling dynamics along the three test directions (same colors as in the left panel). Right panel: irreversible work done by pulling forces along the test directions (same colors as in the left panel)

trajectory. Among the three different pathways, the third direction clearly encountered lower energy barriers by the N-SH2 loops that close the +3–+5 position, thus ensuring faster convergence of configurational space in each "umbrella window."

5. The way in which this direction is restrained needs to be discussed. When an internal distance (r) is considered, a correction term is needed to make the unbound region flat, equal to $2RT\ln(r)$, deriving from the Jacobian of the coordinate conversion; this term can be seen also as an entropic contribution deriving from larger volumes being accessible at higher r values. The term is not needed if the position along a Cartesian axis is chosen as the collective coordinate, as in the example discussed here [39].

6. There are several WHAM implementations available to the scientific community. Besides the GROMACS *wham* module, used in the present protocol, this calculation can be performed also with standalone software like PyWham [68] and Grossfield's C implementation [69].

7. To ensure that all points in the PMF are calculated with similar statistics, before applying WHAM, one should check that histograms are, as much as possible, uniformly distributed along the reaction coordinate with a non-negligible overlap. Ideally, an overlap of about 5–10% between consecutive histograms could be considered as acceptable. Nonetheless, this condition may be difficult to obtain and additional "umbrella windows" may be needed to ensure the correct overlap between histograms and allow the calculation of PMF with acceptable errors.

   The histograms shown in Fig. 3 suggest that the scheme used for the "umbrella windows" ensures an acceptable distribution of histograms along the reaction coordinate. Here, Fig. 6 shows instead the case of nonuniformly distributed histograms: some values of the reaction coordinate are not covered by histograms (e.g., at ξ values equal to 1.2 nm and 2.3 nm). In the same regions, the resulting PMF is poorly defined (*see* Fig. 6, right panel).

8. As discussed in the introduction, a seminal work by Hub and coworkers [37] showed that an alternative approach to "window" refinement in PMF calculations is to perform different independent replicates of US simulations on predetermined US "windows" disregarding the presence of significative gaps between histograms. The statistical errors can then be calculated over the set of independent runs, with a significant improvement in the estimation of errors themselves. Nonetheless, this approach can become highly demanding in terms of computational costs when calculations on large systems have to be performed.

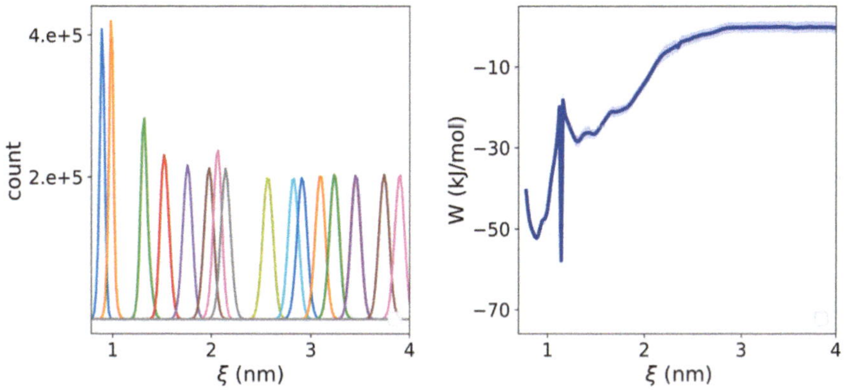

Fig. 6 Incomplete coverage of reaction coordinate values. Left panel: histograms of values of the reaction coordinate explored in each "umbrella window." Right panel: PMF obtained by applying the WHAM analysis to data from "umbrella windows." Errors are reported as blue shadow and they have been estimated by the bootstrap method

References

1. Stumpf M, Thorne T, de Silva E et al (2008) Estimating the size of the human interactome. Proc Natl Acad Sci 105(19):6959–6964
2. Azzarito V, Long K, Murphy N, Wilson A (2013) Inhibition of α-helix-mediated protein–protein interactions using designed molecules. Nat Chem 5(3):161–173
3. Laraia L, McKenzie G, Spring D et al (2015) Overcoming chemical, biological, and computational challenges in the development of inhibitors targeting protein-protein interactions. Chem Biol 22(6):689–703
4. Arkin M, Tang Y, Wells J (2014) Small-molecule inhibitors of protein-protein interactions: progressing toward the reality. Chem Biol 21(9):1102–1114
5. Muttenthaler M, King G, Adams D, Alewood P (2021) Trends in peptide drug discovery. Nat Rev Drug Discov 20(4):309–325
6. Bullock B, Jochim A, Arora P (2011) Assessing helical protein interfaces for inhibitor design. J Am Chem Soc 133(36):14220–14223
7. Siegert T, Bird M, Makwana K, Kritzer J (2016) Analysis of loops that mediate protein–protein interactions and translation into submicromolar inhibitors. J Am Chem Soc 138(39):12876–12884
8. Seo M, Kim P (2018) The present and the future of motif-mediated protein–protein interactions. Curr Opin Struct Biol 50:162–170
9. Dar K, Bhat A, Amin S et al (2019) Exploring proteomic drug targets, therapeutic strategies and protein – protein interactions in cancer: mechanistic view. Curr Cancer Drug Targets 19(6):430–448
10. Liu BA, Jablonowski K, Raina M et al (2006) The human and mouse complement of SH2 domain proteins—establishing the boundaries of phosphotyrosine signaling. Mol Cell 22(6): 851–868
11. Bradshaw JM, Waksman G (2002) Molecular recognition by SH2 domains. Adv Protein Chem 61:161–210
12. Tutone M, Almerico A (2021) Computational approaches: drug discovery and design in medicinal chemistry and bioinformatics. Molecules 26(24):7500
13. Tartaglia M, Mehler E, Goldberg R et al (2001) Mutations in PTPN11, encoding the protein tyrosine phosphatase SHP-2, cause Noonan syndrome. Nat Genet 29(4):465–468
14. Tartaglia M, Niemeyer C, Fragale A et al (2003) Somatic mutations in PTPN11 in juvenile myelomonocytic leukemia, myelodysplastic syndromes and acute myeloid leukemia. Nat Genet 34(2):148–150
15. Tartaglia M, Martinelli S, Stella L et al (2006) Diversity and functional consequences of germline and somatic PTPN11 mutations in human disease. Am J Hum Genet 78(2): 279–290
16. Bocchinfuso G, Stella L, Martinelli S et al (2007) Structural and functional effects of disease-causing amino acid substitutions affecting residues Ala72 and Glu76 of the protein tyrosine phosphatase SHP-2. Proteins Struct Funct Bioinf 66(4):963–974

17. Martinelli S, Torreri P, Tinti M et al (2008) Diverse driving forces underlie the invariant occurrence of the T42A, E139D, I282V and T468M SHP2 amino acid substitutions causing Noonan and LEOPARD syndromes. Hum Mol Genet 17(13):2018–2029
18. Martinelli S, Nardozza A, Delle Vigne S et al (2012) Counteracting effects operating on Src homology 2 domain-containing protein-tyrosine phosphatase 2 (SHP2) function drive selection of the recurrent Y62D and Y63C substitutions in Noonan syndrome. J Biol Chem 287(32):27066–27077
19. Chen Y, LaMarche M, Chan H et al (2016) Allosteric inhibition of SHP2 phosphatase inhibits cancers driven by receptor tyrosine kinases. Nature 535(7610):148–152
20. Prahallad A, Heynen G, Germano G et al (2015) PTPN11 Is a central node in intrinsic and acquired resistance to targeted cancer drugs. Cell Rep [E] 12(12):1978
21. Okazaki T, Chikuma S, Iwai Y et al (2013) A rheostat for immune responses: the unique properties of PD-1 and their advantages for clinical application. Nat Immunol 14(12): 1212–1218
22. Marasco M, Berteotti A, Weyershaeuser J et al (2020) Molecular mechanism of SHP2 activation by PD-1 stimulation. Sci Adv 6(5): eaay4458
23. Higashi H, Tsutsumi R, Muto S et al (2002) SHP-2 tyrosine phosphatase as an intracellular target of *Helicobacter pylori* CagA protein. Science 295(5555):683–686
24. Hayashi T, Senda M, Suzuki N et al (2017) Differential mechanisms for SHP2 binding and activation are exploited by geographically distinct Helicobacter pylori CagA oncoproteins. Cell Rep 20(12):2876–2890
25. Bobone S, Pannone L, Biondi B et al (2021) Targeting oncogenic Src homology 2 domain-containing phosphatase 2 (SHP2) by inhibiting its protein–protein interactions. J Med Chem 64(21):15973–15990
26. Gilson M, Given J, Bush B, McCammon J (1997) The statistical-thermodynamic basis for computation of binding affinities: a critical review. Biophys J 72(3):1047–1069
27. Charlier L, Nespoulous C, Fiorucci S et al (2007) Binding free energy prediction in strongly hydrophobic biomolecular systems. Phys Chem Chem Phys 9(43):5761
28. Limongelli V (2021) Ligand binding free energy and kinetics calculation in 2020. Wires 10(4):e1455
29. Reif M, Zacharias M (2021) Computational tools for accurate binding free-energy prediction. Methods Mol Biol:255–292
30. Deng N, Cui D, Zhang B et al (2018) Comparing alchemical and physical pathway methods for computing the absolute binding free energy of charged ligands. Phys Chem Chem Phys 20(25):17081–17092
31. Kästner J (2011) Umbrella sampling. WIREs Comput Mol Sci 1(6):932–942
32. Kirkwood JG (1935) Statistical mechanics of fluid mixtures. J Chem Phys 3(5):300–313
33. Torrie G, Valleau J (1977) Nonphysical sampling distributions in Monte Carlo free-energy estimation: Umbrella sampling. J Comput Phys 23(2):187–199
34. Kumar S, Rosenberg JM, Bouzida D et al (1992) The weighted histogram analysis method for free-energy calculations on biomolecules. I. The method. J Comput Chem 13:1011–1021
35. Kästner J (2011) Umbrella sampling. Wiley Interdiscip Rev Comput Mol Sci 1:932–942
36. Souaille M, Roux B (2001) Extension to the weighted histogram analysis method: combining umbrella sampling with free energy calculations. Comput Phys Commun 135(1):40–57
37. Hub JS, De Groot BL, Van Der Spoel D (2010) g_wham – A free weighted histogram analysis implementation including robust error and autocorrelation estimates. J Chem Theory Comput 6(12):3713–3720
38. Roux B (1995) The calculation of the potential of mean force using computer simulations. Comput Phys Commun 91(1–3):275–282
39. Doudou S, Burton NA, Henchman RH (2009) Standard free energy of binding from a one-dimensional potential of mean force. J Chem Theory Comput 5:909–918
40. Meng Y, Roux B (2015) Efficient determination of free energy landscapes in multiple dimensions from biased umbrella sampling simulations using linear regression. J Chem Theory Comput 11(8):3523–3529
41. Woo H, Roux B (2005) Calculation of absolute protein–ligand binding free energy from computer simulations. Proc Natl Acad Sci 102(19): 6825–6830
42. Fu H, Gumbart JC, Chen H et al (2018) BFEE: a user-friendly graphical interface facilitating absolute binding free-energy calculations. J Chem Inf Model 58:556–560
43. Fu H, Chen H, Cai W et al (2021) BFEE2: automated, streamlined, and accurate absolute binding free-energy calculations. J Chem Inf Model 61(5):2116–2123

44. Fu H, Chen H, Blazhynska M et al (2022) Accurate determination of protein: ligand standard binding free energies from molecular dynamics simulations. Nat Protoc 17(4): 1114–1141
45. Anselmi M, Calligari P, Hub J et al (2020) Structural determinants of phosphopeptide binding to the N-terminal Src homology 2 domain of the SHP2 phosphatase. J Chem Inf Model 60:3157–3171
46. Chen W, Deng Y, Russell E et al (2018) Accurate calculation of relative binding free energies between ligands with different net charges. J Chem Theory Comput 14(12):6346–6358
47. Abraham MJ et al (2015) GROMACS: high performance molecular simulations through multi-level parallelism from laptops to supercomputers. SoftwareX 1–2:19–25
48. Pettersen EF, Goddard TD, Huang CC et al (2004) UCSF Chimera—a visualization system for exploratory research and analysis. J Comput Chem 25(13):1605–1612
49. Dunbrack RL Jr (2002) Rotamer libraries in the 21st century. Curr Opin Struct Biol 12(4):431–440
50. Lemkul JA (2019) From proteins to perturbed Hamiltonians: a suite of tutorials for the GROMACS-2018 molecular simulation package [Article v1.0]. Living J Comp Mol Sci 1(1): 5068
51. Homeyer N, Horn AH, Lanig H, Sticht H (2006) AMBER force-field parameters for phosphorylated amino acids in different protonation states: phosphoserine, phosphothreonine, phosphotyrosine, and phosphohistidine. J Mol Model 12(3):281–289
52. Jorgensen WL et al (1983) Comparison of simple potential functions for simulating liquid water. J Chem Phys 79:926
53. Darden T, York D, Pedersen L (1993) Particle mesh Ewald: an N· log (N) method for Ewald sums in large systems. J Chem Phys 98(12): 10089–10092
54. Hess B, Bekker H, Berendsen HJC, Fraaije JGEM (1997) LINCS: a linear constraint solver for molecular simulations. J Comput Chem 18(12):1463–1472
55. Olsson M, Søndergaard C, Rostkowski M, Jensen J (2011) PROPKA3: consistent treatment of internal and surface residues in empirical pKa predictions. J Chem Theory Comput 7(2): 525–537
56. Søndergaard C, Olsson M, Rostkowski M, Jensen J (2011) Improved treatment of ligands and coupling effects in empirical calculation and rationalization of pKa values. J Chem Theory Comput 7(7):2284–2295
57. Anandakrishnan R, Aguilar B, Onufriev AV (2012) H++ 3.0: automating pK prediction and the preparation of biomolecular structures for atomistic molecular modeling and simulations. Nucleic Acids Res 40(W1):W537–W541
58. Bussi G, Donadio D, Parrinello M (2007) Canonical sampling through velocity rescaling. J Chem Phys 126(1):014101
59. Zeller F, Zacharias M (2014) Efficient calculation of relative binding free energies by umbrella sampling perturbation. J Comput Chem 35(31):2256–2262
60. Ngo S, Vu K, Bui L, Vu V (2019) Effective estimation of ligand-binding affinity using biased sampling method. ACS Omega 4(2): 3887–3893
61. Pohorille A, Jarzynski C, Chipot C (2010) Good practices in free-energy calculations. J Phys Chem B 114(32):10235–10253
62. Bocchinfuso G, Bobone S, Mazzuca C et al (2011) Fluorescence spectroscopy and molecular dynamics simulations in studies on the mechanism of membrane destabilization by antimicrobial peptides. Cell Mol Life Sci 68(13):2281–2301
63. Lovell SC, Word JM, Richardson JS, Richardson DC (2000) The penultimate rotamer library. Proteins Struct Funct Bioinf 40(3): 389–408
64. Scouras AD, Daggett V (2011) The Dynameomics rotamer library: amino acid side chain conformations and dynamics from comprehensive molecular dynamics simulations in water. Protein Sci 20(2):341–352
65. Guerois R, Nielsen JE, Serrano L (2002) Predicting changes in the stability of proteins and protein complexes: a study of more than 1000 mutations. J Mol Biol 320(2):369–387
66. Feyfant E, Sali A, Fiser A (2007) Modeling mutations in protein structures. Protein Sci 16(9):2030–2041
67. Eck M, Shoelson S, Harrison S (1993) Recognition of a high-affinity phosphotyrosyl peptide by the Src homology-2 domain of p56lck. Nature 362(6415):87–91
68. Sun L, Noel JK, Sulkowska JI, Levine H, Onuchic JN (2014) Connecting thermal and mechanical protein (un) folding landscapes. Biophys J 107(12):2950–2961
69. Grossfield A WHAM: the weighted histogram analysis method, version 2.0.11. http://membrane.urmc.rochester.edu/wordpress/?page_id=126

Chapter 8

NMR Relaxation Dispersion Experiments to Study Phosphopeptide Recognition by SH2 Domains: The Grb2-SH2–Phosphopeptide Encounter Complex

Fabio C. L. Almeida, Karoline Sanches, Icaro P. Caruso, and Fernando A. Melo

Abstract

Protein interactions are at the essence of life. Proteins evolved not to have stable structures, but rather to be specialized in participating in a network of interactions. Every interaction involving proteins comprises the formation of an encounter complex, which may have two outcomes: (i) the dissociation or (ii) the formation of the final specific complex. Here, we present a methodology to characterize the encounter complex of the Grb2-SH2 domain with a phosphopeptide. This method can be generalized to other protein partners. It consists of the measurement of ^{15}N CPMG relaxation dispersion (RD) profiles of the protein in the free state, which describes the residues that are in conformational exchange. We then acquire the dispersion profiles of the protein at a semisaturated concentration of the ligand. At this condition, the chemical exchange between the free and bound state leads to the observation of dispersion profiles in residues that are not in conformational exchange in the free state. This is due to fuzzy interactions that are typical of the encounter complexes. The transient "touching" of the ligand in the protein partner generates these new relaxation dispersion profiles. For the Grb2-SH2 domain, we observed a wider surface at SH2 for the encounter complex than the phosphopeptide (pY) binding site, which might explain the molecular recognition of remote phosphotyrosine. The Grb2-SH2–pY encounter complex is dominated by electrostatic interactions, which contribute to the fuzziness of the complex, but also have contribution of hydrophobic interactions.

Key words Encounter complex, NMR, Relaxation dispersion, Dynamics, SH2

1 Introduction

Protein interactions are fundamental for cell signaling, where dissociation constants (K_D) between pico and nanomolar are defined as stable binding, whereas interactions from tenths of micromolar to millimolar affinities are characterized as intermediate or transient. The Src homology 2 (SH2) domain's main function is to modulate signal transduction through the recognition of

Teresa Carlomagno and Maja Köhn (eds.), *SH2 Domains: Functional Modules and Evolving Tools in Biology*, Methods in Molecular Biology, vol. 2705, https://doi.org/10.1007/978-1-0716-3393-9_8,

phosphotyrosine (pY) [1, 2]. The high-affinity recognition of phosphotyrosine is essential to the fidelity of the signal transduction pathways regulated by protein recognition. The characterization of the SH2 and phosphopeptide bound state is largely described, especially by X-ray crystallography [1, 3–8]. However, a major challenge in structural biology is the determination of transient events such as the formation of encounter complexes in the pathway for protein recognition. They are intermediates that approximate the ligand to the binder, facilitating the formation of the stereospecific complex. In general, it involves regions larger than the binding site, which are important to be described. NMR spectroscopy is especially powerful in the description of protein dynamics and transient events [9], and it was therefore used to describe the encounter complex between the Grb2-SH2 domain and the endothelial growth factor (EGF) phosphopeptide [10]. Here, we will describe in detail the protocol to characterize the protein regions involved in the formation of an encounter complex through NMR spectroscopy, more specifically using Carr–Purcell–Meiboom–Gill (CPMG) relaxation dispersion. It is important to mention that the structural and dynamical description of an encounter complex is rare [10–13], and new methods to describe them will contribute to the understanding of the biological function of any molecular recognition event.

An important feature for the molecular recognition of the SH2 domain is the presence of two distinct binding sites. Site I is mostly responsible for the direct binding to the phosphotyrosine (pY), while site II is responsible for the binding specificity, recognizing pY + 1, pY + 2, or pY + 3 sequences. Generally, the phosphopeptide binds at the surface of the SH2 in an extended conformation spanning from α-helix 1 to α-helix 2 with W121 in between. Furthermore, Grb2-SH2 has shown its multivalency in being able to sense remote tyrosine. Remote tyrosine phosphorylation, other than the primary recognition site, increases the scope of the interaction between Grb2-SH2 and the Linker for Activation of T cells (LAT) [14], which is pivotal for a protein as Grb2 to recognize its target efficiently in the presence of multiple binding partners inside the cell and thus essential to the efficiency of the signaling pathways. In a recent publication, we showed that the interaction surface at the encounter complex of the Grb2-SH2 and the phosphopeptide is much wider than the pY recognition site and involves the entire pY recognition face. We suggested that the recognition face is wide enough to electrostatically sense remote pY sites [10].

In this chapter, we show the protocol used to characterize the encounter complex of the Grb2-SH2 domain [10] and generalize it for use in other complexes. It is reasonable to think that evolution acted on the formation of both the more elusive encounter complex and the final complex. These days, many structural biology methods are available for structure determination, but just a few methods are able to characterize transient encounter complexes [15, 16]. Here, we show how NMR relaxation dispersion experiments can be employed to discover which protein residues participate in the encounter complex. This is especially important for proteins such as Grb2, which participate in signal transduction pathways. Grb2 evolved to interact with hundreds of cellular partners; therefore, understanding the fuzzy interactions of the encounter complex of Grb2-SH2 with a phosphopeptide contributes to the comprehension of how it evolved as an adaptor protein and what forces are involved in this process [10]. Encounter complexes are dominated by electrostatic interactions, which contribute to the fuzziness of the complex, but few hydrophobic interactions can provide some specificity. Solvation also plays an important role, since each partner of the encounter complex is largely solvated [10].

To identify the encounter complex, we use ^{15}N CPMG NMR relaxation dispersion (RD) experiments. The CPMG experiment was first proposed many years ago as a way to measure transverse relaxation rate [17, 18]. In 1999, Palmer and colleagues revolutionized the field of protein dynamics by proposing the first viable CPMG-RD experiment, which measures the effective transverse relaxation rate (R_2^{eff}) as a function of the frequency of refocusing pulses, described as the CPMG frequency (ν_{CPMG}). With the introduction of the relaxation compensation experiment [19], the problem of evolution of both in-phase and antiphase coherences during CPMG was solved by mixing equal amounts of both coherences during the CPMG constant evolution time. Later on, in 2008, Kay and colleagues developed an efficient way to decouple hydrogen during CPMG evolution, enabling the simplification of the pulse sequence into a single block with the evolution of only in-phase coherences (Fig. 1) [20]. Because of the spin flip-flop effect, the antiphase magnetization displays faster decay than the in-phase. With the application of an efficient ^{1}H-decoupling and using a continuous-wave CPMG (CW-CPMG), it became possible to suppress *J*-coupling evolution during the constant-time CW-CPMG enabling the CPMG evolution with in-phase magnetization. CW-CPMG becomes more sensitive to obtaining the relaxation profiles of proteins.

2 Materials

2.1 Stock Solutions for Nonlabeled and Labeled M9 Minimal Media

1. MilliQ H_2O (ddH_2O).
2. 1 M $CaCl_2$ (autoclave and store at room temperature).
3. TS2 solution (0.2 V%) consisting of 100 mg $ZnSO_4 \times 7H_2O$, 30 mg $MnCl_2 \times 4H_2O$, 300 mg H_3BO_3, 200 mg $CoCl_2 \times 6H_2O$, 20 mg $NiCl_2 \times 6H_2O$, 10 mg $CuCl_2 \times 2H_2O$,

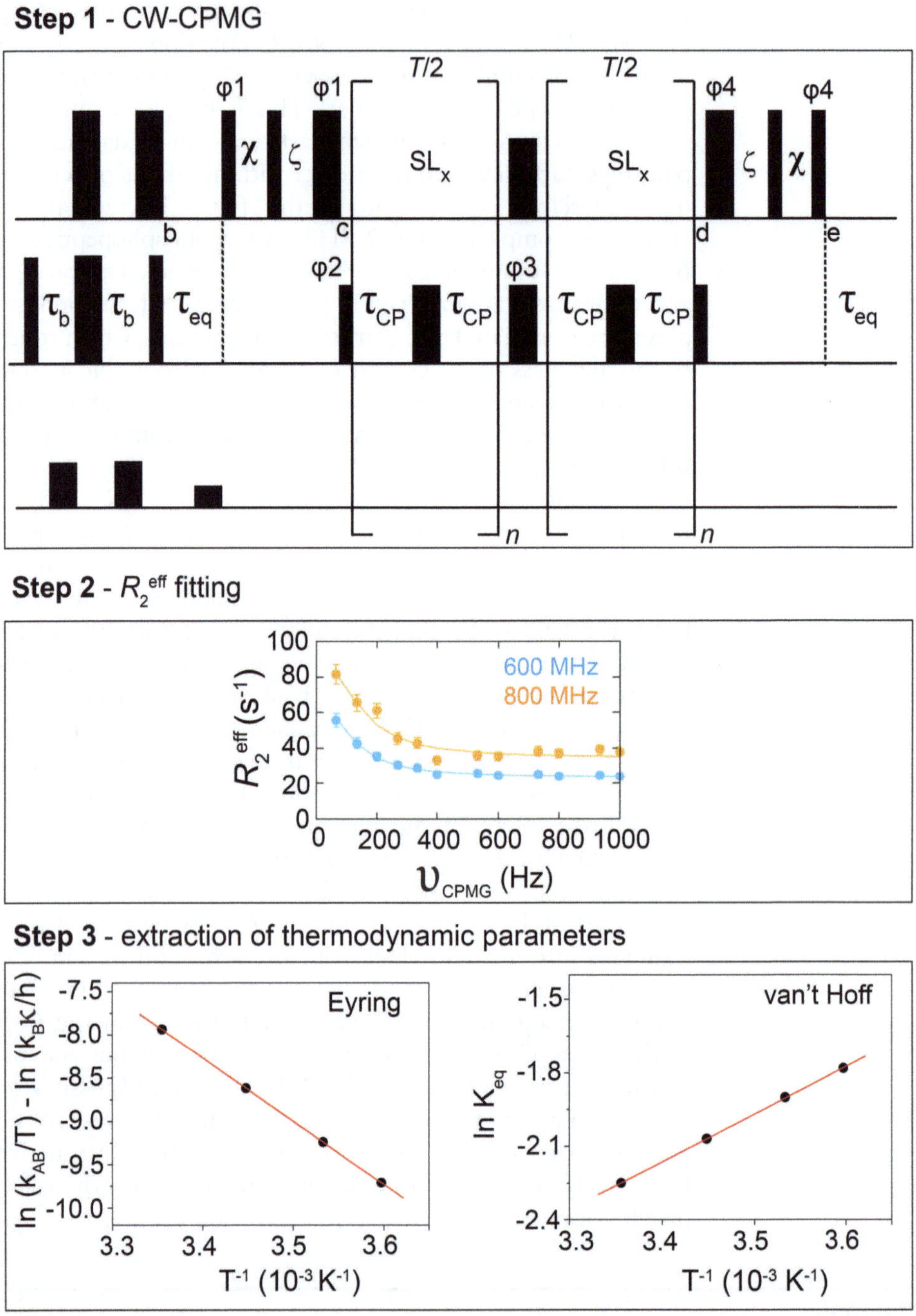

Fig. 1 Steps for characterization of the protein dynamics and thermodynamic profile using CW-CPMG, valid for intermediate exchange regime where $R_2 \ll k_{ex}$ and $p_B \ll 1$. *Step 1* includes recording the CPMG experiments, here we used the CW-CPMG pulse sequence by Kay and colleagues [20] using $T = 30$ ms in the case of Grb2-SH2. Before starting the CPMG pulse train, the magnetization is refocused to the *z*-axis by INEPT ($2I_zS_z$). A spinlock (SL_x) is applied at the *T*/2 CPMG period and a φ_3 refocuses the evolution of water magnetization between the two SL_x. *Step 2* shows the relaxation dispersion decay for two fields (600 and 800 MHz) and the fitting of the data through Carver and Richards equation. After fitting the data, *Step 3* is performed for a thermodynamic profile of the transition state through the Eyring equation and for the excited state using van't Hoff equation

900 mg $Na_2MoO_4 \times 2H_2O$, 20 mg Na_2SeO_3 in 1 L ddH_2O (0.2 μm filtered and store at 4 °C).

4. 10 mM Fe (III) chloride (0.2 μm filtered and store at −20 °C).
5. Antibiotics (kanamycin for our construct).
6. Vitamin cocktail in 20 mM NaPi: 1 mg/mL Biotin, 1 mg/mL Choline Chloride, 1 mg/mL Folic Acid, 1 mg/mL Nicotinamide, 1 mg/mL Pantothenic Acid, 1 mg/mL Pyridoxal Hydrochloride. pH to 7.4, 0.2 μm filtered, and store at −20 °C.
7. 0.1 mg/mL Riboflavin (B2) (0.2 μm filtered and store at −20 °C).
8. 5 mg/mL Thiamine (B1) (0.2 μm filtered and store at −20 °C).
9. 1 M $MgSO_4$ (0.2 μm filtered and store at room temperature).
10. 20% d-glucose stock solution.
11. 5 × M9 salts: 233 mM Na_2HPO_4, 110 mM KH_2PO_4, and 43 mM NaCl. The pH should be ~6.8 but adjust if necessary. Autoclave and store at room temperature.
12. NH_4Cl.

2.2 Recombinantly Expressed and Purified ^{15}N-Labeled Grb2-SH2

1. The plasmid containing the Grb2-SH2 gene (or the gene of interest) with 6× His tag.
2. Antibiotics (kanamycin for our construct).
3. *E. coli* BL21 (DE3) Gold.
4. Luria-Bertani (LB) growth media.
5. 20% d-glucose stock solution.
6. $^{15}NH_4Cl$.
7. Isopropyl-β-d-thiogalactopyranoside (IPTG).
8. Phenylmethanesulfonyl fluoride (PMSF).
9. 0.2 M Co^{2+} for charging column.
10. Lysis buffer (50 mL/L of culture): 50 mM Tris–HCl (pH 8.0) containing 300 mM NaCl, 1 mM EDTA, 1 mM PMSF, and 5 mM of sodium azide.
11. Buffer gradient for cobalt affinity purification: 50 mM Tris–HCl (pH 8.0) with 100 mM NaCl and imidazole ranging from 0.01 to 1 M.
12. Buffer for size-exclusion purification and NMR experiments: 20 mM NaPi (pH 7.0) with 200 mM NaCl, 1 mM EDTA, 1 mM PMSF, and 5 mM of sodium azide.
13. 0.22 μm syringe filters; we use Minisart®.
14. 5 mL metal affinity resin precharged with Co^{2+}; we use HiTrap HP from Cytiva.

15. Size-exclusion chromatography resin; we use Superdex 75 from Cytiva.
16. 15% SDS-PAGE.

2.3 Phosphopeptide Synthesis

1. Phosphopeptide EpYINSQV [21] (recognition epitope from Epidermal Growth Factor receptor) synthesized via Fmoc solid-phase peptide synthesis and purified to >95% purity via reverse-phase HPLC.

2.4 NMR Experiments

1. 5-mm glass tubes for NMR.
2. Deuterium oxide (D_2O).
3. NMR spectrometers operating at 600 and 800 MHz, equipped with ^{15}N, ^{13}C, and ^{1}H TXI probe heads.
4. 4,4-dimethyl-4-silapentane-1-sulfonic acid (DSS) for indirect chemical shift referencing [22].
5. Spectrometer control software TopSpin.

2.5 Data Processing and Analysis

1. Data processing software (nmrPipe) [23].
2. NMR data visualization CCPN [24].
3. nmrPipe [23] for peak integration and fitting exponential decays of peak volumes.
4. Software for fitting the relaxation dispersion curves.

3 Methods

Here, we describe the methodology applied to characterize the conformational dynamics on the Grb2-SH2 domain through ^{15}N CPMG-RD. The methodology applied here can be adapted for any protein-ligand complex to characterize the conformational equilibrium of the protein in its free state and the ligand-binding equilibrium. We start describing sample preparation, e.g., expression and purification, following with the NMR experiments, spectra processing, and data analysis.

3.1 Sample Preparation

For starting with the labeling, we first start with 200 mL nonlabeled minimal media as follows.

3.2 Minimal Media

For 200 mL of M9 minimal media nonlabeled:

1. 50 mL of MiliQ H_2O (ddH_2O).
2. 20 μL of 1 M CaCl2.
3. 400 μL of TS2 solution (0.2 V%).
4. 200 μL of 10 mM Fe (III) chloride.
5. 200 μL of vitamin cocktail.

6. 200 μL of Riboflavin (B2) (0.1 mg/mL stock).
7. 200 μL of thiaminhydrochloride (B1) (5 mg/mL stock).
8. 400 μL of 1 M $MgSO_4$.
9. 2–3 mL of 20% glucose (2–3 g/L).
10. 0.2 g of NH_4Cl (dissolve in water previously).
11. 40 mL of 5 × M9 salts.
12. Complete with ddH_2O to final volume 200 mL.

Adjust the concentrations accordingly to prepare 1 L media for ^{15}N-labeled protein. Please note, 1 L was enough for our experiments, you can adjust depending on your needs and protein expression levels.

3.3 Grb2-SH2 Expression and Purification

1. The Grb2-SH2 construct used for our experiments comprises M55 to D150 (numbered M1 to D96) tagged with $(His)_6$ for cobalt affinity purification. The Grb2-SH2 domain is expressed in *E. coli* BL21 (DE3) Gold.
2. Incubate 1 mL of Grb2-SH2 frozen culture in 25 mL of LB media and 50 μg/mL of kanamycin at 100 RPM, 37 °C overnight. Alternatively, a single colony from agar plate can also be used here.
3. Dilute the 25 mL overnight bacterial culture in 200 mL of M9 nonlabeled minimal media with 50 μg/mL of kanamycin. It is important to add the quantity needed to reach an initial optical density (OD) at 600 nm of 0.2 (in our case was ~25 mL).
4. Grow this starting culture at 37 °C, shaking at 120 RPM until it reaches an $OD_{600} = 0.9$.
5. Harvest the cells by 20 min centrifugation at 20 °C and 4000 × *g* and resuspend the pellet in 1 L M9 minimal media supplemented with 1 g/L $^{15}NH_4Cl$ and 50 μg/mL of kanamycin. Prepare 1 L following **step 3.2.** and adjusting the concentrations accordingly. Please note that for this step we use ^{15}N-labeled ammonium chloride.
6. Incubate at 37 °C shaking at 120 RPM until an $OD_{600} = 0.9$ is reached (around 4 h).
7. Collect a sample (40 μL) for SDS-PAGE analysis (before induction). Induce protein production by adding 0.4 mM IPTG and incubate overnight at 20 °C shaking at 120 RPM.
8. On the next day, collect the sample for SDS-PAGE (after induction) and harvest the cells by centrifugation at 4000 × *g* for 20 min at 4 °C.
9. Resuspend the pellet in lysis buffer (Subheading 2.2, **Item 10**). From now on, it is important to keep the sample cold during all the steps.

10. Lyse the cells by 15 cycles of sonication (2 s ON e 1 s OFF) in ice followed by a centrifugation step at 34,957 × *g* at 10 °C for 50 min to remove cell debris.
11. Filter the soluble material with 0.22 μm filters.
12. Apply the filtrate to a 5 mL metal affinity column precharged with Co^{2+} and pre-equilibrated with 10 mM imidazole in buffer for affinity purification (Subheading 2.2, **Item 10**). We use concentrations of 10, 20, 40, 60, 80, 100, 200 mM, and 1 M imidazole in the gradient of the affinity purification buffer (Subheading 2.2, **Item 11**). You may need to adapt this gradient to your experimental system, decreasing or increasing the gradient steps to remove any undesired protein. The Grb2-SH2 is eluted with 200 mM imidazole.
13. Concentrate the sample using ultra centrifugal filters (3 kDa) to 1.0 mL and apply it to the size-exclusion chromatography column in the size-exclusion purification buffer (Subheading 2.2, **Item 12**).
14. Verify the purity of the protein by 15% SDS-PAGE (*see* **Note 1**).

3.4 Phosphopeptide–Grb2-SH2 Complex

1. Prepare the protein-ligand complex by mixing 300 μM of Grb2-SH2 and a semisaturated concentration of phosphopeptide (2.1 mM) in 20 mM inorganic phosphate and 200 mM NaCl to compete with the phosphopeptide for the interaction with the protein. This is necessary to tune the binding affinity to an appropriate one, to render ^{15}N CPMG-RD informative of the on/off ligand binding (chemical exchange), in addition to the conformational exchange observed in for Grb2-SH2 in the free state. In this environment, the phosphopeptide binds to Grb2-SH2 with millimolar affinity and the chemical exchange can be observed by NMR (*see* **Note 2**).

3.5 CPMG Experiments

3.5.1 CPMG-RD to Characterize the Encounter Complex

1. Measure the CPMG relaxation dispersion profiles for the protein at the free state.
2. Calculate pl ^{1}H hard pulse using zgpr pulse sequence on bruker.
3. Calculate of ^{15}N hard pulse using EMinvp90f3 pulse sequence on bruker.
4. Calculate flip-back pulse.
5. To measure the CPMG-RD experiments, use the pulse sequence "hsqcrexetf3gpsitc3d" on the list of standard Bruker pulse sequences. This is the relaxation compensation version [19]. The CW-CPMG-RD [20] version is not available at Bruker pulse sequence and could either be coded or requested to the authors pulse sequence library.
6. Make sure you have the CPMG frequency lists: vdlist.

Step 1 - CPMG-RD of the protein at the free state:

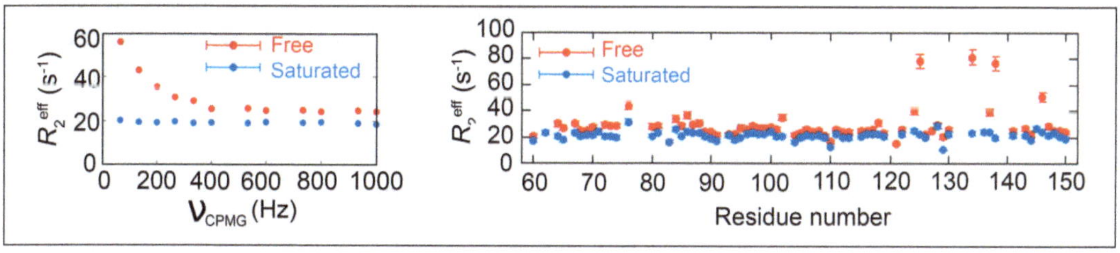

Step 2 - CPMG-RD at semi-saturated condition:

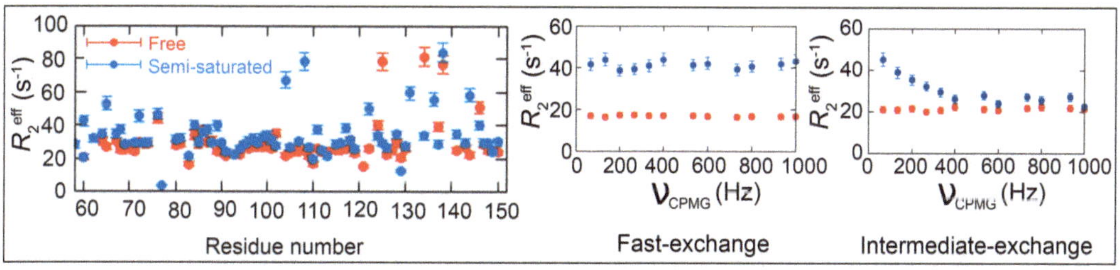

Fig. 2 Steps employed for characterization of the encounter complex. *Step 1*: R_2^{eff} plot shows all the residues displaying elevated R_2^{eff} at free state are quenched at a saturated concentration of phosphopeptide (Step 1, right). On the left, the CPMG-RD curves at free and saturated conditions of a specific residue. At the free state, the residue showed a relaxation dispersion curve typical of the intermediate exchange regime (red, Step 1, left). At a saturated concentration of phosphopeptide, the relaxation dispersion is quenched (blue, Step 1, left). With that, we can investigate residues that already display relaxation dispersion at free state, i.e., prior to semisaturated concentration of phosphopeptide. Thus, it is expected that this (these) residue(s) not be involved in the encounter complex. *Step 2*: The R_2^{eff} plot for all residues at a semisaturated concentration of phosphopeptide shows residues that have not shown any relaxation dispersion at free state (red, Step 2, left), now acquired elevated R_2^{eff} values (blue, Step 2, left). These residues are involved in the encounter complex, and because we are using inorganic phosphate buffer, a K_D of mM is found thus becoming possible to monitor chemical exchange due to the encounter complex formation (Step 2, left and right). Residues with a fast-exchange regime (Step 2, middle) show flat but elevated R_2^{eff}

7. Measure the CPMG relaxation dispersion profiles for the protein with ligand concentrations at semisaturated conditions. It may be necessary to tune for the ideal ligand concentration. Search for a sample condition where K_D is at the millimolar range. Many of the sample conditions can be changed to get to this condition, such as pH, salt concentration, buffer, and temperature. Other biophysical methods such as fluorescence anisotropy, surface plasmon resonance, or thermophoresis could be helpful to tune the sample conditions (*see* **Note 2**).
8. Compare the relaxation dispersion profiles of the free state with the semiconcentrated and check which residues acquired relaxation dispersion. There may also have residues with a flat profile but above the average or above the R_2^{eff} of the free state. This is a typical behavior of fast-exchange regime (*see* **Note 3**), where much larger values of ν_{CPMG} would be necessary to refocus the exchange contribution to $R_{2,\infty}^{eff}$ (Fig. 2).

3.5.2 Data Analysis

Plotting of the CW CPMG-RD Using nmrPipe

This method fits concomitantly the series of two-dimensional spectra acquired as a pseudo-3D. After processing the pseudo-3D spectrum, the data will consist of a series of ^{1}H-^{15}N HSQC-like correlation spectra. The first one is the reference, with relaxation constant time (T) equal to zero. All the others have a defined T ($T = 30$ ms for SH2) and varying CPMG frequencies (ν_{CPMG}). We used NMRbox platform [25], which provides an linux box with preinstalled softwares for NMR spectroscopy, to process all our data.

1. Peak pick the most intense correlation spectrum ($T = 0$) using the peaking mode of nmrDraw (peak → peak detection).
2. Click on "detect" to automatically detect all peaks.
3. On Labels, move from Variables Index to ASS (all the peaks should now be identified as "none").
4. Introduce all your assignments by changing "none" to the respective assignment and save the .tab list.
5. Move the variables from ASS to CLUSTID to adjust clustered peak regions and superposed resonances to the same cluster id. Use different CLUSTERIDs to isolated peaks but keep the same CLUSTID number to the peaks in the same region or overlapping. This is important because peaks defined with the same CLUSTID number will be fitted together by the software.
6. Save the peak list as a file.tab.
7. Close nmrDraw.
8. Rename the fit.com file to your peak_list.tab name and execute ./fit.com.
9. Edit the nlin.tab file, removing all the text before the peak list (number 1) and save as nlin_test.tab.
10. Edit nlin2dsp.com (changing the values according to your experiments, *see* **Note 4**).
11. Execute ./nlin2dps.com.
12. Plot your R_2^{eff} curves using Gnuplot.
13. Repeat the same approach for the experiments under the semi-saturated concentration of phosphopeptide.

The expression for the effective transverse relaxation R_2^{eff} for the ground state is given by:

$$R_2^{eff} = -\frac{1}{T}\ln\frac{I(\nu_{CPMG})}{I(0)}$$

where T is the constant relaxation period with which each spectrum is recorded at each ν_{CPMG}. I and $I(0)$ are the peak intensities for each ν_{CPMG} and at $t = 0$, respectively.

Fit the R_2^{eff} to Calculate the Excited State

The fitting for the R_2^{eff} assuming the simplest two-state exchange model (A and B) is given through the evolution of the magnetization matrix using the Bloch–McConnell equation [26, 27]. For a two-state exchange,

$$A \underset{k_{BA}}{\overset{k_{AB}}{\rightleftharpoons}} B$$

the evolution of the exchanging magnetization A and B assumes the form:

$$\frac{d}{dt}\begin{pmatrix} M_{A\pm}(t) \\ M_{B\pm}(t) \end{pmatrix} = \begin{pmatrix} -R_A - k_{AB} & k_{BA} \\ k_{AB} & \pm i\Delta\omega - R_B - k_{BA} \end{pmatrix}\begin{pmatrix} M_{A\pm}(t) \\ M_{B\pm}(t) \end{pmatrix}$$

where the magnetization of states A and B are complex numbers represented as M_A and M_B. $M_+ = M_x + iM_y$ and $M_- = M_x - iM_y$ k_{AB} and k_{BA} are the exchange rates between the chemical or conformational states A and B, R_A and R_B are the relaxation rates of the transverse magnetization of the states, and $\Delta\omega$ is the difference in the chemical shifts of state A and B and provides structural information on the excited state B.

The equation can also be written as a modified Bloch–McConnell equation as

$$\frac{d}{dt}\vec{\boldsymbol{M}}_{\pm}(t) = \widehat{\boldsymbol{R}}_{\pm}\vec{\boldsymbol{M}}_{\pm}(t)$$

where $\widehat{\boldsymbol{R}}_{\pm}$ is the relaxation matrix and $\vec{\boldsymbol{M}}_{\pm}$ represents the magnetization vectors.

The fitting of a CPMG-RD experiment involves the solution of the following differential equation:

$$\vec{M}(t) = \exp\left\{\widehat{R}\right\}\vec{M}(0)$$

Many available programs can be used for the individual and global fittings of the Bloch–McConnell equation, including in-house academically accessible MATLAB scripts and the software RELAX [28]. Here, we describe the software CPMG_FITD9 coded by Prof. Dmitry Korzhnev (University of Connecticut Health Center) and available at the NMRbox platform [25]. NMRbox is a resource of software packages preconfigured to facilitate biomolecular NMR and computational data analysis. It can be used remotely to access the virtual machine through VNC Viewer. Several examples of CPMG_FIT9 fitting are available at

NMRbox, helping to build the input files for the analysis of relaxation dispersion data.

Briefly, the program reads the relaxation profiles at two magnetic fields, acquired at the same constant-time T (30 ms). After that, the initial approximations for k_{ex}, p_B, R_2 and $\Delta\omega$ are set. It is worth mentioning that, in the fitting scripts, the m-value sets the number of sites in the exchange model and the mode of fitting, which can be either numerical solutions of the Bloch–McConnell equation ($m > 2$) or analytical solutions for two-state exchange ($m = 2$). It is also possible to use $m = 3$ for a two-state exchange model, allowing only k_{ex}^{AB} as a variable and forcing k_{ex}^{BC} and k_{ex}^{AC} equal to zero. For $m = 2$ (two-site exchange model), the data is fitted to the modified Carver and Richards equation.

This software enables either the numeric solution or the use of the analytical solution of a two-state intermediate exchange regime ($R_2 \ll k_{ex}$ and $p_B \ll 1$) described in [29, 30]:

$$R_2^{\text{eff}} = R_2 + \frac{k_{ex}}{2} - \frac{1}{t_{\text{CPMG}}} \cosh^{-1}[D_+ \cosh(\eta) - D_- \cos(\eta)]$$

where:

$$D_\pm = \frac{1}{2}\left(\frac{\varphi + 2\Delta\omega^2}{\sqrt{\varphi^2 + \zeta^2}} \pm 1\right)$$

$$\eta_\pm = \sqrt{2}\tau\sqrt{\sqrt{\varphi^2 + \zeta^2} \pm \varphi}$$

$$\zeta = -2\Delta\omega(p_A - p_B)k_{ex}$$

$$\varphi = ((p_A - p_B)k_{ex})^2 - \Delta\omega^2 + 4p_A p_B k_{ex}^2$$

Therefore, the rate of exchange (k_{ex}), the population of the excited state (p_B), and the chemical shift differences between the excited and ground states ($\Delta\omega$) can be determined, which in turn provide structural information on the "invisible" state.

For the thermodynamics characterization of chemical or conformational equilibria, data must be acquired at multiple temperatures. If we consider an exchange between the two states, A and B, the rate of transition k_{AB} can be related to the Boltzmann distribution:

$$\frac{p_B}{p_A} = \frac{k_{AB}}{k_{BA}} = e^{\left(-\frac{\Delta G}{RT}\right)}$$

where ΔG is the free Gibbs energy for the exchange $A \rightleftharpoons B$. The Boltzmann distribution follows the Eyring equation as:

$$\ln\frac{hk_{AB}}{k_B\kappa T} = -\frac{\Delta H^\ddagger}{RT} + \frac{\Delta S^\ddagger}{R}$$

$$k_{AB} = \frac{k_B\kappa T}{h} e^{\frac{-\Delta H^\ddagger}{RT} + \frac{\Delta S^\ddagger}{R}}$$

where κ is a transmission coefficient, and k_B and h are the Boltzmann and Planck constants, respectively. Knowing that $\ln\frac{k_B\kappa}{h}$ is a constant, one can build the linear plot ($\ln \frac{k_{AB}}{T} - \ln \frac{k_{BK}}{h} \times 1/T$) and have the thermodynamic profile of the transition state (Fig. 1) (*see* **Note 5**).

$$\ln \frac{k_{AB}}{T} - \ln \frac{k_{BK}}{h} = \frac{\Delta S^{\ddagger}}{R} - \frac{\Delta H^{\ddagger}}{RT}$$

For the equilibrium state, we have the Gibbs activation free energy following by the van't Hoff equation, where $K_{eq} = p_B/p_A$

$$\Delta G = -RT \ln K_{eq}$$

$$\Delta G = \Delta H - T\Delta S$$

$$RT \ln(K_{eq}) = -\Delta H + T\Delta S$$

We can build the linear plot ($\ln K_{eq} \times 1/T$) and have the thermodynamic profile for the equilibrium.

$$\ln K_{eq} = -\frac{\Delta H}{R}\frac{1}{T} + \frac{\Delta S}{R}$$

Now that we have briefly described the equations and parameters used in the fitting, let's follow the protocol to analyze the data:

1. Following what you plotted in Subheading 3.3, **step 1**, get all your .dsp files (for each temperature and magnetic field) and move them to a new folder. Inside this folder, you must have directories called: backup, data, fits, input, and logs (use the command MKDIR to create it).
2. Launch the program under NMRbox by typing "cpmg_fitd9" at the Linux terminal. Help commands are available by typing "help" in the software prompt. One can go through any of the commands of the scripts using the helping tools. For instance, to go through the "read" flags, just type "help read."
3. You also will need scripts to prepare the input files in an adequate format. Seven examples are available at the NMRbox platform (/usr/software/cpmg_fitd9/examples) [25].
4. Through global or individual fitting of relaxation profiles, extract k_{ex} ($k_{ex} = k_{AB} + k_{BA}$) (conformational or chemical exchange rate), p_B (population of the excited state), and $\Delta\omega$ (chemical shift difference between the ground (A) and excited state (B)) for each residue of a protein. There are examples available for the global fitting of the relaxation data assuming that k_{ex} and p_B are global variables, meaning that they have the same value for all residues of the protein. One can also globally fit data at multiple temperatures.
5. Validate the model through the chi-square (χ^2) test provided by the program. The chi-square value, the number of degrees of freedom (F), and the p-value are given as output of the test.

The closer the p-value to unity, the higher the probability of the dynamical model reflecting the experimental data. It happens when the chi-square value is approximately equal to the number of degrees of freedom.

6. Analyze the data of the free protein to obtain relaxation dispersion curves (Fig. 1) and compare with the semisaturated to check which residues acquired relaxation dispersion (Fig. 2) (*see* **Note 5**).

4 Notes

1. The size exclusion profile of Grb2-SH2 is generally composed of two peaks because of the different oligomerization states of the domain. It is important to collect it separately because although in SDS-PAGE they appear identical, the NMR spectra are completely different. To determine the encounter complex, it is important to work with the monomeric Grb2-SH2 state.
2. To characterize the encounter complex by NMR, one must take advantage of the on/off equilibrium between the ligand and the receptor and tune it to render intermediate chemical exchange relaxation dispersion profiles. To achieve this condition, k_{off} should be between 400 and 1500 s^{-1}. The on rate constant for protein-protein interaction is in the timescale of supra-τ_c dynamics [31], typically <50 μs. The relaxation dispersion profiles are not sensitive to the on-rates, and depend only on k_{off}. Using a sample condition where K_D is in the millimolar range, k_{off} is frequently in the correct window to render intermediate on/off chemical exchange. Therefore, the residues involved in the encounter complex present relaxation dispersion curves because of the on/off chemical exchange.

   To characterize the encounter complex of the Grb2-SH2 domain, we used a semisaturated concentration (2.1 mM) of ligand in the inorganic phosphate buffer. The buffer competes with the phosphopeptide for binding to R86 while the NaCl shields electrostatic interactions. The phosphopeptide binds with K_D ~ mM enabling the chemical exchange due to the "binding intermediates" to be investigated through NMR.
3. There may also be residues with a flat profile but above the average or above the R_2^{eff} of the free state. This is typical behavior of fast-exchange regime, where much larger values of ν_{CPMG} would be necessary to refocus the exchange contribution to $R_{2,\infty}^{eff}$. In our case, we measured ^{15}N CPMG-RD which shows that some residues with no relaxation dispersion in the free state acquire dispersion profiles at a semisaturated concentration. This is due to fuzzy interactions between the phosphopeptide and the protein, which lead to chemical exchange processes observable by NMR. The encounter

complex engages a wider region of Grb2-SH2 than the stabilized complex. The residues in conformational exchange in the free state which are also involved in the formation of the encounter complex are more difficult to recognize, but they will frequently show changed relaxation dispersion curves upon the addition of the ligand in semisaturated conditions.

4. Script to convert file nlin to files amenable to CPMG_FITD9

```
#!/bin/csh
set assign=('awk '{print $23}' nlin_test.tab')
  awk '{gsub(/[[:alpha:]]+/,"  ");sub(/^[  ]*/,x);NF-=2}1'
nlin_test.tab &gt; nlin_wo_characters

echo $assign
set j=1
while ( $j <= ${#assign} )
awk -v j=$j 'NR==j {for (i=26 ; i&lt;= NF ; i++) print
-log($i/$26)/0.03}' nlin_test.tab &gt; R2
awk -v j=$j 'NR==j {for (i=26 ; i&lt;= NF ; i++) print $i*
$19/($26*$18*0.03), "#"}' nlin_test.tab &gt; Rerror
awk -v j=$j 'NR==j {for (i=26 ; i&lt;= NF ; i++) print
($23)}' nlin_test.tab &gt; resname
awk -v j=$j 'NR==j {for (i=27 ; i&lt;= NF ; i++) print
($26)}' nlin_wo_characters &gt; resnumber
awk -v j=$j 'NR==j {for (i=26 ; i&lt;= NF ; i++) print
$1}' nlin_test.tab &gt; peaknumber
awk -v j=$j 'NR==j {print $23}' nlin_test.tab > t2
paste peaknumber R2 Rerror vdlist resname resnumber > disp/
{$assign[$j]}_15n_800_25oc.out
#paste vdlist R2 Rerror t2> disp2/test_N.dsp
echo $j $assign[$j]
@ j = $j + 1
end
cat disp/* > dsp.tmp
awk '{if ($2 == 0) gsub("0","-0.00000",$2); print $1,$2,
$3,$4,$5, $6, $7}' dsp.tmp > disp/R2mt_rpib_15-
n_800_25oc_m.dsp
rm nlin_wo_characters R2 Rerror resname resnumber peaknum-
ber dsp.tmp
# create a folder call disp, add vdlist without the initial
zero to PRC.., call the previously created file (nlin.tab)
nlin_test.tab, and delete all the first lines.
```

5. For Grb2-SH2, the increase in the temperature led to a small decrease in p_B and an increase in k_{ex} (Fig. 1). The excited state is enthalpically favorable and entropically unfavorable at 298 K. This profile is observed for conformational changes with

exposure of hydrophobic residues to the solvent, leading to an entropic cost.

References

1. Pascal SM, Yamazaki T, Singer AU et al (1995) Structural and dynamic characterization of the phosphotyrosine binding region of a Src homology 2 domain-phosphopeptide complex by NMR relaxation, proton exchange, and chemical shift approaches. Biochemistry 34: 11353–11362. https://doi.org/10.1021/bi00036a008
2. Liu BA, Engelmann BW, Nash PD (2012) The language of SH2 domain interactions defines phosphotyrosine-mediated signal transduction. FEBS Lett 586:2597–2605. https://doi.org/10.1016/j.febslet.2012.04.054
3. Kessels HWHG, Ward AC, Schumacher TNM (2002) Specificity and affinity motifs for Grb2 SH2-ligand interactions. Proc Natl Acad Sci U S A 99:8524–8529. https://doi.org/10.1073/pnas.142224499
4. Zhou S, Shoelson SE, Chaudhuri M et al (1993) SH2 domains recognize specific phosphopeptide sequences. Cell 72:767–778. https://doi.org/10.1016/0092-8674(93)90404-E
5. McNemar C, Snow ME, Windsor WT et al (1997) Thermodynamic and structural analysis of phosphotyrosine polypeptide binding to Grb2-SH2. Biochemistry 36(33): 10006–10014. https://doi.org/10.1021/bi9704360
6. Nioche P, Liu WQ, Broutin I et al (2002) Crystal structures of the SH2 domain of Grb2: highlight on the binding of a new high-affinity inhibitor. J Mol Biol 315:1167–1177. https://doi.org/10.1006/jmbi.2001.5299
7. Rahuel J, Gay B, Erdmann D et al (1996) Structural basis for specificity of GRB2-SH2 revealed by a novel ligand binding mode. Nat Struct Mol Biol 3:586–589. https://doi.org/10.1038/nsb0796-586
8. Waksman G, Kominos D, Robertson SC et al (1992) Crystal structure of the phosphotyrosine recognition domain SH2 of v-src complexed with tyrosine-phosphorylated peptides. Nature 358:646–653. https://doi.org/10.1038/358646a0
9. Liu Z, Gong Z, Dong X, Tang C (2016) Transient protein–protein interactions visualized by solution NMR. Biochim Biophys Acta 1864: 115–122. https://doi.org/10.1016/j.bbapap.2015.04.009
10. Sanches K, Caruso IP, Almeida FCL, Melo FA (2020) The dynamics of free and phosphopeptide-bound Grb2-SH2 reveals two dynamically independent subdomains and an encounter complex with fuzzy interactions. Sci Rep 10:1–13. https://doi.org/10.1038/s41598-020-70034-w
11. James LC, Tawfik DS (2005) Structure and kinetics of a transient antibody binding intermediate reveal a kinetic discrimination mechanism in antigen recognition. Proc Natl Acad Sci U S A 102:12730–12735. https://doi.org/10.1073/pnas.0500909102
12. Tang C, Iwahara J, Clore GM (2006) Visualization of transient encounter complexes in protein-protein association. Nature 444:383–386. https://doi.org/10.1038/nature05201
13. Bashir Q, Scanu S, Ubbink M (2011) Dynamics in electron transfer protein complexes. FEBS J 278:1391–1400. https://doi.org/10.1111/j.1742-4658.2011.08062.x
14. Huang WYC, Ditlev JA, Chiang HK et al (2017) Allosteric modulation of Grb2 recruitment to the intrinsically disordered scaffold protein, LAT, by remote site phosphorylation. J Am Chem Soc 139:18009–18015. https://doi.org/10.1021/jacs.7b09387
15. Ross PD, Subramanian S (1981) Thermodynamics of protein association reactions: forces contributing to stability. Biochemistry 20: 3096–3102. https://doi.org/10.1021/bi00514a017
16. Ubbink M (2009) The courtship of proteins: understanding the encounter complex. FEBS Lett 583:1060–1066. https://doi.org/10.1016/j.febslet.2009.02.046
17. Meiboom S, Gill D (1958) Modified spin-echo method for measuring nuclear relaxation times. Rev Sci Instrum 29:688–691. https://doi.org/10.1063/1.1716296
18. Carr HY, Purcell EM (1954) Effects of diffusion on free precession in nuclear magnetic resonance experiments. Phys Rev 94:630
19. Loria JP, Rance M, Palmer AG (1999) A relaxation-compensated Carr-Purcell-Meiboom-Gill sequence for characterizing chemical exchange by NMR spectroscopy. J Am Chem Soc 121:2331–2332
20. Hansen DF, Vallurupalli P, Kay LE (2008) An improved ^{15}N relaxation dispersion experiment for the measurement of millisecond time-scale

dynamics in proteins. J Phys Chem B 112: 5898–5904. https://doi.org/10.1021/jp074793o

21. Yuzawa S, Yokochi M, Hatanaka H et al (2001) Solution structure of Grb2 reveals extensive flexibility necessary for target recognition. J Mol Biol 306:527–537. https://doi.org/10.1006/jmbi.2000.4396
22. Wishart DS, Bigama CG, Yao J et al (1995) 13 C and 15 N chemical shift referencing in biomolecular NMR. J Biomol NMR 6:135–140
23. Delaglio F, Grzesiek S, Vuister GW et al (1995) Nmrpipe – a multidimensional spectral processing system based on UNIX pipes. J Biomol NMR 6:277–293
24. Vranken WF, Boucher W, Stevens TJ et al (2005) The CCPN data model for NMR spectroscopy: development of a software pipeline. Proteins 59:687–696
25. Maciejewski MW, Schuyler AD, Gryk MR et al (2017) NMRbox: a resource for biomolecular NMR computation. Biophys J 112:1529–1534. https://doi.org/10.1016/j.bpj.2017.03.011
26. McConnell HM (1958) Reaction rates by nuclear magnetic resonance. J Chem Phys 28: 430–431. https://doi.org/10.1063/1.1744152
27. Hansen DF, Led JJ (2003) Implications of using approximate Bloch-McConnell equations in NMR analyses of chemically exchanging systems: application to the electron self-exchange of plastocyanin. J Magn Reson 163: 215–227. http://discovery.ucl.ac.uk/1308703/
28. Morin S, Linnet TE, Lescanne M et al (2014) relax: the analysis of biomolecular kinetics and thermodynamics using NMR relaxation dispersion data. Bioinformatics 30:2219–2220. https://doi.org/10.1093/BIOINFORMATICS/BTU166
29. Carver J, Richards R (1972) A general two-site solution for the chemical exchange produced dependence of T2 upon the carr-Purcell pulse separation. J Magn Reson 6:89–105. https://doi.org/10.1016/0022-2364(72)90090-X
30. Davis DG, Perlman ME, London RE (1994) Direct measurements of the dissociation-rate constant for inhibitor-enzyme complexes via the T1ρ and T2 (CPMG) methods. J Magn Reson Ser B 104:266–275. https://doi.org/10.1006/jmrb.1994.1084
31. Ban D, Funk M, Gulich R et al (2011) Kinetics of conformational sampling in ubiquitin. Angew Chem Int Ed Engl 50:11437–11440. https://doi.org/10.1002/anie.201105086

Chapter 9

Using Linear Motif Database Resources to Identify SH2 Domain Binders

Hugo Sámano-Sánchez, Toby J. Gibson, and Lucía B. Chemes

Abstract

The SH2-binding phosphotyrosine class of short linear motifs (SLiMs) are key conditional regulatory elements, particularly in signaling protein complexes beneath the cell's plasma membrane. In addition to transmitting cellular signaling information, they can also play roles in cellular hijack by invasive pathogens. Researchers can take advantage of bioinformatics tools and resources to predict the motifs at conserved phosphotyrosine residues in regions of intrinsically disordered protein. A candidate SH2-binding motif can be established and assigned to one or more of the SH2 domain subgroups. It is, however, not so straightforward to predict which SH2 domains are capable of binding the given candidate. This is largely due to the cooperative nature of the binding amino acids which enables poorer binding residues to be tolerated when the other residues are optimal. High-throughput peptide arrays are powerful tools used to derive SH2 domain-binding specificity, but they are unable to capture these cooperative effects and also suffer from other shortcomings. Tissue and cell type expression can help to restrict the list of available interactors: for example, some well-studied SH2 domain proteins are only present in the immune cell lineages. In this article, we provide a table of motif patterns and four bioinformatics strategies that introduce a range of tools that can be used in motif hunting in cellular and pathogen proteins. Experimental followup is essential to determine which SH2 domain/motif-containing proteins are the actual functional partners.

Key words SH2 domain, Short linear motif (SLiM), Regular expression, Binding specificity, Phosphotyrosine, Cell signaling, Pathogen hijack

1 Introduction

The Src homology 2 (SH2) domain is a major protein interaction module that is central to tyrosine kinase signaling. SH2 domains are the main class of phosphotyrosine (pTyr) recognition module [1] and are present in metazoa and close unicellular relatives which have receptor tyrosine kinases (RTKs) as well as soluble tyrosine kinases (TKs) [2, 3]. The human proteome contains roughly 120 types of SH2 domains distributed across over 100 human proteins including protein kinases, phosphatases, adaptors,

Teresa Carlomagno and Maja Köhn (eds.), *SH2 Domains: Functional Modules and Evolving Tools in Biology*, Methods in Molecular Biology, vol. 2705, https://doi.org/10.1007/978-1-0716-3393-9_9,

ubiquitin ligases, transcription factors, and guanine nucleotide exchange factors [1, 3]. SH2 domains act in combination with other modular signaling domains to regulate multiple cellular processes ranging from mitogenic signaling, cytoskeletal reorganization, and cell adhesion to T cell activation. The deregulation of SH2 function is a common feature in diseases such as cancer, diabetes, and immunodeficiencies [4]. The widespread distribution of SH2 domains places them at the core of cell signaling, allowing the formation of dynamic protein interaction networks that direct multiple signaling pathways in response to tyrosine kinase activation [5].

While SH2 domains are broadly expanded in eukaryotes [6], only a few examples of SH2 domain-containing proteins have been described in bacteria; in particular in several *Legionella* species and in some members of the *Coxiellacea* family [7]. Bacterial secreted proteins, also known as effector proteins when they are substrates of bacterial secretion systems, have been shown to commonly contain functional linear motifs that mimic host protein functions in order to interfere with host signaling contributing to pathogenesis (reviewed in [8]). Proteins containing phosphotyrosine EPIYA motifs that bind to the SH2 domain of the C-terminal Src kinase (CSK) have been described in known secreted proteins from several bacterial pathogens including *Helicobacter pylori* [8]. Although pathogenic strains of *Escherichia coli* secrete proteins that can be tyrosine phosphorylated [9], no effector has been described or predicted as containing a CSK-binding motif in any *Escherichia* species.

SH2 domains are small (~100 amino acid) protein modules with an invariant fold made up of a central antiparallel β-sheet formed by three or four β strands flanked by two α helices (Fig. 1a) [4]. The secondary structure elements are connected by variable loops that play a key role in determining pTyr motif-binding specificity [10]. The pTyr peptide binds in an extended conformation perpendicular to the central β sheets (Fig. 1a) [11, 12]. Several SH2 domains present a two-pronged binding mode involving the pTyr plus a second residue C-terminal to the pTyr. The pTyr establishes a bidentate salt bridge with a universally conserved arginine residue (ArgβB5) at the pTyr-binding pocket. In contrast to this conserved binding mode, the interactions with the amino acids surrounding the pTyr are more variable and play a role in determining SH2-ligand specificity and affinity. These amino acids are denoted by their position N-terminal (P−1, P−2, etc.) or C-terminal (P+1, P+2, P+3, etc.) to the pTyr. In Src family kinase SH2 domains, a hydrophobic residue at position P+3 binds to the P +3 pocket, which is molded by the EF and BG loops (Fig. 1b). Residue βD5 in the pocket determines the specificity for the P+3 residue. Thus, TyrβD5 in Src SH2 domains dictates a preference for Ile, Leu, or Val at P+3, while IleβD5 in the p85 subunit of PI3K

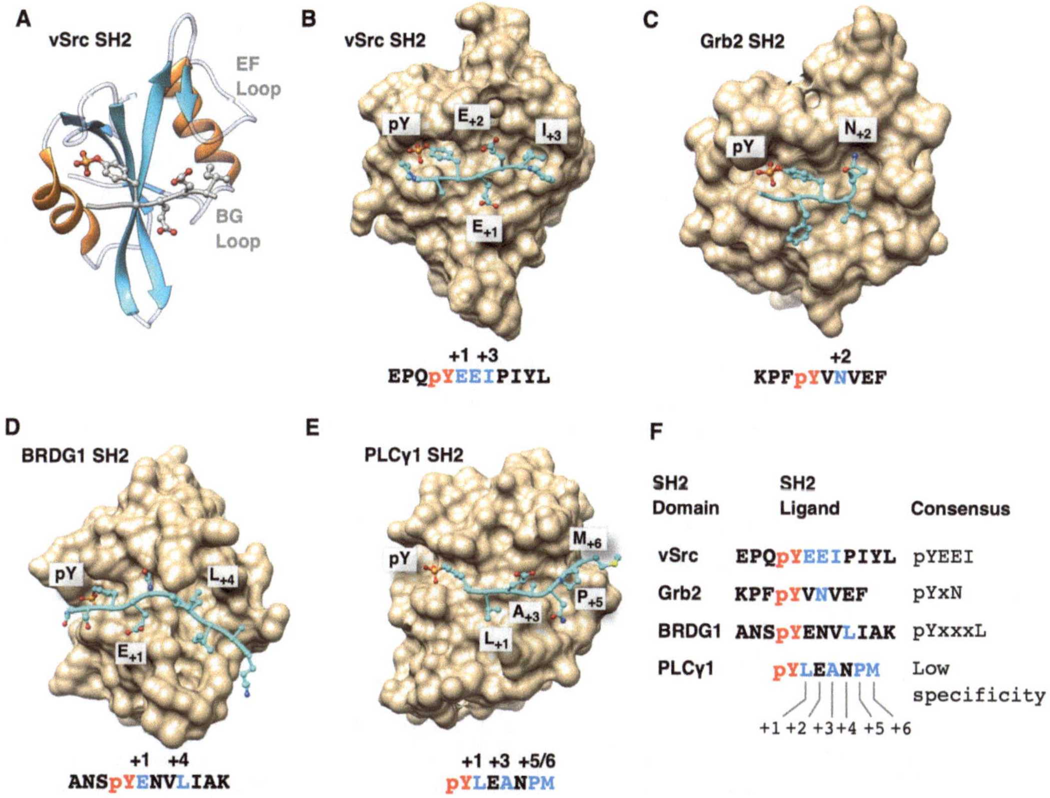

Fig. 1 SH2 domains bind to phosphorylated tyrosine motifs. (**a**) Structure of the vSRC SH2 domain. The SH2 fold is composed of a central antiparallel β-sheet formed by three β strands (cyan) flanked by two α helices (orange). The EF and BG loops determine binding specificity (PDB:1SPS). (**b**) The vSrc SH2 domain bound to a pYEEI peptide represents the canonical two-pronged interaction mode involving pTyr and a hydrophobic residue at P+3 (PDB:1SPS). (**c**) The Grb2 SH2 domain binds peptides containing the pYxN consensus. A hydrophobic residue in the EF loop forces the peptide to adopt a β-turn configuration (PDB:1BMB). (**d**) The BRDG1/STAP-1 SH2 domain has a plugged P+3 pocket and binds peptides with a pYxxxL consensus, harboring a leucine or isoleucine at P+4 (PDB:3MAZ). (**e**) The SH2 domain from phospholipase C-γ1 bound to a high-affinity ligand exemplifies interactions at positions P+5/P+6 (PDB:2PLD). (**f**) Binding motifs and consensus sequence for the different SH2 domains are shown in (**a**–**e**)

makes a deeper groove creating a preference for Met at P+3 [4]. Other binding modes are also possible. In Grb2 SH2 domains, the presence of Trp in loop EF1 occludes the P+3-binding pocket and forces the peptide backbone to adopt a β-turn conformation, allowing an Asn at P+2 to make hydrogen bonds to the SH2 domain (Fig. 1c) [10]. In BRDG1 domains, changes in the EF and BG loops occlude access to the P+3-binding pocket and unplug access to the P+4-binding pocket, which binds a Leu or Ile (Fig. 1d) [10]. Some SH2 domains (e.g., PLCγ1 and SHP2) have extended interactions up to P+5 and P+6 (Fig. 1e), and residues N-terminal to the pTyr can also act as specificity determinants [4, 10] creating a family of related binding motifs (Fig. 1f).

High- and low-throughput approaches revealed different features of SH2 domain–motif interactions. Initial insights were gleaned from low-throughput studies including crystal structures and binding assays performed using isothermal titration calorimetry, fluorescence spectroscopy, and surface plasmon resonance [13, 14]. High-throughput screening (HTS) using peptide microarrays [15, 16] and oriented peptide array libraries (OPAL) [17], also called SPOT arrays, allowed screening of a wide array of SH2 domains under identical experimental conditions and provided a comprehensive picture of SH2 domain-binding specificity [18]. HTS approaches map sequence preferences from positions P−2/P−3 to P+4/P+5 and group related SH2 domains by their binding specificity, which does not strictly follow phylogenetic clustering [19]. SH2 domains can have narrow sequence specificity (e.g., Grb2), facilitating specialized functions that require binding a reduced set of partners, or broader sequence preferences (e.g., Src family) that allow binding a larger set of substrates and promote their functions as signaling hubs [5, 19]. With some exceptions, the binding affinity of an SH2 domain to a pTyr-containing ligand is moderate, ranging from 0.1 μM to 10 μM for equilibrium dissociation constant values (Kd) [20]. The moderate affinity of SH2 domain-mediated interactions enables the dynamic responses required for pTyr-mediated signaling [5].

The sequence preference of SH2 domains can be encoded using regular expressions (Table 1) that define amino acid preferences at positions N- and C-terminal to the pTyr. A summary of the binding selectivity of human SH2 domains derived from OPAL screens in [18] is provided in Table 2. In these screens, pTyr is fixed and all positions except for the one being tested are randomized. The strongest selectivity is at the main specificity-determining positions (N at P+2 for Grb2, hydrophobic residues at P+3 for several SH2s or at P+4 for BRDG1) and identify other positions that influence binding. The selectivity of SH2 domains for physiological pTyr motifs is higher than that derived from the OPAL screens, which reveal moderate selectivity for many SH2 domains (Table 2). This is likely due to context-dependent positional effects and disallowed residues that are not captured by SPOT arrays. SPOT arrays for the same domain can give dramatically different results (Fig. 2a–d) due to changes in the peptide array design or in the experimental conditions [21]. For example, the importance of sequence context is revealed when comparing the canonical SPOT array from [18] to a SPOT array based on a fixed ligand sequence, where only one position is varied (Fig. 2a, b). The results may also change when changing the number of positions being tested or when two positions are fixed instead of one (Fig. 2c, d). The SPOT arrays for Grb2 reveal that proline is disallowed at P+1, where it would prevent the sharp β-turn conformation [21] (Fig. 2a–d). Proline is also disallowed at P+1/P+2 in CRK and

Table 1
Definition of the symbols used in regular expressions for linear motifs[a]

<table>
<tr><th>Character</th><th>Use</th></tr>
<tr><td>X</td><td>Specific amino acid, written in one-letter code</td></tr>
<tr><td>[XZ]</td><td>List of amino acids allowed in a single position of the motif</td></tr>
<tr><td>[^XZ]</td><td>List of amino acids NOT allowed in a single position of the motif</td></tr>
<tr><td>{min, max}</td><td>Specifies the minimum and maximum range allowed for the character to the left of the expression. For example, A{1,4} allows for A, AA, AAA, or AAAA</td></tr>
<tr><td>^</td><td>Indicates the N-terminus of the protein</td></tr>
<tr><td>$</td><td>Indicates the C-terminus of the protein</td></tr>
<tr><td>|</td><td>OR statement used to separate distinct regular expressions in a single motif definition when two or more alternative regular expressions are possible. Instances must match at least one of the regular expressions</td></tr>
<tr><td>(X)</td><td>Bracketing a single character is redundant and can be used to mark positions of interest, e.g., a residue that is covalently modified (such as phosphorylation). Brackets can also be used to group a regular expression</td></tr>
<tr><td>.</td><td>Wildcard. A dot matches any amino acid. In text, wildcards are commonly denoted by a lowercase "x" character in pseudo-regular expressions, as seen in Table 2. This is because they get confused with a full stop, and three consecutive dots are converted by word processor autocorrection into the single character for the ellipsis</td></tr>
</table>

[a]Regular expressions can be used to represent motif models. This table shows the characters and syntax used by ELM to encode and annotate linear motif models which is a subset of the POSIX standard

other SH2 domains [21] (Table 2). Most SH2 domains, exemplified by Grb2 (Fig. 2a, d), show a preference for negative charge with positive charge being disfavored in the positions immediately after pTyr (Table 2). Charge preference is reversed for the CRK domain, where the surface-exposed D91 residue creates a preference for positive charge C-terminal to the pTyr (Table 2 and Fig. 2e).

A functional SH2 domain–motif interaction is only possible if the SH2 domain-containing protein and the pTyr-containing binding partner are present in the same cell type and subcellular compartment and are co-expressed. Spatiotemporal constraints are therefore essential to define functional SH2 domain–motif interactions [5]. While individual SH2-binding motif affinities are often modest, tandem SH2 domains present in signaling proteins may bind to doubly phosphorylated motifs, and SH2/SH3 domain combinations are also frequent. Many scaffolding and adaptor proteins establish such multivalent interactions, which create cooperativity, increasing the specificity and robustness of signaling pathways [5].

Table 2
Regular expression for SH2 domain-binding motifs based on published SPOT arrays data

SH2 group[a]	SH2 domain containing protein	UniProt ID	Function	SH2 motif pattern	Consensus peptide	Relative specificity	Prep quality[b]	Notes	Key references
IA	ABL1	P00519	Non-receptor tyrosine kinase	xx**(Y)**[^KRPW][^KRHW][^KRHW][^KR]	PP(Y)ENVM	Low	G	Similar preferences to other Class I NRtyrK for negative and hydrophobic residues but low specificity	[18]
IA	ABL2	P42684	Non-receptor tyrosine kinase	[^KRSW][^RW]**(Y)**[^KRNPYW][^KRPFYW][^KRHNQGFYW][^KRFYW]	PP(Y)EMPM	Low	G/D	Similar preferences to other Class I NRtyrK for negative and aliphatic residues but low specificity	[18]
IA	BLK	P51451	Non-receptor tyrosine kinase	[^KRHW][^KRW]**(Y)**[Eds][ENd][VLide][De]	DE(Y)EEVD	Medium	A	Low saturation OPAL	[18]
IA	BRK/PTK6	Q13882	Non-receptor tyrosine kinase	[^KR][^KW]**(Y)**[DEstna][ENTMwyifvasqd][VTAPYil][^KRHW]	NP(Y)EEVD	Low	G		[18, 21]
IA	BTK	Q06187	Non-receptor tyrosine kinase	[QVENDpil][ED]**(Y)**[DEsa][EDNH][EVTD]x	QE(Y)DEED	Medium	G/D	Low saturation OPAL. P+3, P+4 reduced signal	[18]
IA	FGR	P09769	Non-receptor tyrosine kinase	xx**(Y)**[EDsvyqta][NEtdmqsag][^KRHNQGW][^KYW]	PE(Y)ENLD	Low	VG	Similar preferences to other SFKs for negative and hydrophobic residues but low specificity	[18]

IA	FRK	P42685	Non-receptor tyrosine kinase	[^KM]x**(Y)** [ESTYdqaivfl] [^KRHGP] [ILVFyedapm] [DEfwyh]	PF(Y) EEFD	Low	G/D	Similar preferences to other SFKs for negative and hydrophobic residues but low specificity	[18]
IA	FYN	P06241	Non-receptor tyrosine kinase	[^KR]x**(Y)** [DEVYfstaq] [^KRHPFW] [ILVPfdeta] [^KRH]	PE(Y) EEID	Low	G/W	Similar preferences to other SFKs for negative and hydrophobic residues but low specificity	[18]
IA	HCK	P08631	Non-receptor tyrosine kinase	[^KRW][^KRW]**(Y)** [EDsta][NE] [PILVr][DEr]	NE(Y) ENPD	Strong	G/W	Low saturation OPAL. When P+3 is optimal (Liu et al PMID:20627867), more residues are tolerated at P+1, P+2 ,and P+4	[18]
IA	ITK	Q08881	Non-receptor tyrosine kinase	[^KRQSFYW] [^KRYW]**(Y)** [^FYW][DNehmk] [^KRHNQSYW] [^KRHN]	EE(Y) EDEM	Low	W	Low saturation OPAL	[18]
IA	LCK	P06239	Non-receptor tyrosine kinase	[^KRM]x**(Y)** [^KRHNGPW] [^KRHP] [^KRHNQG] [^KRQP]	VE(Y) EELE	Low	VG	Similar preferences to other SFKs for negative and hydrophobic residues but low specificity	[18]
IA	LYN	P07948	Non-receptor tyrosine kinase	xx**(Y)**[EDstafyqh] [^KRPI][^KRHNQ] [^KRP]	PE(Y) ENLF	Low	W	High saturation OPAL. Similar preferences to other SFKs for negative and hydrophobic residues but low specificity	[18]
IA	NCK1	P16333	Non-receptor tyrosine kinase	[^KRHW][^KRW]**(Y)** [DEstay] [^KRHGFYW] [PVateslm] [SDtaemg]	PV(Y) DEPS	Medium	G	P+1 and P+3 are strong determinant positions	[18]

(continued)

Table 2 (continued)

SH2 group[a]	SH2 domain containing protein	UniProt ID	Function	SH2 motif pattern	Consensus peptide	Relative specificity	Prep quality[b]	Notes	Key references
IA	NCK2	O43639	Non-receptor tyrosine kinase	[^KRHW][^KRGW]**(Y)** [DESta] [^KRHGPFYW] [PVats][DTsag]	PV(Y) ENVD	Medium	G/W	P+1 and P+3 are strong determinant positions	[18]
IA	SRC	P12931	Non-receptor tyrosine kinase	xx**(Y)**[DEVIstay] [^KRHG] [ILVdepamf]x	PI(Y) EEID	Low	G/W	Similar preferences to other SFKs for negative and hydrophobic residues but low specificity. Low signal at P+4	[18]
IA	SYK-C	P43405	Non-receptor tyrosine kinase	[^W][^W]**(Y)** [EQSTAVnlmfyi] [DESTAnqgmp] [Lpivdestan] [QSTdeaglm]	GA(Y) EELQ	Low	G/DD	Strong L at P+3, acidic and small preferences with other residues often tolerated	[18]
IA	SYK-N	P43405	Non-receptor tyrosine kinase	*Nonspecific, no motif can be defined*	MM(Y) EFLW	–	VG	Low saturation and low specificity. There are no clear preferences for most positions except IL at P+3	[18]
IA	TEC	P42680	Non-receptor tyrosine kinase	xx**(Y)**[DESAnhy] [DENFqstilvm] [EAPVil][DE]	ED(Y) DEVD	Medium	A		[18]
IA	TXK	P42681	Non-receptor tyrosine kinase	*Nonspecific, no motif can be defined*	EE(Y) EEEE	–	G	Preferences for acidic residues but all residues tolerated	[18]
IA	YES	P07947	Non-receptor tyrosine kinase	[^KR]x**(Y)**[^KRGPW] [^KRHW][^KRHNW] [^KR]	PL(Y) DNLD	Low	G	Similar preferences to other SFKs for negative and hydrophobic residues but low specificity	[18]

IA	ZAP70-N	P43403	Non-receptor tyrosine kinase	[^KRW][^KRW]**(Y)** [Edsav][NEdilm] [ILVEdp][De]	EE(Y) ENVD	Medium	VG	Low saturation OPAL	[18]
IB	CRK	P46108	Adaptor signaling	xx**(Y)**[^EPILVFYW] [^HDEFW] [^HDENGYW] [^DEILVFYW]	NR(Y) DTPR	Low	G/D	Unusual allowance for positively charged residues and preference for hydrophobic residues at P+3	[18, 21]
IB	CRKL	P46109	Adaptor signaling	xx**(Y)**[^EGPILVFYW] [KRNQSTMav] [PILVTkram] [^DEILFYW]	NR(Y) DKPR	Low	G	Unusual allowance for positively charged residues and preference for hydrophobic residues at P+3	[18]
IB	MATK	P42679	Non-receptor tyrosine kinase	[^KRW]x**(Y)** [ETAYVqsimwlfd] [^KGPYW] [ATIVMFelg] [^KRHN]	IE(Y) EMAM	Low	W	Clear preferences in all positions but many residues tolerated	[18]
IB	RASA1_N	P20936	RAS GTPase regulator	*Nonspecific, no motif can be defined*	EE(Y) DEEE	–	G/D	Preferences for acidic and aliphatic residues, disfavouring positive and aromatics but very low specificity	[18]
IB	RASA1_C	P20936	RAS GTPase regulator	[^KRHYW][^KRFYW] **(Y)**[Sdenqta] [^KRHFYW][Pv]x	DN(Y) SNPL	Medium	G	Very low saturation OPAL and very low signal at P+4	[18]
IB	SH2D1A	O60880	Adaptor signaling	xx(Y)[^KRGPW] [^RHDGPW] [VILmy]x	TI(Y) FLVG	Medium	G/W	Primarily hydrophobic preference but most positions tolerate other residues. Main specificity at P+3	[18]
IB	SH2D1B	O14796	Adaptor signaling	[^KRHW][^KRW]**(Y)** [^KRHGPW] [^KRHGPW] [Vlade] [DSGTMhae]	TV(Y) STVD	Medium	G	Primarily hydrophobic and small preference but most positions tolerate other residues. Main specificity at P+3	[18]

(continued)

Table 2
(continued)

SH2 group[a]	SH2 domain containing protein	UniProt ID	Function	SH2 motif pattern	Consensus peptide	Relative specificity	Prep quality[b]	Notes	Key references
IB	SHIP1	Q92835	Phosphatidyl-inositol phosphatase	xx**(Y)**x[LFMtviq] [Lviarp] [GLVQFmtap]	TG(Y) SLLG	Medium	VG	Primarily hydrophobic and not much charge	[18]
IB	SHIP2	O15357	Phosphatidyl-inositol phosphatase	[^KRH][^KW]**(Y)** [DEStmn] [Tmvlnqsark] [LIVAPrq] [RGMNsaqth]	PP(Y) DTLR	Medium	VG	Some clear acidic and aliphatic preferences but other residues often tolerated	[18]
IC	BCAR3	O75815	Adaptor signaling	*Nonspecific, no motif can be defined*	LE(Y) DEEE	–	G	Preferences for acidic and hydrophobic residues but almost all residues tolerated	[18]
IC	BMX	P51813	Adaptor signaling	[^KRHDE][^KRW]**(Y)** [DE][Nde][ED] [ED]	PP(Y) ENEE	Strong	G/W	Low saturation OPAL. Preference for DEP at P−1	[18]
IC	CSK	P41240	Non-receptor tyrosine kinase	EP[IL]**(Y)**[TAG]	EPI(Y) Axxx	Strong	–	Unusual for SH2, strong preference at -3 position which is not tested on arrays. This pattern is from the ELM resource	[22]
IC	CSK	P41240	Non-receptor tyrosine kinase	xx**(Y)**[STA]N[VP]x	xx(Y) ANVx	Strong	G/W	Low saturation OPAL. P−2, P−1, and P+4 low to absent signal	[18]
IC	FER	P16591	Non-receptor tyrosine kinase	[^KRHW][^KRW]**(Y)** [Ed][Ned][EYd]d	DE(Y) ENED	Strong	G	Low saturation OPAL. Preference for DE at P−2. Weak signal at P+4	[18]

IC	FES	P07332	Non-receptor tyrosine kinase	[^HW][^KW]**(Y)** [Edimyv] [^KRHTW][^KRHW] [DSGtaemmqp]	DE(Y) ENVD	Low	G/W	N is main preference at P+2 but many residues tolerated	[18]
IC	GADS/ GRAP2	O75791	Adaptor signaling	[^RSW][^DW]**(Y)** [^KRHPFYW][Ne] [^KRHPFYW][^KR]	PE(Y) ENEM	Medium	G/W		[18]
IC	GRAP	Q13588	Adaptor signaling	[^KRGFYW] [^KRHDGFYW]**(Y)** [^KRHDGPW][Ne] [ELVItmd][^KR]	EE(Y) ENEE	Medium	G	P−2 and P−1 are low saturation	[18]
IC	GRB2	P62993	Adaptor signaling	[^W][^D]**(Y)** [^KRHDPW][Neqm] [^KRHW][^kr]	PP(Y) ENEE	Medium	VG	Strong saturation OPAL. When P+2 is fixed at N, P is disallowed at P+3	[18, 21]
IC	GRB2	P62993	Adaptor signaling	xxx**(Y)**[^HDNPW]N [^DP][^d]x	DPS(Y) VNVQ	Medium	–	SPOT based on Shc1 pTyr317. Tolerance of basic residues unlike Huang et al. [18]	[7]
IC	GRB7	Q14451	Adaptor signaling	*Nonspecific, no motif can be defined*	IL(Y) DNEN	–	VG	Preferences for acidic P+1 and P+2 but almost all residues tolerated	[18]
IC	GRB10	Q13322	Adaptor signaling	[^KRH][^KRW]**(Y)** [DE][Ndetqsm] [^KRHW] [DEnmstagq]	NE(Y) ENVD	Medium	VG		[18]
IC	GRB14	Q14449	Adaptor signaling	[^W][^KW]**(Y)** [DEsa][^KRHW] [^KRHW] [^KRHFYW]	DN(Y) ENVD	Low	G		[18]
IC	HSH2D	Q96JZ2	Adaptor signaling	[Milve][Mle]**(Y)** [Ednsm]N [Mfativ]x	MM(Y) ENMM	Medium	G/W	Low saturation OPAL. Positions P−2 and P+4 have very weak signal	[18]

(continued)

Table 2
(continued)

SH2 group[a]	SH2 domain containing protein	UniProt ID	Function	SH2 motif pattern	Consensus peptide	Relative specificity	Prep quality[b]	Notes	Key references
IC	SH2D2A	Q9NP31	Adaptor signaling	`[^KR][^KRFYW]`**`(Y)`**`[DEsatv]Nxx`	DH(Y)DNLx	Medium	–	Low saturation OPAL. P+3 and P+4 low to absent signal	[18]
IC	SH3BP2	P78314	Adaptor signaling	`x[EIVndq]`**`(Y)`**`E[Nvi][eqv]x`	NV(Y)ENEx	Strong	G	Low saturation OPAL. P+3 and P+4 low to absent signal	[18]
IC	SRMS	Q9H3Y6	Non-receptor tyrosine kinase	`[^KRGW][^KRNGMW]`**`(Y)`**`[EVFYsa]N[Vlf]x`	PP(Y)ENVS	Strong	G	Low saturation OPAL. P+4 low to absent signal	[18]
IC	TENC1 (TNS2)	Q63HR2	pTyr/lipid phosphatase	`xx`**`(Y)`**`[DEry][Nyrgfm][^K][RHDEN]`	NP(Y)DNVR	Medium	G	Rare tolerance of R in all positions. Preference for N at P+2 but most residues are tolerated	[18]
IC	TNS1 (TENS1)	Q9HBL0	pTyr/lipid phosphatase	`[^KRSW][^KRW]`**`(Y)`**`[EDstvfl][^KRHP][^KRH][^KR]`	VE(Y)ENEE	Low	G	Acid preference also aliphatic and small. Basic residues excluded	[18]
IC	TNS4 (CTEN)	Q8IZW8	Adaptor signaling	`[^KR][^KDW]`**`(Y)`**`[DE][Ntfylm][LVFIPt][^KRW]`	FG(Y)DNLF	Medium	G/W	Low saturation OPAL	[18]
IIA	PLCg1_C	P19174	Phospholipase signaling	*Nonspecific, no motif can be defined*	EE(Y)VEEE	–	G	Preferences for acidic and hydrophobic residues but almost all residues tolerated	[18]
IIA	PLCg2_C	P16885	Phospholipase signaling	*Nonspecific, no motif can be defined*	YP(Y)EEDE	–	VG	Preferences for acidic and aromatic residues but all residues tolerated	[18]

IIA	PIK3R1_N/ p85A	P27986	PI3Kinase regulatory subunit	[^HW][^W]**(Y)** [^KRHNGPW] [^KRHYW] [^KRHGIYW] [DSAGemtq]	DH(Y) EEMD	Low	G/W	Weak preferences for acid and aliphatic and N at P+2	[18]
IIA	PIK3R1_C/ p85A	P27986	PI3Kinase regulatory subunit	[^KRHFYW] [^KRFYW]**(Y)** [EVmi][Nmil] [Mlvt]x	DD(Y) ENMx	Strong	G	Low saturation OPAL. P+4 absent signal	[18]
IIA	PIK3R2_N/ p85B	O00459	PI3Kinase regulatory subunit	x[^K]**(Y)** [^KRHNQGPW] [^KRHW][^KRW] [^KRNPILVF]	NE(Y) ENMD	Low	G	Very weak preferences for acid and aliphatic and N at P+2	[18]
IIA	PIK3R2_C/ p85B	O00459	PI3Kinase regulatory subunit	[^W][^W]**(Y)** [^KRHDNGPW] [^KRHYW] [^KRHDENGPYW] [^KRENFYW]	DE(Y) VNIM	Low	G	Very weak preferences for aliphatic and N at P+2	[18]
IIA	PIK3R3_N/ p55G	Q92569	PI3Kinase regulatory subunit	[^KRHW][^KYW]**(Y)** [VEIflmt] [NElmtp] [Melfpa][DTAG]	DE(Y) VNMD	Medium	G	Reduced preference for N at P +2	[18]
IIA	PIK3R3_C/ p55G	Q92569	PI3Kinase regulatory subunit	[^H]x**(Y)** [^KRHDNGPW] [^KRHYW] [^KRHDGYW] [^KRENVFYW]	DH(Y) ENMM	Low	G/D	Very weak preferences for aliphatic and E at P+1, N at P+2	[18]
IIA	PTPN6_C	P29350	Tyrosine phosphatase	*Nonspecific, no motif can be defined*	PE(Y) IEEE	–	G	Some preferences visible but all residues tolerated. Air bubble on P+3 P+4 Val	[18]
IIA	PTPN11_N	Q06124	Tyrosine phosphatase	*Nonspecific, no motif can be defined*	LE(Y) VEEE	–	G	Some preferences visible but all residues tolerated	[18]

(continued)

Table 2 (continued)

SH2 group[a]	SH2 domain containing protein	UniProt ID	Function	SH2 motif pattern	Consensus peptide	Relative specificity	Prep quality[b]	Notes	Key references
IIA	PTPN11_C	Q06124	Tyrosine phosphatase	`[^RYW][^RW]`**(Y)**`[^KRHQPFYW][^KRHGPILVYW][^KRHNQSPYW][^KRHNY]`	EM(Y)EDIM	Low	G/D	Shows preferences but many residues tolerated, mainly disfavouring positives and aromatics	[18]
IIA	VAV1	P15498	Signaling GTPase	`[^HW][^W]`**(Y)**`[ELIVMfaqy][^KRHPFYW][EPad][Destagm]`	DL(Y)LNED	Low	W	Preference for acidic, aliphatic, P+2 and small at P+4	[18]
IIA	VAV2	P52735	Signaling GTPase	`[^KR][^KW]`**(Y)**`[EVilmtqd][TMvenqsag][EPdsaqt][EDnqstma]`	NP(Y)ETPD	Low	VG	Preference for acidic, small and aliphatic (P+1)	[18]
IIA	VAV3	Q9UKW4	Signaling GTPase	`[^RHW]x`**(Y)**`[ELVMIfy][^KRHGPYW][DEPyas][Dest]`	DE(Y)ENED	Low	G	Preference for acidic, also some aliphatic and small	[18]
IIB	APS/SH2B2	O14492	Adaptor signaling	*Nonspecific, no motif can be defined*	EE(Y)IELE	–	G	Acidic, aliphatic preferences but all residues appear to be tolerated	[18]
IIB	BLNK	Q8WV28	Non-receptor tyrosine kinase	`xx`**(Y)**`[Der][Der]xx`	DE(Y)DDRE	Medium	VG	Low saturation OPAL	[18]

IIB	SH2B	Q9NRF2	Adaptor signaling	[^KR][^KRGM]**(Y)** [^KRHNPM] [FYWLedn] [FYLedwvi] [Defylw]	PI(Y) YFFD	Medium	VG	Preferences for aromatic, acidic and more weakly aliphatic residues	[18]
IIB	SH2B3 (LNK, SLNK)	Q9UQQ2	Adaptor signaling	xx**(Y)**[DEY] [DFkwerhy] [EIYWlvmf] [^KRH]	EE(Y) YDEE	Medium	VG	P+1 strong DYE preference but all residues tolerated. P−2, P+3 weak saturation. Acidic and aromatic preference but often tolerating other residues	[18]
IIB	SHB	Q15464	Adaptor signaling	[^KRHW][^KRHW]**(Y)** [DE][Edt][Led] [DE]	NE(Y) DELE	Strong	VG/D	Low saturation OPAL	[18]
IIB	SHC1	P29353	Adaptor signaling	[^W][^KW]**(Y)** [EDSlmyqagi] [NTMqsalvi] [ILVmats]x	TH(Y) ENLS	Medium	G	Low saturation at P+4	[18]
IIB	SHC2	P98077	Adaptor signaling	[^HW][^W]**(Y)** [EILSdavm] [^RHDGPW] [LVIYfmat]x	DE(Y) ELLT	Low	G		[18]
IIB	SHC3	Q92529	Adaptor signaling	[^KRHFYW] [^KRFYW]**(Y)** [Ets][Ntmqsal] [LIvm]x	PH(Y) ENLx	Medium	G	Low saturation OPAL, signal absent at P+4	[18]
IIB	SHC4	Q6S5L8	Adaptor signaling	xx**(Y)**[^KPW]x [LIVYfmatwr]x	DR(Y) ELLW	Low	G		[18]
IIB	SHD	Q96IW2	Adaptor signaling	xx**(Y)**[EDk] [Nktilvme][LYi] x	kk(Y) ENLk	Medium	VG	Low saturation OPAL	[18]

(continued)

Table 2
(continued)

SH2 group[a]	SH2 domain containing protein	UniProt ID	Function	SH2 motif pattern	Consensus peptide	Relative specificity	Prep quality[b]	Notes	Key references
IIB	SHE	Q5VZ18	Adaptor signaling	*Nonspecific, no motif can be defined*	EE(Y) ENTM	–	VG	P+1 and P+2 have preference for EN but most positions accept all residues	[18]
IIB	SHF	Q7M4L6	Adaptor signaling	[De]x**(Y)**[FYed] [Edkf][Lyde] [De]	DE(Y) YELD	Strong	G/D	Low saturation OPAL. Specificity is likely to be less strict	[18]
IIC	BRDG1/ STAP1	Q9ULZ2	Adaptor signaling	xx**(Y)**[EDStafy] [YTLFesmvi] [YTEDSf][ILFv]	EE(Y) EFYI	Medium	VG	Has a known preference for hydrophobic residues at P+4	[18]
IID	SUPT6H	Q7KZ85	Transcription elongation factor	[^FW][^FY]**(Y)**xx [KRHEdpm][KMP]	KK(Y) FPKM	Low	G/DD	Rare positive charge preferences. Weak signal in P+4	[18]
Not class ified	RavO Legionella	Q5ZWF6	Bacterial effector protein	xx[^K] **(Y)**[ILVtf] [^KHDEW] [^KHDENW][^HD]x	PS(Y) INVQ	Low	–	SPOT array based on the Shc1 pTyr317 peptide	[7]

Regular expression for SH2 domain-binding motifs based on SPOT array experiments. Upper case letters indicate strongly preferred residues and lower case letters indicate weaker preferences

[a]Classification based on Huang et al. [18], group I domains harbor an aromatic residue (Tyr or Phe) at βD5, group II domains harbor a hydrophobic residue (Ile, Leu, Val, Cys, Met) at βD5, and group III domains (STAT-family) harbor a hydrophilic residue (Glu, Gln, or Lys) at βD5

[b]*VG* very good signal, *G* good signal, *D* degradation, *DD* very strong degradation, *W* weak signal, *A* absent band, – not performed

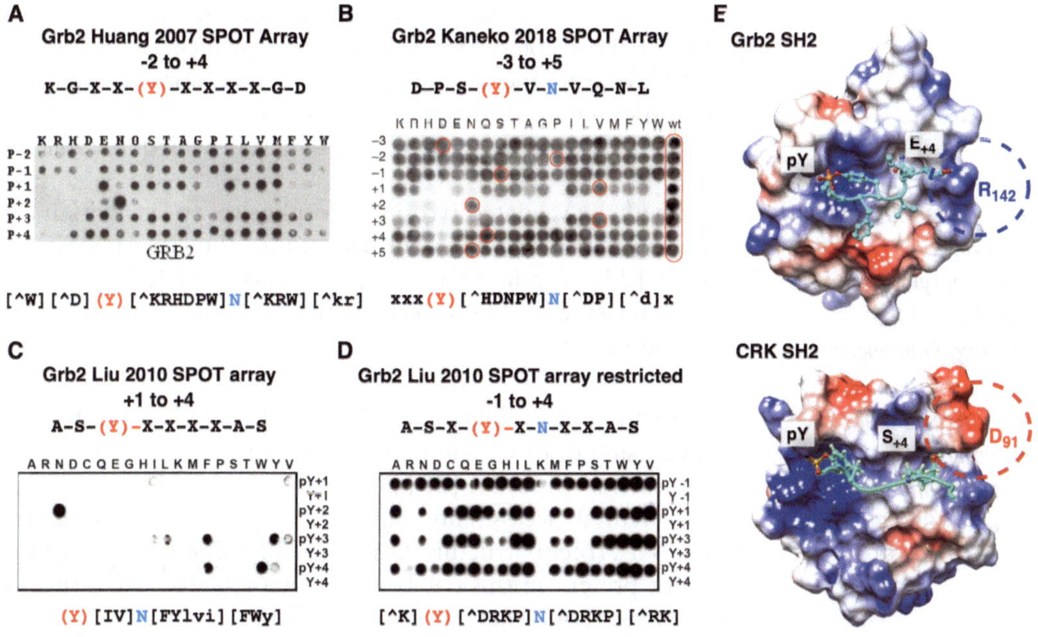

Fig. 2 Sequence preference of SH2 domains revealed by SPOT arrays. (**a–d**) SPOT array for the Grb2 domain performed by different groups and using different peptide array designs. Each SPOT array is labeled with the group that performed the work, the residues being tested, with numbering referenced to the central pTyr, and the design of the array. Fixed positions indicate the amino acid present, and positions being randomly varied are labeled with an "X." pTyr is indicated by the red (Y) symbol. The regular expression derived from each SPOT array is shown below each panel. (**e**) Surface charge distribution for the Grb2 (PDB:1BMB) and CRK (PDB:1JU5) SH2 domains. Dashed circle: region containing opposite charges. This explains the opposing preferences for charged residues of Grb2 and CRK domains. (Panel (**a**) is adapted from Huang et al. [18], panel (**b**) is adapted from Kaneko et al. [7], and panels (**c**, **d**) are adapted from Liu et al. [21]. The creative commons license under which the cited articles are distributed can be found at https:/creativecommons.org/licenses/by/4.0/)

Here, we provide protocols for identifying SH2-binding motifs in human and pathogenic bacterial proteins. The starting point might be the known selectivity of an SH2 domain of interest (Method 1), a candidate protein containing a pTyr (Method 2), a group of bacterial effector proteins (Method 3), or an entire bacterial proteome (Method 4). We highlight the strengths and shortcomings of each method and the need to perform experimental validations.

2 Materials

2.1 SH2 Domain Specificity

We derive the specificity of a dataset of SH2 domains from Huang et al. [18] from OPAL screens. We assess the amino acid preferences at each position by visual inspection of the SPOT arrays and construct regular expressions, when possible. We record the relative specificity of the regular expressions along with the preparation quality of the recombinant SH2 domains, which we assess from SDS-PAGE experiments. Some regular expressions derived from other experiments or from the eukaryotic linear motif (ELM) database [22] are also included. The data is provided in Table 2.

2.2 Determination of Tyrosine Phosphorylation in Candidate SH2-Binding Sequences

Experimentally validated tyrosine phosphosites (pTyr) within candidate proteins can be identified using the PhosphoSitePlus v6.6.0.4 [23], PeptideAtlas [24], Phospho.ELM [25], and UniProt [26] databases. Links to access the databases and software are presented in Table 3.

2.3 Analysis of Structural Accessibility of Putative SH2 Domain-Binding Motifs

The structural accessibility of candidate SH2-binding motifs can be assessed using disorder annotation (DisProt) [27] or disorder predictors such as MobiDB [28] and IUPred3 [29]. For several species, structural accessibility can also be assessed by using the AlphaFold Protein Structure Database [30]. The structure of proteins not present in this database can be predicted using the web server ColabFold that runs AlphaFold2 using an accelerated homology search [31]. Links to the tools are provided in Table 3.

2.4 Extract Protein Annotations

To check for the expression of a protein in a particular cell type, the Human Protein Atlas offers RNA expression and immunohistochemistry evidence [32]. UniProt includes information regarding observed cellular compartment localization of a protein [26]. Links to the tools are provided in Table 3.

2.5 Construction and Visualization of Multiple Sequence Alignments for a Set of Homologous Proteins Containing Candidate SH2-Binding Motifs

Sequences for the construction of an alignment for a protein containing a candidate SH2-binding motif can be collected using TreeFam V9 [33], BLAST, or other methods. Multiple sequence alignments can then be constructed and visualized by opening the fasta file in the Jalview program [34, 35]. ProViz [36] hosts precomputed alignments for the complete proteomes of human and several model organisms and pathogens. Links to the tools are provided in Table 3.

2.6 Searching for Validated and Predicted Candidate SH2 Domain-Binding Motifs

Experimentally validated SH2 domain-binding motifs can be searched using the ELM database [22]. SH2 domain-binding motifs can also be predicted using the ELM database "prediction" tab and the Scansite 4.0 server [37]. SlimFinder [38] predicts motif models as regular expressions and is able to capture semi-conserved positions. Links to the tools are provided in Table 3.

2.7 Determination of SH2 Domain-Binding Specificity

The ModPepInt Server [39] contains data on the specificity of a dataset of 51 SH2 domains and can be used to assess the predicted SH2 domain specificity of a sequence containing a putative SH2-binding motif. The table provided in this chapter (Table 2)

Table 3
Bioinformatics resources[a]

Resource	Method or database description	Use/notes	URL	References
Protein phosphorylation analysis				
PhosphoSitePlus	Database of experimentally verified phosphosites in proteins	Phosphosites can be identified through low-throughput (LTP) or high-throughput (HTP) experiments. The presence of a higher number and multiple types of experimental validation evidence increases the confidence for the annotation. Search filters allow searching by cell line, tissue, or disease	https://www.phosphosite.org/homeAction.action	[23]
PeptideAtlas	Database of peptides and phosphosites identified in tandem mass spectrometry experiments from several organisms	Phosphosites are identified through high-throughput mass spectrometry experiments and filtering tools are used to assess the reliability of each site	http://www.peptideatlas.org/	[24]
PhosphoELM	Database of experimentally verified serine, threonine, and tyrosine phosphorylation sites in eukaryotic proteins	Each annotation identifies the source of experimental evidence (HTP or LTP) and the publication containing the experimental evidence and knowledge of the modifying kinase and/or binding domain if available	http://phospho.elm.eu.org/	[25]
Protein sequence and structure databases/resources (annotated and predicted)				
UniProt	Database of annotated protein sequences	Sequence and functional annotations with numerous links to, e.g., domain/family, posttranslational modification and structural databases as well as GO annotations. UniProtKB contains highly curated reviewed (Swiss-Prot) and unreviewed (TrEMBL) entries	https://www.uniprot.org/	[26]

(continued)

Table 3 (continued)

Resource	Method or database description	Use/notes	URL	References
AlphaFold database	Database of structural models for proteins made using the AlphaFold algorithm	Contains structural models for full-length proteins <2700 residues in the human proteome and a growing number of proteomes of interest	https://alphafold.ebi.ac.uk/	[30]
IUPred3	Web server for protein intrinsic disorder prediction	Allows prediction of disorder propensity for any sequence using the UniProt ID or also by directly pasting the sequence. Disorder propensity is displayed on a scale of 0 (low disorder propensity) to 1 (highest disorder propensity)	https://iupred3.elte.hu/	[29]
MobiDB	Database and metapredictor of protein intrinsic disorder and mobility	Metapredictor tool that allows prediction of disorder propensity and harbors annotations from structural (X-ray and NMR) databases	https://mobidb.bio.unipd.it/	[28]
DisProt	Database of experimentally annotated intrinsically disordered proteins (IDPs) and protein regions (IDRs)	Contains annotations of disorder defined based on experimental evidence. Defines disordered regions and associated structural transitions and molecular interactions and functions annotated according to PSI-MI standards and with links to the publications	https://disprot.org/	[27]
ColabFold	Web server to run AlphaFold2 with an accelerated homology search based on MMseq2	For protein sequences not present in AlphaFold database, a fast de novo AlphaFold prediction can be performed using a web server	https://colab.research.google.com/github/sokrypton/ColabFold/blob/main/AlphaFold2.ipynb	[31]

Human Protein Atlas	Database of protein localization in human cells	Contains annotated evidence for the localization of proteins and mRNAs in tissues, cell types, and diseases, including subcellular localization	https://www.proteinatlas.org/	[32]
Pfam	Database of protein domain families	Protein domain families are encoded using hidden markov models (HMMs) and protein alignments with functional annotation and links to domain structures and phylogenetic distribution	https://pfam.xfam.org/	[54]
Linear motif databases and SH2 domain specificity predictors				
Scansite 4.0	Web server that searches for motifs within proteins likely to be phosphorylated by specific protein kinases or bind to SH2, 14-3-3, or PDZ domains	Identifies motifs and provides a confidence score as well as cross-references to other databases such as PhosphoELM and PhosphoSitePlus	https://scansite4.mit.edu/#home	[37]
ModPepInt	Web server for predicting the binding partners of SH2, SH3, and PDZ domains	Comprises three tools: SH2PepInt, SH3PepInt, and PDZPepInt. SH2 Pepint is a predictor based on semi-supervised machine learning for 51 SH2 domains. It allows input of a query sequence and identifies SH2 domain binder candidates	https://modpepint.informatik.uni-freiburg.de/	[39]
SLiMFinder	Web server for de novo prediction of linear motifs shared by a collection of protein/peptide sequences	Predicts motif models as regular expressions and is able to capture semi-conserved positions. Will not be able to find motifs that are not present in a large proportion of the proteins in the dataset	http://www.slimsuite.unsw.edu.au/servers/slimfinder.php	[38]

(continued)

Table 3 (continued)

Resource	Method or database description	Use/notes	URL	References
SLiMSearch4	Short linear motif discovery tool	SLiMSearch searches a proteome using a motif consensus (regular expression) to discover putative novel motif instances. The instances are annotated with conservation and structural accessibility scores	http://slim.icr.ac.uk/slimsearch/	[40]
ELM Resource	Database of eukaryotic linear motifs	Contains experimentally validated linear motifs from eukaryotic cells and viral and bacterial pathogens infecting eukaryotes. It is not a comprehensive resource, but data is high quality. Motifs are defined by a motif class, and individual instances in proteins are annotated with reference publications and types of experimental evidence available. Includes several SH2 class motifs. Motifs are encoded as regular expressions allowing prediction of motifs for a sequence of interest using the "Prediction" tab	http://elm.eu.org/	[22]
Protein homolog family databases and multiple sequence alignments and visualization tools				
TreeFam V9	Database of phylogenetic trees inferred from animal genomes	Useful for quickly collecting aligned orthologous and paralogous protein sequences. These can be viewed in Jalview. It is best to remove low-quality sequences containing many X characters from the set	http://www.treefam.org/	[33]

Jalview	Program for interactive editing, analysis, and visualization of multiple sequence alignments (MSAs)	Allows the search of linear motifs using regular expressions, making it easy to find and analyze linear motif sequence conservation patterns. Individual motifs can be defined and marked, allowing the analysis of multiple motifs in one MSA	http://www.jalview.org/getdown/release/	[34]
ProViz	Interactive exploration tool for investigating the functional and evolutionary features of proteins	Allows the exploration of linear motif candidates in a protein of interest providing pre-built sequence alignments. Sequence alignments are annotated with conservation, protein domains, post-translational modifications, disorder prediction, and known linear motifs	http://slim.icr.ac.uk/proviz/	[36]
Tools for the prediction of bacterial effector proteins				
BastionHub	Database of manually curated bacterial effector proteins	To reduce the number of false positives, only proteins with evidence of secretion can be used to predict interactions with host proteins	https://bastionhub.erc.monash.edu/	[41]
SignalP	To predict signal peptides in proteins	When little evidence exists that a bacterial protein is secreted, the presence of a signal peptide can be used as evidence that the protein is at least localized to the plasma membrane where several secretion systems might later translocate it	https://services.healthtech.dtu.dk/service.php?SignalP	[42]
BioCyc	Genome annotation database that includes transcription unit architecture information	In bacteria, proteins that belong to the same pathway tend to be located in the same genomic regions, have the same trancription orientation, and even be transcribed simultaneously. This can be used to predict new effector proteins	https://biocyc.org/	[43]

[a] Databases and web servers useful in the process of annotation for putative SH2 domain-binding motifs

additionally summarizes SH2 domain-binding specificities encoded as regular expressions, which are derived from OPAL screens performed in [18]. A link to ModPepInt is provided in Table 3.

2.8 Search for Candidate SH2 Domain-Binding Motifs at the Proteome Level Using Regular Expressions

The SlimSearch4 server [40] can be used to search for proteins harboring candidate SLiMs that conform to a defined regular expression. The regular expressions can be derived from experiments performed on a specific SH2 domain or from Table 2, which contains the results of a high-throughput screen across 76 SH2 domains plus data for an additional SH2 domain and a motif class from bacterial effector proteins. A link to SlimSearch4 is provided in Table 3.

2.9 Materials for Bacterial Effector Identification

BastionHub [41] is a manually curated database of effector proteins from gram-negative bacteria. In particular, the substrates of secretion systems III, IV, and VI can be directly translocated into the cytosol of the infected cell, locating the effector in the correct cellular compartment where interactions with SH2 domains may take place. Other translocated bacterial proteins can be secreted after a signal peptide is detected and cleaved within the bacteria. UniProt annotation or the prediction software SignalP [42] can be used to check the presence of the signal peptide. The operon organization of bacterial genes can be used to associate genes of unclear function with secretion and pathogenesis if they are located within the same operon. BioCyc [43] is a database with about 20,000 bacterial genomes with predicted or manually curated transcription unit structures. Links to these resources are provided in Table 3.

3 Methods

Identifying functionally relevant SH2 domain–motif interactions is not an easy task. The partially overlapping specificities of SH2 domains often make it unclear which SH2 domains will bind to a new candidate SH2-binding motif. The sequence preferences of a collection of SH2 domains obtained from OPAL screens (*see* **Note 1**) can be used to make an initial assessment of SH2 domain specificity (Table 2). However, motifs derived from different methodologies often differ (*see* **Note 2**), and even when the same method is used, assessments from different laboratories might disagree (Fig. 2). This can be due to changes in the experimental conditions (*see* **Note 3**) or the peptide array design (*see* **Note 4**), since specificity is strongly determined by the sequence context (*see* **Note 5**). The quality of high-throughput data is influenced by recombinant protein preparation quality and by the saturation of the SPOT arrays (Table 2). For some SH2 domains, knowledge of sequence preferences is limited or absent (i.e., for STAT domains) due to poor solubility of the constructs (*see* **Note 6**). Problems with the saturation of SPOT arrays also hamper the accurate determination of SH2 domain-binding preferences (*see* **Note 7**).

Narrowing down to a single SH2 domain given an SH2-binding motif candidate might be difficult, and some pTyr motifs might be functionally bound by multiple SH2 domains, for example, as part of a sequential signaling process. Ultimately, bioinformatics tools are useful to guide the initial search for SH2 domain/motif candidates, but temporal and spatial colocalization should also be evaluated and followed by validation through in vitro binding assays and direct in cell binding experiments to confirm a functional interaction.

Here, we provide four protocols for the identification of SH2 domain binders in human and bacterial proteins. The methods can have different starting points, of which we provide examples (Fig. 3).

Predicting SH2 domain-binding motifs in human proteins.

3.1 Using SLiMSearch to Find Candidate SH2 Interacting Motifs Using the BTK SH2 Domain Pattern as the Query

1. Load the SLiMSearch4 page http://slim.icr.ac.uk/slimsearch/index.php.
2. Paste the motif pattern `[QVENDpil][ED](Y)[DEsa][EDNH][EVTD]x` for BTK (*see* **Note 8**) from Table 2 into the SLiMSearch box. Note the strong acidic preference for the residues surrounding the pTyr.
3. Replace the x with a "." (dot character) in the pattern `[QVENDpil][ED](Y)[DEsa][EDNH][EVTD].` to denote a wildcard residue (*see* **Note 9**) and form the correct regular expression. For more information on how to interpret the motif pattern, also *see* Table 1.
4. Keep the default settings for species (*Homo sapiens*), intrinsic disorder score (0.4), and flank length (5). *See* **Note 10** for more information on the meaning of the disorder score.
5. Submit the search. The output will appear in 1–2 min. If you attempt a SLIMSearch for a motif pattern from Table 2 and no predicted candidates are obtained, consider revising your regular expression (*see* **Note 11**) (Fig. 3).
6. Now go to the SLiMSearch help to learn about the displayed results, how to re-rank them, and the available links to other resources.
7. Re-rank the list using the down arrow in the Metazoan column so that the probabilities in this column are ranked from best to worst (note that smaller values correspond to best probabilities). Figure 4 shows the top hits in the search.
8. Examine the hit for HCLS1 which is expressed in the hematopoietic lineage. Is the motif match (positions 376–382 of HCLS1) in a region of predicted intrinsically disordered protein (IDP)? For this, interpret the disorder score as explained in

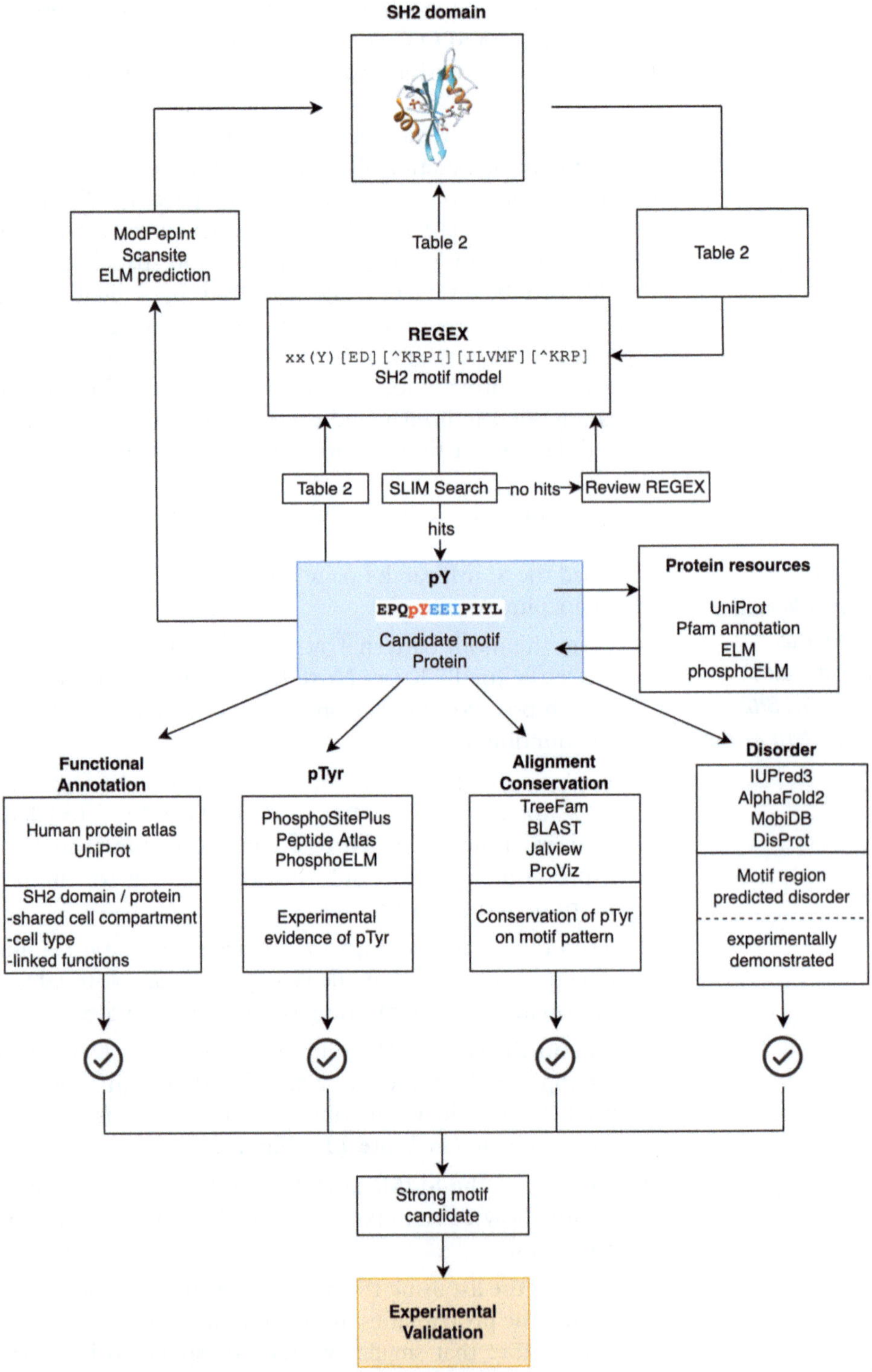

Fig. 3 Pipeline for identifying SH2 domain-binding motifs. Different methods can be used to identify a candidate SH2-binding motif in a protein of interest. The start point might be the known specificity of a given SH2 domain (central blue box), a candidate SH2-interacting protein, or a region of a protein containing a reported pY residue. In every case, bioinformatics tools can help reduce the search space, but experimental validations need to be performed to assign the SH2 domain specificity of a given pY motif

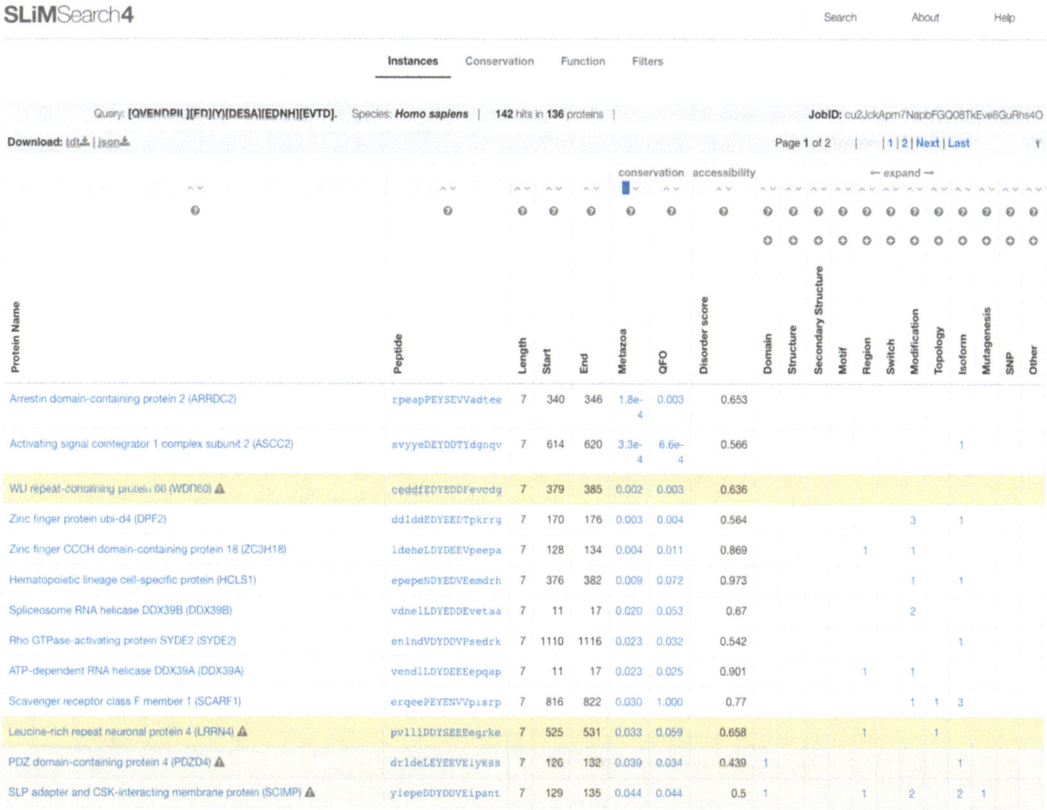

Protein Name	Peptide	Length	Start	End	Metazoa	QFO	Disorder score	Domain	Structure	Secondary Structure	Motif	Region	Switch	Modification	Topology	Isoform	Mutagenesis	SNP	Other
Arrestin domain-containing protein 2 (ARRDC2)	rpeapPEYSEVVadtee	7	340	346	1.8e-4	0.003	0.653												
Activating signal cointegrator 1 complex subunit 2 (ASCC2)	svyyeDEYDDTYdgnqv	7	614	620	3.3e-4	6.6e-4	0.566									1			
WD repeat-containing protein 60 (WDR60) ⚠	ceddfEDYEDDFevcdg	7	379	385	0.002	0.003	0.636												
Zinc finger protein ubi-d4 (DPF2)	ddlddEDYEEDTpkrrg	7	170	176	0.003	0.004	0.564							3		1			
Zinc finger CCCH domain-containing protein 18 (ZC3H18)	ldeheLDYDEEVpeepa	7	128	134	0.004	0.011	0.869					1		1					
Hematopoietic lineage cell-specific protein (HCLS1)	epepeNDYEDVEemdrh	7	376	382	0.009	0.072	0.973							1		1			
Spliceosome RNA helicase DDX39B (DDX39B)	vdnelLDYEDDEvetaa	7	11	17	0.020	0.053	0.67							2					
Rho GTPase-activating protein SYDE2 (SYDE2)	enlndVDYDDVPsedrk	7	1110	1116	0.023	0.032	0.542									1			
ATP-dependent RNA helicase DDX39A (DDX39A)	vendlLDYDEEEepqap	7	11	17	0.023	0.025	0.901					1		1					
Scavenger receptor class F member 1 (SCARF1)	erqeePEYENVVpisrp	7	816	822	0.030	1.000	0.77							1	1	3			
Leucine-rich repeat neuronal protein 4 (LRRN4) ⚠	pvlllDDYSEEEegrke	7	525	531	0.033	0.059	0.658					1			1				
PDZ domain-containing protein 4 (PDZD4) ⚠	drldeLEYEEVElykss	7	126	132	0.039	0.034	0.439	1								1			
SLP adapter and CSK-interacting membrane protein (SCIMP) ⚠	yiepeDDYDDVEipant	7	129	135	0.044	0.044	0.5	1				1		2		2	1		

Fig. 4 Screenshot of the SLiMSearch output page showing the top hits for the BTK motif after sorting for the scores in Metazoa. These hits can be followed up for contextual information augmenting or undermining the likelihood that the match is genuine. The 6th hit is HCLS1 expressed in the hematopoietic lineage which strongly suggests that it will be in the same cells as BTK

Note 10. Clicking on its name opens the UniProt entry. Check the "PTM/Processing" section in the UniProt entry to see if Y378 is known to be phosphorylated.

9. Scroll down to the "PTM Databases" section and click on the PhosphoSitePlus link (for this you must click on the region containing the UniProt ID P14317). Is there both high-throughput (HTP) and low-throughput (LTP) evidence for Y378 phosphorylation? Check the help for the type of experiments in these categories. *See* **Notes 12** and **13** for more information on how to interpret annotations of pTyr data.
10. Click on the Human Protein Atlas link in the expression section of the UniProt entry. Does the entry report strong HCLS1 expression in bone marrow and lymphoid tissues?
11. On the SLiMSearch output, clicking on the peptide sequence opens the ProViz alignment browser in a new window. It centers on the matched motif sequence. Below the alignment

Fig. 5 Screenshot of the ProViz HCLS1 output page showing the alignment for mammalia. The candidate motif is highly conserved with an acidic residue preference in mammals, and the tyrosine is marked as phosphorylated. IUPred indicates that this region is strongly predicted to be intrinsically disordered. Indels are observed on either side of the motif but not in the motif itself. These properties suggest that the motif is functional but do not establish which partner(s) will bind. Two other phosphorylated tyrosine motifs are also highly conserved

are rows of metadata—are there any SLiMs shown? Is the IDP prediction shown? The output is highly customizable so read the help before continuing (*see* **Note 14**).

12. Set the alignment group to "mammalia" as shown in Fig. 5. Is the Y378 shown to be phosphorylated? Is the candidate motif conserved in mammalian species?

13. After conducting these steps, the potential BTK SH2-interacting SLiM is revealed as a strong candidate. It is in the appropriate tissue/cell types, it is in a region of protein disorder, it is conserved, and it is known to be phosphorylated. The next steps would be experimental, firstly to assay binding affinity in vitro and secondly to assay proximity of the proteins in cell. Since BTK is itself a tyrosine kinase, it would be of interest to know whether other conserved tyrosine residues in HCLS1 are its substrates for phosphorylation.

3.2 Determining the Binding Specificity of a Candidate SH2-Binding Motif in the ACE2 Human Protein: A Receptor for SARS-CoV2

1. Use the human angiotensin-converting enzyme 2 (ACE2) protein as a starting point. For this, go to the UniProt resource (https://www.uniprot.org/) and type ACE2_HUMAN in the search window (UniProt:Q9BYF1). The protein has a length of 805 amino acids. Click on the entry and download the ACE2 sequence in fasta format by going to the "Sequence and isoform" tab and selecting the download button. Save the file as a .txt or .fasta file.
2. Identify predicted disordered regions in ACE2. Go to the IUPred3 server (https://iupred3.elte.hu/), and type the UniProt accession in the box that requests the SWISS-PROT/TrEMBL identifier. Submit with default parameters (IUPred3 long disorder and Medium smoothing). The output will be displayed. Unclick the "masking" option on the top right of the graph. The entire prediction will show as a red line. The C-terminal tail of ACE2 (residues 770–805) is predicted to be disordered by IUPred3 (*see* **Note 10**).
3. Identify experimentally annotated disordered regions in ACE2. At the bottom of the IUPred output, you can see that the C-terminal tail is experimentally annotated as disordered (EXP DIS annotation below the IUPred3 graph). If you hover over the red box with the "EXP DIS" annotation, you will see it corresponds to residues 769–805 of ACE2 annotated in the DisProt database. To verify this annotation, go to DisProt (https://disprot.org/), and type the UniProt accession Q9BYF1 in the search tab. Click on entry DP02854 corresponding to ACE2. What experimental technique was used to verify disorder in the ACE2 tail? After Subheadings 3.2, **step 2** and **3**, the ACE2 C-terminal tail can be established as a disordered region.
4. Identify phosphotyrosines in ACE2. Go to the PTM/Processing tab in the UniProt entry and check the C-terminal tail. Residue Y781 is marked as a phosphosite. Scroll down to the PTM databases section and click on the PhosphoSitePlus link. What kind of evidence is available? How much evidence is there compared to other phosphosites in ACE2? The number and type of evidence are important for establishing the confidence of a given phosphosite (*see* **Notes 12** and **13**). After the analysis, Y781 appears as a strong pTyr site.
5. Identify a candidate SH2-binding motif in ACE2. Inspect the sequence surrounding pY781. For this, you can use the fasta sequence downloaded in Subheading 3.2, **step 1**, and build your own alignment using the tools from Table 3. A more straightforward method is to go to ProViz (http://slim.icr.ac.uk/proviz/), and search for the ACE2_HUMAN protein. Select the first hit and click on it. This will open a browser page with the ACE2 alignment. On the top right of the page go to the "Alignments" tab, which is set to "qfo," and select

"vertebrates" from the drop down menu. Go to the "Proteins" tab on the left of the browser page, and click on it to expand the information. This allows choosing which proteins to display. The options are "show models only," "show all," or "hide all." Choose the "show all" option, to display all vertebrate ACE2 sequences. This will expand the alignment. Then, close the "Proteins" Tab and return to the alignment. Scrolling to the protein C-terminus, you will find a highly hydrophobic stretch (residues 741–761), which is annotated as a transmembrane helix (Fig. 6a). Y781 is conserved across virtually all sequences, and the sequence $_{779}$NPYASID$_{785}$ in ACE2 has features of an SH2-binding motif, with a conserved hydrophobic residue [IVMA] at P+3 and a conserved acidic residue at P+4 (Fig. 6a). The pattern derived from the conservation observed in the alignment can be expressed as `NP(Y)xS[IVma]D` (*see* **Note 15**).

6. Determine the SH2 specificity of the ACE2 $_{779}$NPYASID$_{785}$ motif candidate (*see* **Note 16**). Compare your motif pattern from Subheading 3.2, **step 5** to the major known specificity groups (Figs. 1, 2 and Table 2). The motif is excluded as a binder of most Grb2-family group IC domains (Fig. 1c) as it lacks N at P+2. It is excluded as a group IIC BRDG1 domain binder since it does not contain a hydrophobic residue at P+4 (Fig. 1e), and it is also excluded as a CRK/CRKL domain binder since it has an aspartic acid at P+4, instead of the preferred positive charge (Fig. 2c). Matching the human ACE2 `NP(Y)ASID` sequence to the regular expressions in Table 2 reveals it maps the specificity pattern of most Group IA SH2 domains, which includes several Src family kinases (SFKs) (Fig. 6b).
7. Go to the ModPepInt server (https://modpepint.informatik.uni-freiburg.de/) and click on the SH2PepInt link. Input the motif sequence in fasta format in the "Protein/Peptide FASTA" box:

```
>ACE2_HUMAN_SH2_Motif
NPYASID
```

Click down and select "all" SH2 domains. This will search over the 51 models present in this web server (*see* **Note 17**). Leave all other fields as default and click start. The results will be ready in 2–5 min and will return the following SH2 domains:

```
ABL1, ABL2, BLK, E109111, E185634, EAT2, FES,
FGR, FRK, GRB10, GRB14, HCK, LCK, MIST, PTK6,
SOCS2, SOCS5, SRC, TEC
```

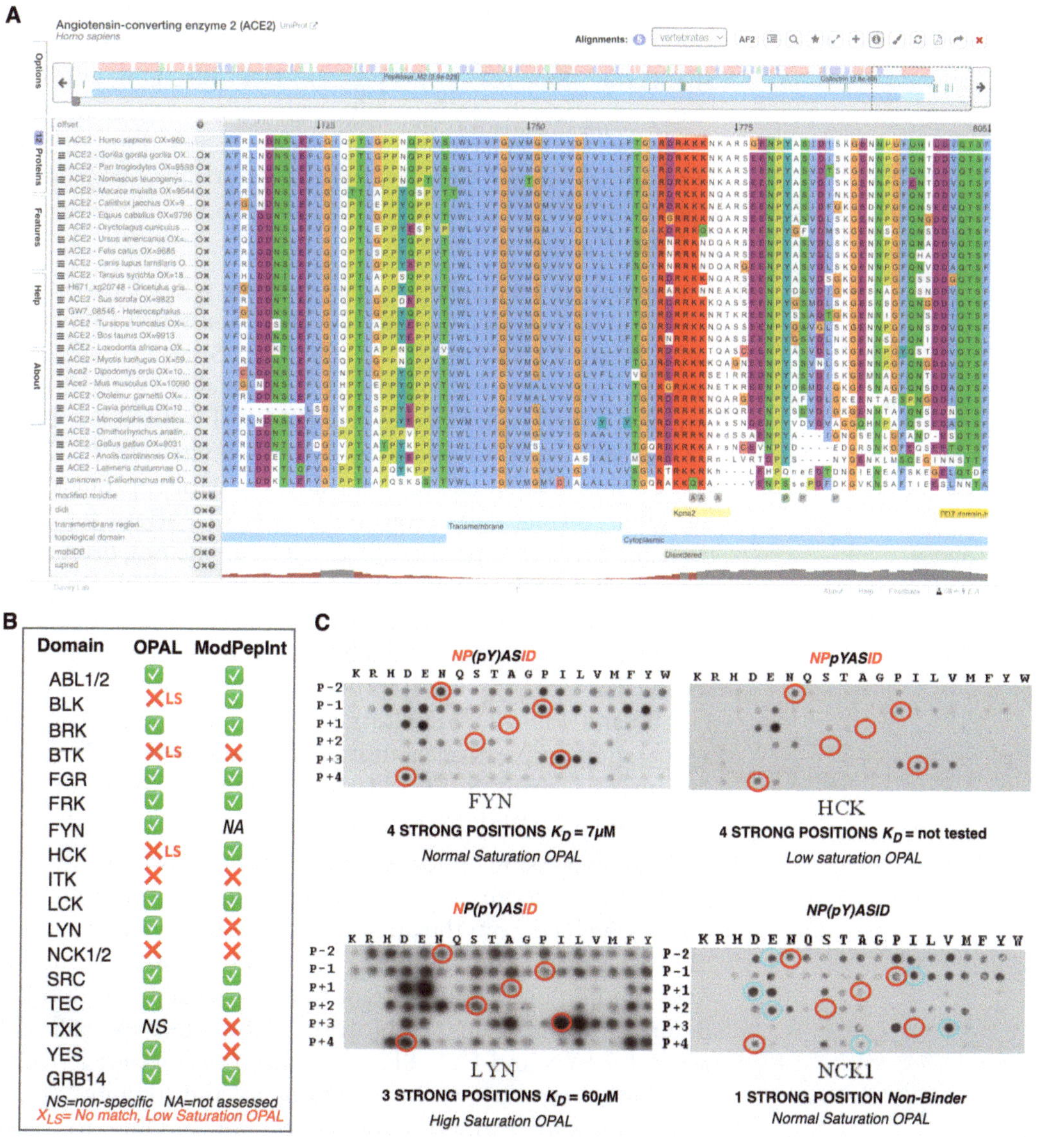

Fig. 6 SH2 domain-binding specificity for the ACE2 motif. (**a**) Alignment of vertebrate ACE2 sequences visualized in ProViz. The hydrophobic stretch from residues 741–761 corresponds to a transmembrane region. It is followed by the cytosolic tail. The tyrosine of the candidate SH2-binding motif (pY781) is highly conserved and marked as phosphorylated. The candidate motif is predicted by IUPred to be located in a highly conserved disordered region. The consensus sequence `NP(Y)xS[IVma]D` derived from the substitution pattern in the alignment broadly matches SFK-family SH2 domain specificity. (**b**) SH2 domain-binding specificity for the `NP(Y)ASID` motif assessed with regular expressions derived from OPAL screens (Table 2) [18] or with ModPepInt. Several SFK-family SH2 domains match the motif pattern. (**c**) Specificity of four SH2 domains for the `NP(Y)ASID` motif derived from direct inspection of SPOT array. (SPOT arrays are adapted from Huang et al. [18]). The motif sequence is marked by red circles in each SPOT array. The number of strong positions is scored and shown as red letters in the motif sequence above each SPOT array. A strong spot is defined as having one of the strongest intensities at a given position. For NCK1, cyan circles mark the amino acids of the strong `EI(Y)DEVA` Tir motif from *Enteropathogenic Escherichia coli*. For each SH2 domain, experimental measurements of binding affinity are reported when available. (The creative commons license under which the cited articles are distributed can be found at https:/creativecommons.org/licenses/by/4.0/)

Inspecting the output of ModPepInt and the group definitions of Table 2 reveals many SH2 domains belong to the SFK-family (Src, Yes, Fyn, Fgr, Lck, Hck, Blk, Lyn, and Frk) Group IA specificity group. This is in broad agreement with the results of the regular expression-based analysis done in Subheading 3.2, **step 6**. The two sets are not fully overlapping, since ModPepInt has some SH2 domain models that are absent in Table 2 and vice versa.

8. A summary of the results obtained with the regex-based search (Subheading 3.2, **step 6**) and the ModPepInt server (Subheading 3.2, **step 7**) is shown in Fig. 6b. Many SFKs are selected as positive hits with both methods. The quality of the SPOT arrays and the smaller size of the ModPepInt panel (*see* **Note 17**) are important for interpreting these results. The BLK and HCK domains are predicted as positives by ModPepInt but not by Table 2 (Fig. 6b). However, the SPOT arrays for both domains have low saturation (e.g., HCK in Fig. 6c), decreasing the reliability of the models derived from them (*see* **Notes 7** and **18**). Therefore, these domains can still be considered plausible candidates. Some positive hits belong to other specificity classes, such as Grb14 (Class IC). For some domains (e.g., TXK in Fig. 6b and Table 2), SPOT arrays show a nonspecific pattern, and a regular expression cannot be derived (*see* **Note 7**). From this analysis, many SFKs and some additional SH2 domains appear as likely candidates to bind the `NP(Y)ASID` motif.
9. Regular expressions encode the residues allowed at certain positions, but more information on the likelihood of binding can be obtained from the direct inspection of spot intensities in SPOT arrays. The number of strong positions in an SH2-binding motif is important to assess its ability to bind to different SH2 domains (*see* **Note 19**). The inspection of SPOT arrays for FYN and LYN SH2 domains reveals 4 and 3 strong positions, respectively, in addition to the required pTyr (Fig. 6c), supporting the prediction for binding to these domains. The prediction is supported by in vitro binding experiments [44] revealing micromolar-affinity binding of the ACE2 SH2-binding motif candidate to both SH2 domains.
10. The 2022 release of ELM predicted binding of the ACE2 `NP(Y)ASID` motif to NCK1/2 SH2 domains [22]. The regular expression used by ELM is

```
ELM Regex:
(Y)[DESTNA][^GWFY][VPAI][DENQSTAGYFP]
```

However, both Table 2 and ModPepInt predict NCK1/2 as non-binders (Fig. 6b). Compare the regular expression for NCK1 from Huang et al. [18] (Table 2) to the ELM regex:

```
Table 2 Regex:
[^KRHW][^KRW](Y)[DEstavy][^KRHGFYW][PVateslm]
[SDtaemg]
```

Check position P+3 in both regular expressions. The ELM expression allows Ile at P+3, making the ACE2 match. This raises an important point about the limitations of prediction methods for SH2 domain binding. The ELM regex pattern was built from structures and binding data from known NCK1/2 SH2 domain binders, such as the strong `EI(Y)DEVA` sequence from the *Enteropathogenic E. coli* (EPEC) Tir protein [45]. Inspecting the NCK1 SPOT array (Fig. 6c) reveals that the `NP(Y)ASID` ACE2 motif has weak amino acids Ala and Ile at positions P+1 and P+3, while the `EI(Y)DEVA` Tir motif has strong amino acids Asp and Val at the same positions. This is relevant, because P+1 and P+3 are two main binding determinants for interactions with the Src SH2 domain (Fig. 1a, b). In the Tir motif, a substitution to a weaker Ile is tolerated at P+3, likely because P+1 has a strong Asp residue [45]. However, the lack of a strong amino acid at either P+1 or P+3 decreased the binding likelihood for the ACE2 sequence (*see* **Note 5**). In accordance with this expectation, in vitro binding experiments show that the NCK1 SH2 domain does not interact with the pY781 ACE2 motif [44].

11. As shown in Fig. 6b, several candidate SH2 domains have been identified, and bioinformatics analysis alone cannot fully distinguish stronger from weaker binders. At this point, the candidate interactions need to be tested experimentally.

Predicting SH2 domain-binding motifs in bacterial effector proteins.

The following two protocols aim to identify new mechanisms of pathogenesis in bacteria.

3.3 Identifying SH2 Domain-Binding Motifs in Bacterial Proteins Using the Tir Proteins from the Genus Escherichia as a Model

Translocated intimin receptor (Tir) protein from pathogenic *E. coli* strains (including EPEC) is a secreted protein that resides in the plasma membrane of eukaryotic target cells. The extracellular side binds to the intimin protein that anchors the bacterium to the cell surface. This is bounded by two transmembrane helices, so that natively disordered N- and C-termini of Tir reside in the host cell cytosol. EPEC Tir contains a tyrosine residue in position 474 that, when phosphorylated, binds to the SH2 domain of NCK [46]. Its homolog in *Enterohemorrhagic E. coli* does not contain this tyrosine, and no phosphorylation has been reported, so it is assumed to use a different infection mode. All Tir proteins contain an NPY motif (positions 452–454, Fig. 7) that mediates the binding with membrane-curving I-BAR domains [47].

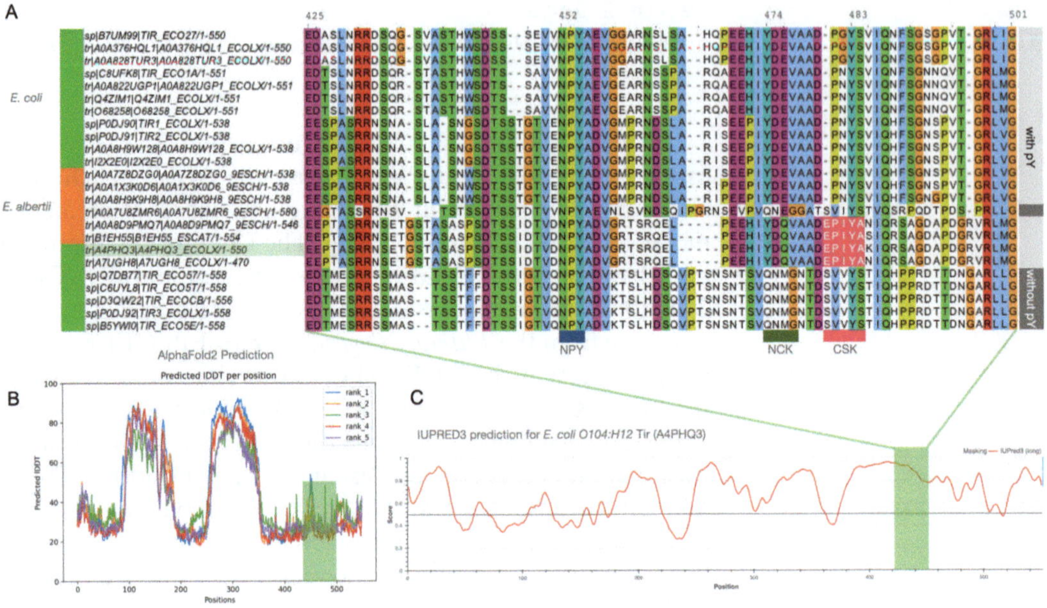

Fig. 7 Predicting SH2-binding motifs in bacterial proteins. (**a**) Alignment of representative sequences obtained after using an *Enteropathogenic Escherichia coli* Tir protein as reference (B7UM99), the numbering at the top of the alignment corresponds to this sequence. Four different sequences show the presence of a CSK SH2 domain-binding motif (pink) in *E. coli* (green) and *E. albertii* (orange) species. The position of the previously described NPY and NCK SH2 domain-binding motifs are marked by a blue and a green rectangle, respectively. The subset of sequences that contain the validated phosphotyrosine are coloured in light gray. (**b**) Predicted local distance difference test plot indicating the per-residue confidence of the AlphaFold2 result on *E. coli* O104:H12 Tir protein (UniProt:A4PHQ3, highlighted in green in the alignment). (**c**) IUPred3 disorder prediction for the same Tir protein. The region shown in A of the sequence alignment is highlighted in green in (**b**, **c**). Both predictors mark this region as disordered

1. Use the Tir protein from EPEC *O127:H6* (strain E2348/69) (https://www.uniprot.org/uniprotkb/B7UM99/) as a starting point and reference sequence (first sequence in Fig. 7a).
2. Run a BLAST by clicking the BLAST icon at the top Entry section in the UniProt page. Choose UniProtKB as the target database, narrow the search by restricting the taxonomy to *Escherichia* (Escherichia[561]), and leave the default number of hits on 250, since we want to cover more *Escherichia* species than *coli*.
3. Select all obtained hits and save them in "My Basket."
4. Open "My Basket," select the last 250 entries, or all of them if you do not have previously saved sequences, and press download. The file does not need to be compressed as it will not take more than 250 KB.
5. Do a first cleaning of the sequences. Open the file in Jalview (http://www.jalview.org/), go to File -> Input Alignment -> From file, and select the file with the 250 sequences. The reference sequence has 550 amino acids, and most retrieved

sequences are shorter than 1100 amino acids; therefore, you can remove sequences that are longer than 1100 (which is about one standard deviation of the mean of the width of the sequences). Similarly, very short sequences (shorter than ~400 bp) can be removed since they are probably fragments of proteins. You can scroll through the sequences and find the long and short ones or sort the sequences using Calculate -> Sort -> By Length. If you select a sequence and press Delete on the keyboard, the sequence will be deleted. After deleting the too long or too short sequences, you should have about 200 sequences left.

6. Perform a multiple sequence alignment. Within Jalview, go to Web Service -> Alignment -> ClustalO -> with Defaults (*see* **Note 20**).
7. Explore the alignment and recognize two groups, one with a conserved Proline-rich region in positions 17–23 of the original Tir protein. Most of the sequences in the second group are named as Short-chain fatty acid transporter or Tail fiber protein suggesting they are artifacts of the BLAST search. You can get rid of these sequences by manually selecting them. In addition, you can quickly look for sequences that have in their fasta header (their description) the term "Short-chain fatty acid" by going to Select -> Select all, and then Select -> Find, type "Short-chain fatty acid" sequences, select the box that says Include Description, press Select all, and remove them. At this point, you will have about 100 sequences in your alignment.
8. There are now several columns that only contain gaps. They are not informative and need to be removed: go to Edit -> Remove Empty Columns.
9. By exploring the alignment, you can find that several residues are not properly aligned; this is because the sequences that we just eliminated affected the precision of the alignment. Remove all gaps by selecting Edit -> Remove All Gaps and do a new alignment (Subheading 3.3, **step 6**); this time it will be faster.
10. Identify functional sites. Based on the information from the UniProt entry of the initial Tir protein, the first transmembrane region is formed by residues 234–254 and the second by 363–383, which are characterized by hydrophobic residues. The region towards the C-terminus of the sequences (positions 384–550) should correspond to the second intracellular region which is, therefore, in potential contact with tyrosine kinases and SH2 domain-containing host proteins (Fig. 7a). Positions 452–454 of the reference Tir protein contain the fully conserved NPY motif. Position 474 of the reference Tir protein is the Tyr residue that binds to NCK [46]; note that it is not conserved across all the gathered sequences (Fig. 7a).

11. Identify sequence groups. Based on the last Tyr residue identified in Subheading 3.3, **step 10**, all sequences can be split and rearranged into two groups. You can first perform an automatic sorting. For this, go to Calculate -> Sort -> By Pairwise Identity. Then, select the sequences that do not have a Tyr residue in the same position as in the reference Tir protein, and use the down arrow of the keyboard to pull them to the bottom of the alignment (*see* **Note 21**).
12. Identify potential CSK-binding motifs. The alignment shows several other Tyr, not always conserved across the whole set of sequences nor across a single group, which might be explained by the limited number of sequences that we have collected. From the ELM database, retrieve the regular expression that has been defined as the binder of the SH2 domain of CSK (http://elm.eu.org/elms/LIG_CSK_EPIYA_1), select it, and perform a Regular Expression search in Jalview: Select -> Select all and then Select -> Find and paste the regular expression from ELM `EP[IL]Y[TAG]`, and press Find all. A few sequences should be identified as containing a putative CSK domain-binding motif. Place the mouse on top of the names of the sequences to quickly read the species name and strain. You can also double-click on any sequence name to open the link to the UniProt entry. These sequences correspond to *E. coli O104:H12*, a bovine pathogen, and *E. albertii* strain TW07627, a human emerging enteropathogen (Fig. 7a).
13. Verify that the putative CSK-binding motif is in a disordered region. AlphaFold2 can be used to assess protein disorder (*see* **Note 22**). Input any of the sequences with the putative CSK SH2-binding motif to the ColabFold web server. The predicted lDDT score for the region containing the `EPIYA` sequence is lower than 50 confirming that AlphaFold2 predicts disorder in this region (Fig. 7b and *see* **Note 22**). IUPred can also be used to directly predict disorder, and input the candidate protein in IUPred3 (*see* **Note 10**). The region containing the sequence matching the motif (`EPIYA`) shows scores above 0.5, also confirming disorder prediction (Fig. 7c).

3.4 Predicting CSK SH2 Domain-Binding Motifs in a Bacterial Proteome

The following protocol predicts motifs that bind to the CSK SH2 domain in a bacterial proteome:

1. Download the proteome of the opportunistic human pathogen *Chromobacterium violaceum* strain ATCC 12472 (https://www.uniprot.org/proteomes/UP000001424) from Uniprot; it contains about 4400 proteins. Save it in a file called proteome.fasta by going to Components -> Download -> Compressed -> No -> Download. The file is only 2.3 MB.

2. To identify proteins that contain motifs that potentially bind to the CSK SH2 domain, run the following Python3 script in a terminal (*see* **Note 23**), and note that it already includes the regular expression defined in ELM for the LIG_CSK_EPIYA_1 motif (same as in Subheading 3.3, **step 12**):

```
import re
proteome = input("Input full path of the pro-
teome: ")
proteome = open(proteome,'r')
RegExp = 'EP[IL]Y[TAG]'
dic = {}
for line in proteome:
 if re.search('>', line):
 key = line
 dic[key] = ''
 else:
 sequence = dic[key]
 dic[key] = sequence + line
proteome.close()
matches = []
for key in dic.keys():
 if re.search(RegExp, str(dic[key])):
 name = re.findall("\|(.+)\|", str(key))
 matches.append(name)
print(matches)
```

The output is a list of UniProt IDs with matches with the regular expression saved in the RegExp variable. Three proteins had a match with the CSK SH2 domain-binding motif: Q7NWX9, Q7NZE8, and Q7P118.

3. Explore the annotation of the candidates. Input each ID into UniProt. The second protein is named "Probable tyrosine phosphatase" which gives a hint into tyrosine phosphorylation-mediated signaling. A quick exploration of the sequence reveals that regions 463–467 and 473–477 contain EPIYA sequences, which match the ELM regular expression for the SH2-binding motif. Moreover, UniProt shows that the region 466–489 is predicted as disordered, and IUPred3 gives a score higher than 0.5 for both motif matches (*see* **Note 10**).

4. You can explore the operon architecture of this gene in BioCyc (https://biocyc.org/GCF_000007705/NEW-IMAGE?type=OPERON&object=TU26SV-540). It forms an independent transcription unit, but it is surrounded by two genes following the same transcription orientation: one of them is CV_RS04755, a type III secretion system chaperone, giving additional suggestions for a function as an effector and a role in pathogenesis.

4 Notes

1. The sequence specificity for different SH2 domains can be elucidated by using SPOT peptide arrays [17, 18]. One of these technologies is the oriented peptide array library (OPAL), which uncovers favorable and unfavorable residues at different positions in the SH2 ligand. For OPAL experiments, an activated nitrocellulose membrane support is spotted with synthetic peptides which are extended using solid-phase synthesis methods. This method allows the incorporation of unnatural and modified (i.e., phosphorylated) amino acids as building blocks in the synthesis. After synthesis of the nitrocellulose-based SPOT array, usually multiple SH2 domains are recombinantly expressed and purified and confronted with the membrane. Binding peptides can then be detected using tags fused to the SH2 domains (e.g., His, GST). Peptides that do not bind or bind very weakly are also detected, revealing non-preferred sequences. To study SH2 domain specificity, the pTyr position is fixed at 0, and amino acids at positions preceding (P−1, P−2, P−3) or following (P+1, P+2, P+3) the pTyr are varied, while using a randomized mixture of amino acids at all other positions (usually except for Cys). The results of the array can then be used to build position-specific scoring matrices (PSSMs) or regular expressions that reflect SH2 domain-binding specificity (Table 2).
2. *Strengths and limitations of high- and low-throughput methods used to test SH2 domain specificity.* OPAL screens are a powerful method for providing an "unbiased" representation of the binding specificity of a given domain, and they also allow large-scale screening of SH2 domains. However, these screens also have limitations. For example, the number of experimental replicates and experimental conditions tested is often limited by the cost of the assays, and SPOT arrays can have too high or too low signal saturation (Table 2) leading to difficulties in assessing specificity and/or deriving a regular expression. Microarrays are also high-throughput methods, but the immobilization of GST-tagged proteins to the chip surface can cause artifacts and differences with respect to SPOT arrays [15]. Other techniques such as structural studies and in vitro binding assays provide more accurate details of the molecular interactions and allow to quantify the binding affinity, but they are able to test a limited set of sequences leading to the effects of mutations being assessed only in a specific sequence background.
3. *Variation of experimental conditions.* The specificity of an individual SH2 domain reported by different groups using related SPOT array techniques can reveal different preferences at some

positions. One striking example is provided in Fig. 2a, b where charge preference is inverted at several positions following the pTyr [7, 18]. One origin for these differences could be changes in the experimental conditions (buffer, salt concentration, pH, and additives) used in different experiments, which can affect binding specificities in particular for charged and ionizable amino acids (e.g., Asp, Glu, Lys, Arg, and His). Differences in this case may also arise from the different design of the SPOT arrays used in each case (Fig. 2).

4. *Variations of OPAL libraries.* Instead of using randomized amino acids at the non-restricted positions (randomized library), the SPOT array can be constructed by using a specific known SH2 ligand as the reference sequence [7] (nonrandomized library), and all amino acid replacements (except Cys) can be assayed at each position (Fig. 2b). Randomized and nonrandomized arrays often reveal a different specificity pattern. Also, peptide arrays can include a different number of residues N-terminal and C-terminal to the pTyr (Fig. 2), and the length of the peptide used in the array can also modify the output.

5. *Effects of sequence context on SH2 domain specificity.* Often, binding of SH2 and other short ligands to their cognate domains shows cooperative effects between different positions in the motif. For example, a weak amino acid might be tolerated at one position only if a strongly preferred amino acid is present at another position in the ligand. Because each position is assessed independently in SPOT arrays, this information is often lost. This can be revealed by comparing SPOT arrays where only the pTyr is fixed to SPOT arrays where pTyr and a second position are fixed. Figure 2a, d shows one such example for the Grb2 domain, where Pro is disallowed at P+3, when P+2 is fixed to Asn. A second way to reveal these effects is when comparing SPOT arrays to in vitro binding experiments using point mutants. The latter experiments are usually performed with high-affinity ligands, and therefore, these ligands can show higher tolerance for weaker amino acids at other positions, compared with the SPOT arrays.

6. *Limitations with solubility and purity of SH2 domains.* Specificity studies rely on the identification of sequences that bind to specific SH2 domains. However, the in vitro production of SH2 domains, often performed as GST-tagged proteins, is not straightforward and is limited by the solubility and stability of the constructs, for example, STAT1-6, JAK1-3, and SOCS1/2/4/5/7 SH2 domains have very low solubility, limiting the analysis of their binding specificity [19]. The purity obtained in the preparations [18] also affects the outcome of

high-throughput assays to assess SH2 domain specificities. The purity of all SH2 domains used in the Huang et al. [18] study is annotated in Table 2.

7. *Problems arising from low and high saturation of SPOT arrays.* To elucidate the binding preferences of SH2 domains, SPOT arrays are confronted with His- or GST-tagged SH2 domains and the results visualized by incubation with fluorescently labeled or horseradish peroxidase (HRP)-tagged antibodies and revealed using an imaging technique. Protein quality and variations in the incubation times and washing procedures can lead to over- and undersaturation of the SPOT arrays. This produces arrays where too many or too few spots reveal as positive (Fig. 6c). Low and high saturation can preclude a correct assessment of the binding preferences for the SH2 domain and to an uneven comparison to other SH2 domain preferences. When deriving a regular expression from SPOT arrays, low saturation can lead to an apparent high specificity, while high saturation can obscure the determination of sequence preferences at each position. For some motifs, the SPOT arrays don't show specificity, and therefore a motif pattern cannot be built.
8. Bruton's tyrosine kinase (BTK) is expressed in B lymphocytes and other cells of the hematopoietic lineage. It is essential for B cell maturation and antibody production. Proteins not expressed in these lineages cannot interact with BTK. It is predominantly cytosolic but may also have a role in the nucleus.
9. When using regular expressions for searching databases, the correct character for the wildcard position is a dot character ".", while, in text, wildcards are commonly denoted by a lowercase "x" character, as seen in Table 2. This is because they get confused with a full stop, and three consecutive dots are converted by word processor autocorrection into the single character, the horizontal ellipsis. *See* Table 1 for the definition of characters used in regular expressions.
10. *Predicting protein disorder with IUPred3.* The output of IUPred3 is a plot with a smoothed prediction score ranging from 0 to 1.0, which is a measure of the propensity for each position in the protein sequence to be disordered. In the default criterion, a value higher than 0.5 is taken as a strong indication that the region is intrinsically disordered. However, lower cutoffs such as 0.4 are often used.
11. *Revising a regular expression.* Regular expressions can encode for strong, medium, or low specificity, depending on the degeneracy of the motif pattern (Table 2). When a search using a regular expression produces no results, you may consider making the search more relaxed. This can be done by

being more permissive at certain positions (e.g., positions that are less important for binding), by allowing a wildcard at that position, or by allowing more residues at that position. Conversely, if the search has very low specificity and returns too many hits, you may consider making the regular expression more restrictive, which can be done, for example, by allowing only the most preferred amino acids at that position. Highly preferred residues are denoted by capital letters, and the weaker amino acids are denoted by lower case letters in Table 2. Also, some positions described by a "NOT" statement (*see* Table 1) may be converted into a position described by the main preferences at that position assessed by the spot intensities, which is described in the "Notes" section of Table 2 or can be visually assessed directly from the SPOT arrays [18].

12. *Techniques for enrichment of pTyr peptides using SH2 domains as baits*

    High-throughput methods such as mass spectrometry coupled to enrichment methods such as immunodetection [48] or affinity-based enrichment of SH2 domain binders [49] can be used to identify pTyr in specific proteomes and cell types. Determining all pTyr in a particular proteome/cell type is not straightforward. The sensitivity of experimental assays and the low abundance of tyrosine phosphorylation limit the detected pTyr residues, and assays can also produce a sizable proportion of false-positive results [50]. Moreover, phosphorylation is a highly dynamic process that varies with cell state and cell type, precluding the identification of all possible phosphosites in a single experiment. Recent developments have used phage display technology to create SH2 domains with tailored selectivity and higher affinity binding to pTyr, improving pTyr enrichment for high-throughput analysis [51]. Low-throughput methods used to identify phosphotyrosine include detection with anti-pTyr antibodies using western blot assays.

13. *Identification of pTyr sites in candidate proteins.* Several databases including PhosphositePlus, PeptideAltas, PhosphoELM, and UniProt are available to search for pTyr sites in a protein of interest (*see* Subheading 2.2). The number, type, and quality of experimental evidence will increase or decrease the confidence in the pTyr being examined. Because of the possibility of false positives [50], it is always advisable to complement the identification of a particular pTyr site with analysis of evolutionary conservation of the pTyr site, functional annotation, and correct spatiotemporal localization. Because annotations are not complete, the failure to find the particular pTyr residue in one of the databases does not indicate a lack of phosphorylation.

14. We did not explore the filters in SLiMSearch. Use the help to train yourself. The filters can be useful in restricting the size of the output to work through. For example, selecting the cytosol as localization would exclude all matched proteins that are not annotated for this cell compartment. The filter is likely to enrich for candidates that might bind a cytosolic SH2 domain. However, do note that filtering is dependent on annotation quality and so some good candidates may also be missed.
15. *Building a regular expression from an alignment of SH2-binding motifs*. In order to build a regular expression, use the characters explained in Table 1 to model the fixed positions and the wildcard positions. Fixed positions can allow a single amino acid, or several amino acids, as shown in Table 2. Also, among allowed amino acids, some may be more strongly preferred than others. This can be known from the intensity of each amino acid in a SPOT array (Fig. 2) or from the frequency of an amino acid at a given motif position in a multiple sequence alignments. The most preferred amino acids are encoded by capital letters, and the less preferred amino acids can be encoded by lower case letters (Table 2).
16. *SH2 domain classification*. SH2-binding motifs have been classified according to their binding specificity based largely on [18] but integrating other experimental results. SH2 domains were classified into three groups, based on the identity of the βD5 residue (nomenclature based on the Src SH2 domain) and their binding specificity. βD5 contacts the P+1 and P+3 residues in many SH2 ligands and is therefore a major specificity determinant. Group I domains harbor an aromatic residue (Tyr or Phe) at βD5, group II domains harbor a hydrophobic residue (Ile, Leu, Val, Cys, Met) at βD5, and group III domains (STAT-family) harbor a hydrophilic residue (Glu, Gln, or Lys) at βD5.
17. *ModPepInt*. ModPepInt [39] integrated multiple high-throughput datasets and developed SH2 specificity prediction models for 51 SH2 domains using semi-supervised machine learning approaches.
18. *Influence of SPOT saturation on determining SH2 domain specificity*. Low saturation of the SPOT arrays leads to some amino acids being scored as disallowed at a certain position when in reality they may be tolerated, but the signal is too weak to make them detectable. This can lead to a regular expression that is stronger (more discriminating) than in reality. For example, the regular expression for HCK derived from SPOT arrays does not allow Ser at P+2, rejecting the `NP(Y)ASID` sequence (Fig. 6c), but a close inspection of the low-saturation SPOT array image reveals a weak signal for Ser at P+2.

19. *Intensity SPOT Arrays.* SPOT arrays reveal which residues make the strongest binders (higher-intensity spots) and which are the weakest (lower-intensity spots) at a specific position (Fig. 6c). If a motif candidate matches many strong positions, it is expected to be a stronger binder. For SH2 domains, at least 2–3 strong positions in addition to the pTyr are usually required for stable binding. Having more than 3 strong positions will likely make a high-affinity binder.
20. *Running alignments in Jalview.* Jalview includes a connection with Clustal Omega, a multiple sequence alignment software which is good at dealing with modular proteins.
21. *A second group of E. coli pathogenic strains.* The sequences that you moved actually correspond to *Enterohemorrhagic E. coli* strains, another pathogen that mainly uses the `NPY` motif to mediate actin polymerization [52].
22. *Using AlphaFold2 as a disorder predictor.* The predicted local difference distance test (lDDT) is a per-residue confidence score used by AlphaFold2. When the lDDT values are below 50, the score can be used as an indirect disorder predictor [53].
23. *Running the Regular Expression matcher script.* You can save a Python script in any location of your computer. Add a .py as the extension. Open a Terminal (in Windows operating system is called cmd), move to the directory where the Python script is located, and type python3 followed by the name you gave to the script. The program will ask you for the name of the proteome; you need to first add the path to the directory where you saved the proteome.

Acknowledgments

L.B.C. is a National Research Council Investigator (CONICET, Argentina) and has received funding from Agencia Nacional de Promocion Cientifica y Tecnológica (ANPCyT) Grant #PICT-2017/1924 and #PICT-2019/02119. L.B.C. and T.J.G. received support from the European Union's Horizon 2020 Marie Skłodowska-Curie action #778247 (IDPfun).

References

1. Liu BA, Jablonowski K, Raina M et al (2006) The human and mouse complement of SH2 domain proteins-establishing the boundaries of phosphotyrosine signaling. Mol Cell 22(6): 851–868
2. Suga H, Dacre M, de Mendoza A et al (2012) Genomic survey of premetazoans shows deep conservation of cytoplasmic tyrosine kinases and multiple radiations of receptor tyrosine kinases. Sci Signal 5(222):ra35
3. Liu BA, Shah E, Jablonowski K et al (2011) The SH2 domain-containing proteins in 21 species establish the provenance and scope of phosphotyrosine signaling in eukaryotes. Sci Signal 4(202):ra83

4. Marasco M, Carlomagno T (2020) Specificity and regulation of phosphotyrosine signaling through SH2 domains. J Struct Biol X 4: 100026
5. Pawson T (2004) Specificity in signal transduction: from phosphotyrosine-SH2 domain interactions to complex cellular systems. Cell 116(2):191–203
6. Liu BA, Nash PD (2012) Evolution of SH2 domains and phosphotyrosine signalling networks. Philos Trans R Soc Lond Ser B Biol Sci 367(1602):2556–2573
7. Kaneko T, Stogios PJ, Ruan X et al (2018) Identification and characterization of a large family of superbinding bacterial SH2 domains. Nat Commun 9(1):4549
8. Samano-Sanchez H, Gibson TJ (2020) Mimicry of Short Linear Motifs by bacterial pathogens: a drugging opportunity. Trends Biochem Sci 45(6):526–544
9. Phillips N, Hayward RD, Koronakis V (2004) Phosphorylation of the enteropathogenic E. coli receptor by the Src-family kinase c-Fyn triggers actin pedestal formation. Nat Cell Biol 6(7):618–625
10. Kaneko T, Huang H, Zhao B et al (2010) Loops govern SH2 domain specificity by controlling access to binding pockets. Sci Signal 3(120):ra34
11. Waksman G, Shoelson SE, Pant N et al (1993) Binding of a high affinity phosphotyrosyl peptide to the Src SH2 domain: crystal structures of the complexed and peptide-free forms. Cell 72(5):779–790
12. Waksman G, Kominos D, Robertson SC et al (1992) Crystal structure of the phosphotyrosine recognition domain SH2 of v-src complexed with tyrosine-phosphorylated peptides. Nature 358(6388):646–653
13. Machida K, Liu B (2017) Binding assays using recombinant SH2 domains: far-Western, pull-down, and fluorescence polarization. Methods Mol Biol 1555:307–330
14. Ladbury JE, Lemmon MA, Zhou M et al (1995) Measurement of the binding of tyrosyl phosphopeptides to SH2 domains: a reappraisal. Proc Natl Acad Sci U S A 92(8): 3199–3203
15. Tinti M, Panni S, Cesareni G (2017) Profiling Phosphopeptide-binding domain recognition specificity using peptide microarrays. Methods Mol Biol 1518:177–193
16. Tinti M, Kiemer L, Costa S et al (2013) The SH2 domain interaction landscape. Cell Rep 3(4):1293–1305
17. Liu BA (2017) Characterizing SH2 domain specificity and network interactions using SPOT peptide arrays. Methods Mol Biol 1555:357–373
18. Huang H, Li L, Wu C et al (2008) Defining the specificity space of the human SRC homology 2 domain. Mol Cell Proteomics 7(4):768–784
19. Machida K, Thompson CM, Dierck K et al (2007) High-throughput phosphotyrosine profiling using SH2 domains. Mol Cell 26(6): 899–915
20. Kaneko T, Joshi R, Feller SM, Li SS (2012) Phosphotyrosine recognition domains: the typical, the atypical and the versatile. Cell Commun Signal 10(1):32
21. Liu BA, Jablonowski K, Shah EE et al (2010) SH2 domains recognize contextual peptide sequence information to determine selectivity. Mol Cell Proteomics 9(11):2391–2404
22. Kumar M, Michael S, Alvarado-Valverde J et al (2022) The Eukaryotic Linear Motif resource: 2022 release. Nucleic Acids Res 50(D1): D497–D508
23. Hornbeck PV, Kornhauser JM, Latham V et al (2019) 15 years of PhosphoSitePlus(R): integrating post-translationally modified sites, disease variants and isoforms. Nucleic Acids Res 47(D1):D433–DD41
24. Farrah T, Deutsch EW, Hoopmann MR et al (2013) The state of the human proteome in 2012 as viewed through PeptideAtlas. J Proteome Res 12(1):162–171
25. Dinkel H, Chica C, Via A et al (2011) Phospho.ELM: a database of phosphorylation sites--update 2011. Nucleic Acids Res 39 (Database issue):D261–7
26. UniProt C (2021) UniProt: the universal protein knowledgebase in 2021. Nucleic Acids Res 49(D1):D480–D4D9
27. Quaglia F, Meszaros B, Salladini E et al (2022) DisProt in 2022: improved quality and accessibility of protein intrinsic disorder annotation. Nucleic Acids Res 50(D1):D480–D4D7
28. Piovesan D, Necci M, Escobedo N et al (2021) MobiDB: intrinsically disordered proteins in 2021. Nucleic Acids Res 49(D1):D361–D3D7
29. Erdos G, Pajkos M, Dosztanyi Z (2021) IUPred3: prediction of protein disorder enhanced with unambiguous experimental annotation and visualization of evolutionary conservation. Nucleic Acids Res 49(W1): W297–W303
30. Jumper J, Evans R, Pritzel A et al (2021) Highly accurate protein structure prediction with AlphaFold. Nature 596(7873):583–589
31. Mirdita M, Schutze K, Moriwaki Y et al (2022) ColabFold: making protein folding accessible to all. Nat Methods 19(6):679–682

32. Uhlen M, Fagerberg L, Hallstrom BM et al (2015) Proteomics. Tissue-based map of the human proteome. Science 347(6220): 1260419
33. Schreiber F, Patricio M, Muffato M et al (2014) TreeFam v9: a new website, more species and orthology-on-the-fly. Nucleic Acids Res 42(Database issue):D922–5
34. Waterhouse AM, Procter JB, Martin DM et al (2009) Jalview Version 2--a multiple sequence alignment editor and analysis workbench. Bioinformatics 25(9):1189–1191
35. Procter JB, Carstairs GM, Soares B et al (2021) Alignment of biological sequences with Jalview. Methods Mol Biol 2231:203–224
36. Jehl P, Manguy J, Shields DC et al (2016) ProViz-a web-based visualization tool to investigate the functional and evolutionary features of protein sequences. Nucleic Acids Res 44 (W1):W11–W15
37. Obenauer JC, Cantley LC, Yaffe MB (2003) Scansite 2.0: proteome-wide prediction of cell signaling interactions using short sequence motifs. Nucleic Acids Res 31(13):3635–3641
38. Davey NE, Haslam NJ, Shields DC, Edwards RJ (2010) SLiMFinder: a web server to find novel, significantly over-represented, short protein motifs. Nucleic Acids Res 38(Web Server issue):W534–9
39. Kundu K, Mann M, Costa F, Backofen R (2014) MoDPepInt: an interactive web server for prediction of modular domain-peptide interactions. Bioinformatics 30(18): 2668–2669
40. Krystkowiak I, Davey NE (2017) SLiMSearch: a framework for proteome-wide discovery and annotation of functional modules in intrinsically disordered regions. Nucleic Acids Res 45 (W1):W464–W4W9
41. Wang J, Li J, Hou Y et al (2021) BastionHub: a universal platform for integrating and analyzing substrates secreted by Gram-negative bacteria. Nucleic Acids Res 49(D1):D651–D6D9
42. Teufel F, Almagro Armenteros JJ et al (2022) SignalP 6.0 predicts all five types of signal peptides using protein language models. Nat Biotechnol 40(7):1023–1025
43. Karp PD, Billington R, Caspi R et al (2019) The BioCyc collection of microbial genomes and metabolic pathways. Brief Bioinform 20(4):1085–1093
44. Kliche J, Kuss H, Ali M, Ivarsson Y (2021) Cytoplasmic short linear motifs in ACE2 and integrin beta3 link SARS-CoV-2 host cell receptors to mediators of endocytosis and autophagy. Sci Signal 14(665):eabf1117
45. Frese S, Schubert WD, Findeis AC et al (2006) The phosphotyrosine peptide binding specificity of Nck1 and Nck2 Src homology 2 domains. J Biol Chem 281(26):18236–18245
46. Campellone KG, Giese A, Tipper DJ, Leong JM (2002) A tyrosine-phosphorylated 12-amino-acid sequence of enteropathogenic Escherichia coli Tir binds the host adaptor protein Nck and is required for Nck localization to actin pedestals. Mol Microbiol 43(5): 1227–1241
47. de Groot JC, Schluter K, Carius Y et al (2011) Structural basis for complex formation between human IRSp53 and the translocated intimin receptor Tir of enterohemorrhagic E. coli. Structure 19(9):1294–1306
48. Lind SB, Artemenko KA, Pettersson U (2012) A strategy for identification of protein tyrosine phosphorylation. Methods 56(2):275–283
49. Ke M, Chu B, Lin L, Tian R (2017) SH2 domains as affinity reagents for Phosphotyrosine protein enrichment and proteomic analysis. Methods Mol Biol 1555:395–406
50. Kalyuzhnyy A, Eyers PA, Eyers CE et al (2022) Profiling the human Phosphoproteome to estimate the true extent of protein phosphorylation. J Proteome Res 21(6):1510–1524
51. Martyn GD, Veggiani G, Kusebauch U et al (2022) Engineered SH2 domains for targeted Phosphoproteomics. ACS Chem Biol 17(6): 1472–1484
52. Weiss SM, Ladwein M, Schmidt D et al (2009) IRSp53 links the enterohemorrhagic E. coli effectors Tir and EspFU for actin pedestal formation. Cell Host Microbe 5(3):244–258
53. Tunyasuvunakool K, Adler J, Wu Z et al (2021) Highly accurate protein structure prediction for the human proteome. Nature 596(7873): 590–596
54. Mistry J, Chuguransky S, Williams L et al (2021) Pfam: the protein families database in 2021. Nucleic Acids Res 49(D1):D412–D4D9

Chapter 10

Using Surface Plasmon Resonance to Study SH2 Domain–Peptide Interactions

Gabrielle M. Watson, Menachem J. Gunzburg, and Jacqueline A. Wilce

Abstract

Biosensor measurement using surface plasmon resonance enables precise evaluation of peptide–protein interactions. It is a sensitive technique that provides kinetic and affinity data with very little sample and without the need for analyte labels. Here, we describe its application for the analysis of peptide interactions with the Grb7-SH2 domain prepared with a GST-tag for tethering to the chip surface. This has been successfully and reliably used for direct comparison of a range of peptides under different solution conditions as well as direct comparison of peptides flowed over different related SH2 domains in real time. We have used the BIAcore system and describe both the methodology for data collection and analysis, with principles also applicable to other biosensor platforms.

Key words Surface plasmon resonance, Growth receptor-bound protein 7, GST fusion protein expression, Sensor chip preparation, Steady-state analysis

1 Introduction

The characterization of SH2 domain interactions with their peptide ligands provides important insight into their biological role and for the development of potential therapeutic inhibitors or research tools. Alongside structural analysis, affinity data is invaluable for understanding the basis of molecular recognition. It is vital for assessing the specificity of a peptide for a particular SH2 domain, the potential for inhibitor peptides to outcompete the natural ligand or for understanding the effects of different solution conditions on the interaction. There are many methodologies suitable for the measurement of protein–peptide interactions in solution [1]. Isothermal titration calorimetry (ITC) is one of the more traditional methods for the determination of binding affinity, but biosensor measurement using surface plasmon resonance (SPR) has distinct advantages—including high sensitivity, precision, the need

Teresa Carlomagno and Maja Köhn (eds.), *SH2 Domains: Functional Modules and Evolving Tools in Biology*, Methods in Molecular Biology, vol. 2705, https://doi.org/10.1007/978-1-0716-3393-9_10,

for only very small quantities of material, and insight into not only the affinity, but also the kinetics of the interaction.

Surface plasmon resonance (SPR) is an optical technique that detects the interaction of a mobile molecule (analyte) with a binding partner immobilized on the gold surface of the biosensor chip. The interaction increases the refractive index at the surface of the chip that is detected by a change in the angle of incident light that is effectively absorbed rather than reflected on the opposite side of the chip. This response is detected in real time and with great sensitivity so that, association and disassociation can be observed when the analyte is injected or depleted respectively, providing kinetic information. If the interaction reaches a steady state, then the response at equilibrium can be used to determine the affinity of the interaction by steady-state analysis. Here, it is important, as with all binding affinity measurements, that the response is measured for an appropriate range of analyte concentrations for the particular affinity being measured [2].

In our studies of growth receptor-bound protein 7 (Grb7), we were interested to develop a lead peptide with high affinity and specificity for the Grb7-SH2 domain for potential use as an inhibitor. Our starting point was the G7-18NATE peptide that was discovered using phage display and subsequently cyclized using thioether chemistry for enhanced affinity [3]. Initial binding measurements were undertaken using ITC [4] but we subsequently explored the use of SPR for gaining further insight into the interaction [5]. We investigated several SPR strategies for measuring the Grb7-SH2 domain–peptide interaction. Usually it is advantageous for the analyte to be the higher molecular weight binding partner, since this results in the greatest refractive index change upon binding. However, tethering the peptide to the chip surface and flowing the Grb7-SH2 domain across as the analyte were not ideal, as the SH2 domain tended to bind irreversibly to the chip surface. Tethering biotinylated Grb7-SH2 domain to a streptavidin coated chip yielded high quality data, despite the relatively small change in refractive index by peptide binding. Our preferred strategy, ultimately, was to tether the GST-fused form of Grb7-SH2 to the surface of the chip via GST antibodies. Despite this being an even higher molecular weight complex at the surface of the chip, this also provided high quality data and had the added advantage of directly using the GST-fused SH2 domain without the need for cleavage.

This SPR strategy has facilitated many investigations of Grb7-SH2 domain–peptide interactions. These include the direct comparison of G7-18NATE binding to different SH2 family members [6, 7]. The four independent flow cells of the Biacore T100 biosensor chip allow for simultaneous binding experiments to be undertaken [8]. With one flow cell required for assessment of the blank response, three different SH2 domains can be compared. The advantage of this is that the peptide analyte solution flowed across all four cells is essentially identical at each peptide concentration administered, allowing for rigorous comparison. The same chip surface can also be used multiple times to investigate the different effects of the solution conditions. In this way, it was discovered that phosphate enhances the binding of G7-18NATE to the Grb7-SH2 domain [6]. Upon development of peptides incorporating phosphotyrosine mimetics, the SPR data also clearly demonstrated that phosphate was a competitor in the SH2 domain peptide binding site [7, 9].

Overall, SPR has proved to be a robust method for investigating SH2–peptide interactions. While there are many available methodological instructions for applying SPR to characterize molecular interactions, here we provide a detailed protocol for the way this has been applied in our laboratory for the measurement of peptide binding to GST-fused SH2 domains. This includes the method we have used for expression and purification of GST-fused Grb7, as well as the steady-state method we used for deriving binding curves, rather than kinetic analysis. While individual proteins can present unexpected complications, it is likely that any GST-SH2 domain construct would be amenable to this method.

2 Materials

2.1 Buffers

1. Lysis buffer: comprising phosphate buffered saline (1× PBS) at pH 7.4, 2 mM ethylenediamine tetra-acetic acid (EDTA), 0.5% (v/v) Triton-X-100, and 1 mM DTT.
2. Wash buffer: 1× PBS, 1 mM Dithiothreitol (DTT).
3. Elution buffer: 10 mM reduced glutathione (GSH) (Sigma Aldrich) in 1× PBS, 1 mM DTT.
4. SEC buffer: 50 mM MES (pH 6.6), 100 mM NaCl, 1 mM DTT.
5. SPR buffer: 50 mM sodium phosphate (pH 7.4), 150 mM NaCl, and 1 mM DTT.

2.2 GST Fusion Protein Expression and Purification

1. SH2 domain of interest cloned into a bacterial expression vector that encodes a GST expression tag. For example, the pGEX system (Cytiva).
2. Luria-Bertani (LB) media comprising of 1% (w/v) bactotryptone, 0.5% (w/v) yeast extract, and 1% (w/v) NaCl, sterilized by autoclaving.
3. LB agar plates comprising LB media supplemented with 1.5% (w/v) agar. Add the agar prior to autoclave sterilization and pour plates under sterile conditions with the appropriate antibiotics.

4. Shaking Incubator (for 800 mL volumes of cell culture).
5. 1 M stock solution of 1-thio-β-D-galactopyranoside (IPTG).
6. Centrifuge (for pelleting of cell culture).
7. Sonicator (MSE Soniprep 150 Plus).
8. Econo Pump (BioRad) peristaltic pump.
9. 5 mL Protino GST/4B column (Macherey Nagel) or a 5 mL GS Trap FF column (Cytiva).
10. Superdex 200 16/60 column (Cytiva).
11. Äkta Purifier FPLC (Cytiva).
12. NanoDrop spectrophotometer.

2.3 Sensor Chip Preparation and Immobilization

1. BIAcore CM5 series S sensor chip (Cytiva) at room temperature.
2. BIAcore T100 (Cytiva).
3. BIAnormalizing solution (70%) (Cytiva).
4. Amine Coupling Kit (we use Cytiva). Reconstitute the provided 1-ethyl-3-(3-dimethylaminopropyl)carbodiimide hydrochloride (EDC) in water to 0.4 M concentration. Reconstitute the provided N-hydroxysuccinimide (NHS) in water to 0.1 M concentration. Filter the EDC and NHS solutions through 0.2 μm membrane, divide into 0.1 mL volume aliquots and store at −20 °C. Once an aliquot has been thawed, it cannot be refrozen.
5. Polyclonal anti-GST antibody (at 16–60 μg/mL; we use GE Life Science) diluted in 10 mM Na acetate pH 4.0 or 5.0.

2.4 Steady-State Assay

1. 0.2 μm 0.5 mL spin filter.
2. 96-well plate.
3. BIAcore microplate foil.

2.5 Data Analysis

Biacore Analysis Software (Cytiva) or Scrubber 2 (BioLogic Software).

3 Methods

Prepare all solutions using type 1 ultrapure water. Degas SEC and SPR buffers by stirring under vacuum for 30 min (*see* **Note 1**).

3.1 GST Fusion Protein Expression and Purification

This protocol has been optimized for expression and purification of the SH2 domains of Grb7, Grb10, and Grb2, expressed using the pGex2T expression vector. Figure 1 demonstrates an example of the steps involved in the purification of GST-tagged Grb7-SH2.

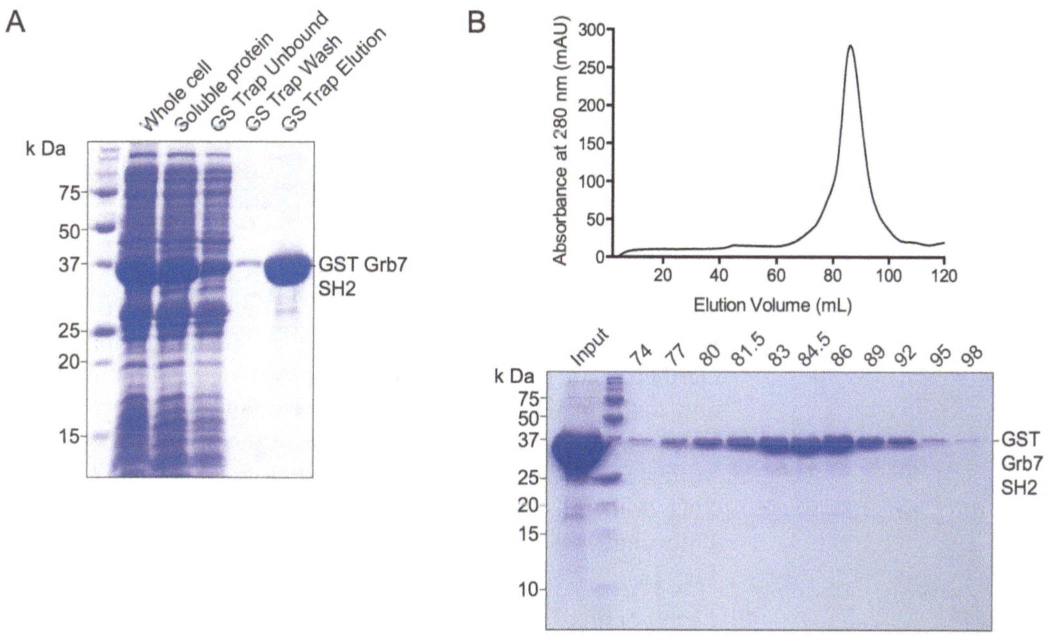

Fig. 1 Purification of GST-tagged Grb7-SH2. (**a**) 15% SDS-PAGE analysis of the steps involved in the GST affinity purification. (**b**) Size-exclusion chromatography of the GST-tagged Grb7-SH2. Chromatogram of the size-exclusion chromatography profile with absorbance at 280 nm displayed (top) and 15% SDS-PAGE of the resultant fractions (bottom)

1. Transform 100 ng of purified plasmid DNA into BL21 (DE3) pLysS competent cells and plate on LB agar plates containing 100 μg/mL ampicillin and 25 μg/mL chloramphenicol. Incubate overnight at 37 °C.
2. Prepare a saturated *E. coli* starter culture by inoculating a scraping of a single bacterial colony from the transformation plate in 50 mL of LB media (containing appropriate antibiotics, here 100 μg/mL ampicillin and 25 μg/mL chloramphenicol). Incubate overnight at 37 °C with shaking at 200–220 rpm for aeration.
3. Inoculate 800 mL of sterilized LB (here containing 100 μg/mL ampicillin and 25 μg/mL chloramphenicol) with 1% v/v starter culture. Incubate at 37 °C with shaking at 200–220 rpm until an optical density at 600 nm (OD600) of 0.6–0.8 is reached. At this optimal cell density, rest the cultures without shaking for 30 min at room temperature. Induce protein expression by adding IPTG to a final concentration of 0.4 mM and supplement the bacterial cultures with a further 100 μg/mL of ampicillin and 25 μg/mL chloramphenicol. Incubate bacterial cultures at 25 °C for 4 h with aeration.

4. Harvest bacterial pellets by centrifugation at 4000 rpm for 20 min. Bacterial pellets can be stored at −80 °C until required.
5. Resuspend bacterial pellet in 30 mL of lysis buffer and lyse cells by sonication (whole cell sample in Fig. 1a).
6. Separate soluble fraction from cell debris via centrifugation at 4 °C for 30–45 min at 15,000 rpm (soluble protein sample in Fig. 1a).
7. Equilibrate a 5 mL Protino GST/4B column (Macherey Nagel) or a 5 mL GS Trap FF (Cytiva) with 5- to 10-column volumes (CV) of lysis buffer using a peristaltic pump. Pass clarified lysate through a syringe filter with a 0.2 μm membrane and load on to the pre-equilibrated column at ~1 mL/min. Wash bound protein with 1- to 3-column volume (CV) of lysis buffer, 5–10 CV of wash buffer, and elute off the GST-tagged SH2 domain of interest with 5 CV of elution buffer (GS Trap Unbound, GS Trap Wash and GS Trap Elution samples in Fig. 1a).
8. Dialyze eluted protein overnight into SEC buffer, concentrate to 2 mL and load on pre-equilibrated Superdex 200 16/60 column (we use Cytiva) on a Äkta Purifier FPLC (we use Cytiva) (Chromatogram example in Fig. 1b).
9. Analyze protein fractions by SDS-PAGE (example shown in Fig. 1b).
10. Pool fractions that are >95% pure. Typically, the GST-SH2 domain does not need to be concentrated for use in the SPR experiments. If a higher concentration is desired, we recommend concentrating with 3K molecular weight cut off concentrators (we use Amicon).
11. Determine Protein concentration (c) using the Beer–Lambert Law, $A = \varepsilon.c.l$, whereby A is the absorbance measured at 280 nm (we use a NanoDrop spectrophotometer), ε is the extinction coefficient (L/mol.cm) predicted by the *ProtParam* server [10], and l is the path length (cm).
12. Store at 4 °C and use within 1 week. Store at −80 °C for long-term storage.

3.2 Peptide Preparation and Quantification

1. Weigh out 1 mg of mass spectrometry verified peptide using a balance suitable for weighing small volumes.
2. Resuspend the peptide in 500 μL of SPR buffer.
3. Filter the resuspended peptide through a 0.2 μm membrane.
4. Measure the absorbance at 280 nm using a NanoDrop spectrophotometer or equivalent instrument.
5. Calculate the extinction coefficient of the peptide using the *ProtParam* web server.

6. Determine the peptide concentration using the Beer–Lambert Law (*see* **Notes 2–4**).
7. Dilute the peptide to a stock concentration of 1 mM in SPR buffer.

3.3 Sensor Chip Preparation and Immobilization

The day before beginning the SPR experiment, we recommend conducting a desorb procedure.

1. Bring a BIAcore CM5 series S sensor chip to room temperature.
2. Dock room temperature sensor chip into the BIAcore T100.
3. Prime the BIAcore T100 with SPR buffer.
4. Start a new sensorgram and run a manual run at 30 μL/min over all flow cells until the observed drift is less than 1 response unit (RU)/minute.

The following can also be achieved using a manual run; however, for convenience, a program can also be generated.

In the program, 0.4 M 1-ethyl-3-(3-dimethylaminopropyl)carbodiimide hydrochloride (EDC) and 0.1 M N-hydroxysuccinimide (NHS) are premixed and flowed over all flow cells for 7 min at 5 μL/min leading to the formation of reactive succinimide esters. Polyclonal anti-GST antibody (at 16–60 μg/mL; GE Life Science) in 10 mM Na Acetate (either pH 4.0 or 5.0) is passed over the chip surface at 5 μL/min for 7 min, leading to the covalent attachment of the antibody to the chip surface. The remaining reactive ester groups are blocked by flowing over 1 M ethanolamine-HCl (pH 8.5) for 7 min at 5 μL/min.

1. Open a wizard template.
2. Select the desired number of flow cells to be coupled, minimum of 2 is recommended with flow cell 1 serving as a reference flow cell.
3. Specify a contact time of 420 s, and a flow rate of 5 μL/min.
4. Include a normalization step in the program.
6. Defrost EDC and NHS aliquots immediately prior to use.
7. Prepare tubes containing BIAnormalizing solution (70%), EDC, NHS, and ethanolamine-HCl.
5. Prepare tubes containing polyclonal anti-GST antibody (at 16–60 μg/mL) diluted in 10 mM Na acetate. The concentration to use depends on the age of the antibody stock solution.
6. Visually inspect all tubes to ensure there are no air bubbles at the bottom of the tube.
7. Eject the sample rack, add samples to the allocated positions and start the program.

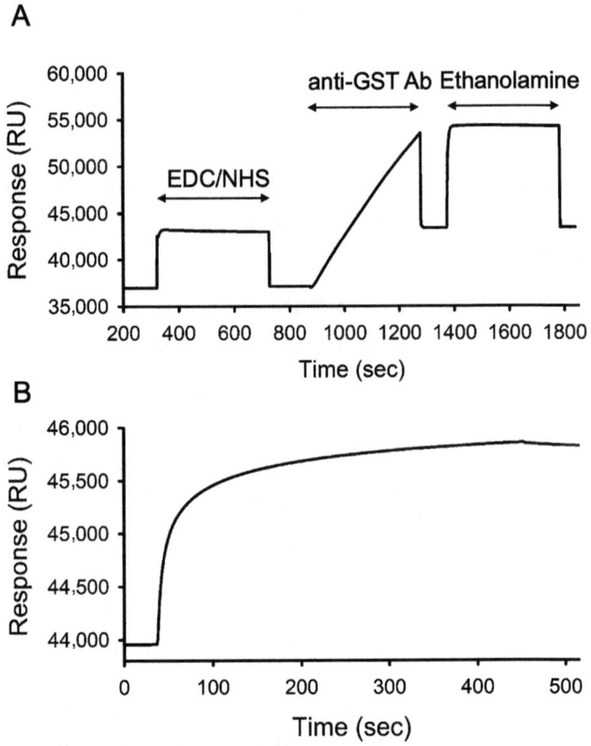

Fig. 2 Anti-GST antibody immobilization and GST-Grb7-SH2 domain capture. (**a**) Sensorgram of Anti-GST antibody capture on a CM5 chip, showing injection of EDC/NHS followed by anti-GST-antibody and ethanolamine. (**b**) Sensorgram of GST-Grb7-SH2 capture on an anti-GST antibody immobilized surface

8. Always visually inspect the sensorgrams following the program. EDC/NHS immobilization should be between 180 and 200 response units (RU). If lower than this, we recommend preparing fresh solutions of EDC and NHS for further experiments. Antibody immobilization is normally between 4000 and 5000 RU. If lower than this, we recommend increasing the concentration of antibody or lowering the pH of the Na acetate solution.

An example sensorgram of GST antibody immobilization is provided in Fig. 2a.

The following can either be set up as a program or as a manual run.

1. Set the flow rate at 5 μL/min.
2. Prepare GST fusion of interest by diluting to 0.7 μM in SPR buffer.
3. Specify flow cell to be immobilized by each protein. We highly recommended to immobilize GST alone on flow cell 1, or an irrelevant GST fusion protein control.

4. Inject the diluted GST-SH2 over the relevant flow cell until approximately 2000 RU of GST-SH2 immobilized, or 800 RU for GST alone.
5. Run a manual run at 30 μL/min until the baseline has stabilized to less than 1 RU/min drift.

An example sensorgram of GST-Grb7-SH2 immobilization is provided in Fig. 2b.

3.4 Steady-State Assay

1. Filter the peptide stock solution through a 0.2 μm spin filter for 2 min at 14,000 g.
2. Prepare a serial dilution of the peptide using SPR buffer. If the approximate K_D of the protein/peptide is known, ensure the concentration range extends to 10× above and below the expected K_D. A minimum of six concentrations should be selected, ideally eight to ten. Ensure that you have at least two buffer blank samples per concentration series, and start the run with three to eight blank startup cycles with buffer blank.
3. Prepare analysis program using the following parameters (*see* **Note 5**):
   (a) 30 μL/min.
   (b) 60 s injection.
   (c) 600 s dissociation.
   (d) 25 °C assay temperature.
4. Transfer the indicated volume (plus an additional 10 μL to account for pipetting errors) to a 96-well plate. Include triplicates of each peptide concentration. Include buffer only controls that will be used during data analysis.
5. Cover the 96-well plate with microplate foil to prevent sample evaporation during the length of the experiment.
6. Add 96-well plate to corresponding tray and insert into the BIAcore T100.
7. Ensure there is enough SPR buffer on the left hand buffer inlet, water on the right hand inlet, and the waste bottle has space.
8. Start program.

An example of the multicycle concentration series of the G7-18NATE peptide binding to GST-Grb7 SH2 is shown in Fig. 3a, and the corresponding binding curve in Fig. 3b.

9. Before proceeding to data analysis, it is important to inspect the sensorgrams for signs of nonspecific binding, and to ensure the assay parameters are appropriate for the binding equation to be conducted. For example, (1) if the responses on flow cell 1 do not return to preinjection levels, then this indicates nonspecific

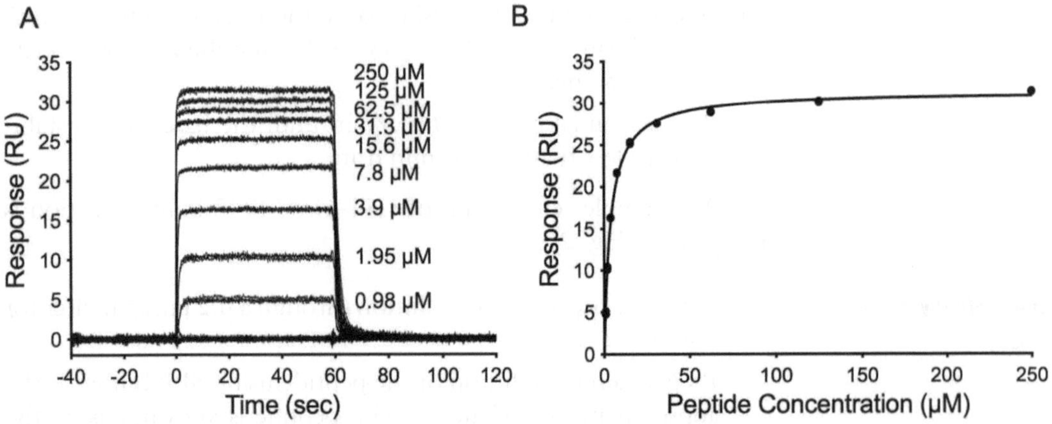

Fig. 3 Example steady-state affinity experiment of the cyclic peptide inhibitor G7-18NATE binding to immobilized GST-Grb7 SH2 using the described method. (**a**) Sensorgram of nine concentrations (0.98–250 μM) of G7-18NATE binding to GST-Grb7 SH2 once the responses from buffer only injections as well as flow cell 1 (GST alone) had been subtracted (double referencing). (**b**) Corresponding equilibrium binding curve

binding to the reference flow cell (*see* **Note 6**), and (2) if an equilibrium binding model is to be used (described in Subheading 3.5), the injection length needs to be long enough for the binding equilibrium to be reached.

3.5 Data Analysis

3.5.1 Buffer and Reference Subtraction and Other Corrections

The assay data need to be reference subtracted in order to correct for the contribution of buffer components and the peptide to the refractive index of the sample. In this process, the response of the reference flow cell (with anti-GST antibody and GST or irrelevant GST fusion protein immobilized) is subtracted from the response of the active flow cell (with anti-GST antibody and the GST-SH2 immobilized) and will result in data where the corrected response is directly related to binding. The two most common programs available for analyzing SPR data, Biacore Evaluation Software, and Scrubber 2, both allow reference subtraction to be performed.

The assay data also require zeroing to y-adjust the sensorgrams such that the baseline response before injection is zero for all the injections. Alignment of the data (*x*-adjustment) must be performed to ensure that the start of all injections is set to time = 0. Double referencing of the data is also required to remove systematic noise and drift from the data. This is performed by subtracting the reference subtracted blank injection from the reference subtracted peptide injection. Either the average of the blanks or the nearest blank to the sample can be used. All these data corrections can be performed using either Biacore Evaluation Software or Scrubber 2.

3.5.2 Fitting Data to Binding Equation

In most cases for peptides binding to SH2 domain, where the dissociation constant is 1 μM or above, the dissociation rate will be >1 s^{-1} and therefore be too fast to accurately fit to kinetic rate equations. The fast kinetics of these interactions allows the binding to reach equilibrium rapidly and therefore the data can be fit to an equilibrium binding model. To facilitate this, a binding curve of equilibrium binding response vs. concentration is required. For each injection, a window is selected after the binding has reached equilibrium, and the data is averaged across this window to give the equilibrium binding response (R_{eq}) for that concentration (C). It is important for the binding to have reached equilibrium before the window used to determine the R_{eq}, otherwise it is invalid to fit the data to an equilibrium binding model. Biacore Evaluation Software, and Scrubber 2 facilitate the calculation of this binding curve from the injection sensorgram data. The binding curve is fit to Eq. 1 binding equation using either Biacore Evaluation Software, and Scrubber or curve fitting software.

$$R_{eq} = \frac{R_{max} C}{C + K_D} \quad (1)$$

In order to obtain reliable fitting parameters, the fit model must visually be close to the actual data, and the residuals should have only small amounts of nonrandom deviations. For the fitted K_D to be reliable, there needs to be sufficient curvature in the binding data. Generally, this is achieved if the maximum concentration of the data is greater than 2 time the K_D of the fit. It is also important for the R_{max} of the fit to be reasonable. This can be checked by comparing the R_{max} to the theoretical R_{max}, that is calculated by Eq. 2, where R_I is the immobilization level of the SH2 domain, and S is the stoichiometry of binding.

$$\text{Theoretical } R_{max} = \frac{\text{MW peptide}}{\text{MW SH2 construct}} \times R_I \times S \quad (2)$$

If the theoretical R_{max} is much greater than the R_{max} of the fit, then the SH2 domain may have low binding activity, while if the fit R_{max} is much greater than the theoretical R_{max}, it suggests super-stoichiometric binding or other artifacts which may affect the reliability for the determined K_D.

4 Notes

1. Following addition of DTT, buffers should be used within 1 day.
2. To obtain accurate affinity measurements by SPR, it is critical to obtain accurate measurements of the peptide concentration.

3. If the peptide does not contain a tryptophan, consider the DirectDetect (Millipore) system to measure peptide concentration.
4. If the lyophilized peptide is known to be highly pure, the concentration can be determined by dividing the mass weighed out by the volume of SPR analysis buffer it was resuspended in. However, we do not recommend this method of quantification if the concentration of free salt in the lyophilized peptide is unknown.
5. We do not typically require a regeneration solution for removal of bound peptide following completion of the dissociation step. If required, we recommend manually testing out different regeneration solutions such as 2 M NaCl. The required regeneration solution will vary between experiments.
6. The choice of SPR buffer may require optimization. We have successfully removed nonspecific binding (observed on reference cell 1) by increasing the NaCl concentration from 150 to 300 mM. Including a detergent (e.g., 0.005% P20) and/or 1–5% bovine serum albumin (BSA) can also remove nonspecific binding events.

References

1. Walport LJ, Low JKK, Matthews JM, Mackay JP (2021) The characterization of protein interactions – what, how and how much? Chem Soc Rev 50:12292–12307
2. Pollard TD (2010) A guide to simple and informative binding assays. Mol Biol Cell 21: 4061–4067
3. Pero SC, Oligino L, Daly RJ et al (2002) Identification of novel non-phosphorylated ligands, which bind selectively to the SH2 domain of Grb7. J Biol Chem 277:11918–11926
4. Porter CJ, Matthews JM, Mackay JP et al (2007) Grb7 SH2 domain structure and interactions with a cyclic peptide inhibitor of cancer cell migration and proliferation. BMC Struct Biol 7:58
5. Gunzburg MJ, Ambaye ND, Hertzog JT et al (2010) Use of SPR to study the interaction of G7-18NATE peptide with the Grb7-SH2 domain. Int J Pept Res Ther 16:177–184
6. Gunzburg MJ, Ambaye ND, Del Borgo MP et al (2012) Interaction of the non-phosphorylated peptide G7-18NATE with Grb7-SH2 domain requires phosphate for enhanced affinity and specificity. J Mol Recognit 25:57–67
7. Watson GM, Gunzburg MJ, Ambaye ND et al (2015) Cyclic peptides incorporating phosphotyrosine mimetics as potent and specific inhibitors of the Grb7 breast cancer target. J Med Chem 58:7707–7718
8. Rich RL, Myszka DG (2007) Higher-throughput, label-free, real-time molecular interaction analysis. Anal Biochem 361:1–6
9. Watson GM, Kulkarni K, Sang J et al (2017) Discovery, development, and cellular delivery of potent and selective bicyclic peptide inhibitors of Grb7 cancer target. J Med Chem 60: 9349–9359
10. Gasteiger E, Hoogland C, Gattiker A et al (2005) Protein identification and analysis tools on the ExPASy server. In: Walker JM (ed) The proteomics protocols handbook. Humana Press, pp 571–607

Part III

Small-Molecule Binders and Inhibitors of SH2 Domains

Chapter 11

Inhibitor Library Screening of SH2 Domains Through Denaturation-Based Assays

Elvin D. de Araujo, Anna Orlova, Qirat F. Ashraf, Richard Moriggl, and Patrick T. Gunning

Abstract

Screening of inhibitor libraries for candidate ligands is an important step in the drug discovery process. Thermal denaturation-based screening strategies are built on the premise that a protein–ligand complex has an altered stability profile compared to the protein alone. As such, these assays provide an accessible and rapid methodology for stratifying ligands that directly engage with the protein target of interest. Here, we describe three denaturation-based strategies for examining protein–inhibitor binding, in the context of SH2 domains. This includes conventional dye-based Thermal Shift Assays (TSA), nonconventional labeled ligand-based TSA, and Cellular Thermal Shift Assays (CETSA). Conventional dye-based TSA reports on the fluorescence of an external hydrophobic dye as it interacts with heat-exposed nonpolar protein surfaces as the temperature is incrementally increased. By contrast, nonconventional-labeled ligand TSA involves a fluorescence-tagged probe (phosphopeptide for SH2 domains) that is quenched as it dissociates from the protein during the denaturation process. CETSA involves monitoring the presence of the protein via Western blotting as the temperature is increased. In all three approaches, performing the assay in the presence of a candidate ligand can alter the melting profile of the protein. These assays offer primary screening tools to examine SH2 domain inhibitors libraries with varying chemical motifs, and a subset of the advantages and limitations of each approach is also discussed.

Key words Thermal shift assay, Cellular thermal shift assay, Fluorescence, SH2 domain, Phosphopeptide, Inhibitor

1 Introduction

Small molecule inhibitor screening and identification is a complex process that can involve a range of chemical scaffolds with unknown target engagement profiles. One of the early-stage assays for exploring and identifying candidate ligands involves heat-induced macromolecular denaturation. This is largely due to amenability of these assays to high-throughput formats, broad applicability, and limited experimental optimization. The most frequently used assay for inhibitor library screening (fluorescence-based thermal shift assays

Teresa Carlomagno and Maja Köhn (eds.), *SH2 Domains: Functional Modules and Evolving Tools in Biology*, Methods in Molecular Biology, vol. 2705, https://doi.org/10.1007/978-1-0716-3393-9_11,

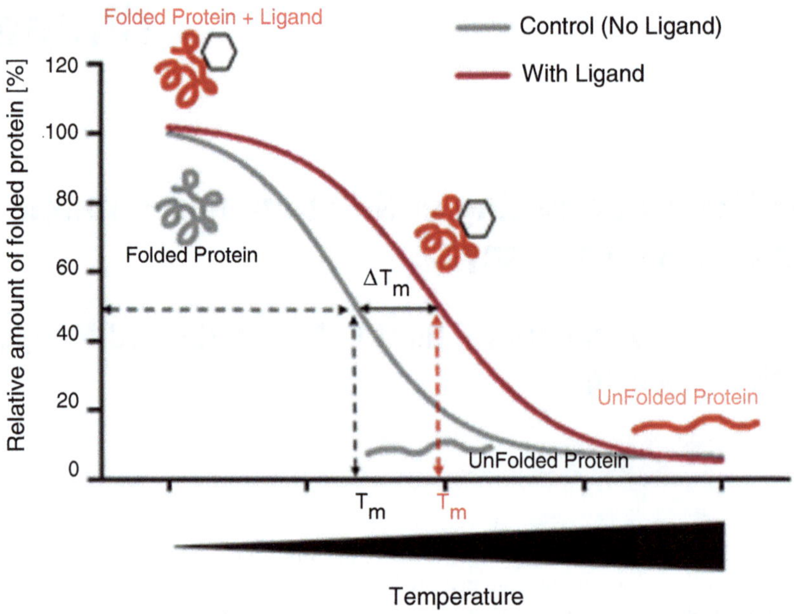

Fig. 1 Schematic explanation of thermal shift assay

or TSA [1]) is built on the classical premise that a protein–ligand complex demonstrates higher stability than the protein alone [2]. This increased stability translates to greater resistance to thermal denaturation, which can be observed by subjecting the protein–ligand mixture to incrementally increasing temperatures in the presence of an environmentally-sensitive (hydrophobic) fluorophore. As the protein unfolds due to increased temperatures, it exposes hydrophobic residues that can interact with the fluorescent dye and increase its quantum yield. By observing fluorescence as a function of temperature, the melting point (T_m) of the protein (or protein–ligand complex) can be determined from the mid-point of the curve (Fig. 1). This is more accurately evaluated from the inflection point of a first-derivative plot. Since this assay does not require modification, immobilization, or excessive quantities of the protein or ligand, TSA is a routinely employed technique, especially at early stages in drug discovery pipelines [3].

Similar to all assays, specific requirements are necessary to ensure appropriate TSA evaluation. For instance, the denaturation profile requires a well-defined cooperative (two-state) transition as multistate or linear unfolding curves present challenges in interpretation of the data. Similarly, protein–ligand complexes are also assumed to maintain a 1:1 stoichiometry. Finally, hydrophobic surfaces on the protein, inhibitor, or detergents in the buffer can lead to interactions with the fluorescent dye, drastically increasing the background signal and reducing the sensitivity or feasibility of the assay.

Finally, a critical aspect in TSA involves the energetics of ligand binding which determine the magnitude of the T_m shift. For example, two ligands with the same K_D for a protein target can result in different T_m values, as a result of differences in the enthalpic (ΔH) and entropic (ΔS) contributions to binding. TSA is heavily biased toward ligands with entropically-driven binding, as the enthalpic components are often masked due to the heat of unfolding [4]. Therefore, a ligand with high (negative) $\Delta H_{binding}$ but low $\Delta S_{binding}$ can result in a dampened T_m shift and therefore may be unintentionally filtered out of a TSA library screen.

Despite these considerations, thermal shift assays are routinely employed for SH2 domain inhibitor screening. SH2 domains are modular domains that are comprised of two alpha helices separated by a central β sheet, creating a pY and pY + 3 pocket [5]. SH2 domains selectively engage with phosphorylated tyrosine-containing peptides which triggers highly specific protein–protein interactions. As such, tyrosine phosphorylation and SH2 domain interactions are typically responsible for altered cellular localization and recruitment of the biomolecule. Conventional inhibitors for SH2 domains exploit the unique features of the pY and pY + 3 pockets and often employ phospho-tyrosine mimetics [6]. However, there are additional druggable sites within the SH2 domain, and thermal shift screening can help identify new chemical moieties. Protocols for thermal shift screening are recurrently highlighted in the literature [7, 8]. By using STAT3 as a representative SH2 domain-containing protein, which is also the highest interspecies conserved STAT transcription factor family member [9], we outline potential thermal shift screening strategies with an emphasis on their applications in SH2 domains. Initially, we describe the assay conditions for the typical TSA experimental set-up, after determining whether a target protein will be amenable to this approach. Additionally, we offer an alternative approach that employs the cognate phospho-tyrosine peptide as a probe to overcome challenges in situations with nonideal TSA profiles or ligands incompatible with the assay conditions/fluorescent dye.

CETSA (cellular thermal shift assay) is a variation of the TSA assay adapted for evaluating protein–ligand interactions in living cells or cell lysates [10]. By observing the melting curve of a target protein in comparison to a melting curve of the target protein incubated with a ligand, it is possible to draw conclusions about target engagement *in cellulo*. The method provides simple readout of protein–ligand interactions in cell culture and can be adapted to primary biopsies or organs [11]. CETSA relies on a few crucial steps: first, incubation of the compound with cells, second, heating of the cells, and lastly cell lysis and readout. This chapter focuses on quantitative Western blot as a readout of CETSA; however, other readouts like mass spectrometry or AlphaScreen are also possible to enhance throughput [11].

2 Materials

2.1 Equipment

1. Real-time PCR instrument, such as the C1000 Touch ThermoCycler equipped with a CFX96 Real-time optical unit.
2. Gradient PCR instrument, such as Eppendorf Mastercycler X59a.
3. Single channel micropipettes (1000 μL, 200 μL, 20 μL).
4. Multichannel micropipettes (300 μL, 20 μL).
5. Centrifuge with a swinging-bucket rotor and adapters for 96-well plates.
6. SDS-PAGE/Western blot equipment.

2.2 Materials and Reagents

1. Clear, U-bottom, 96-well plate.
2. PCR strip tubes.
3. White, thin wall, skirted, 96-well PCR plate.
4. Optically clear, adhesive microseals for 96-well plates.
5. Micropipette tips (1000 μL, 300 μL, 200 μL, 20 μL); 1.5 mL; 2 mL microfuge tubes.
6. Cell culture flasks T25, T75.
7. RPMI 1640 media supplemented with 10% FBS, 1% PenStrep, and 2 mM L-Glutamine.
8. PBS, pH ~7.4.
9. cOmplete, EDTA-free protease inhibitors.
10. Reducing loading dye 6×: 1.6% SDS; 20 mM Tris–HCl, pH 6.8; 16% glycerol; 0.24 g/mL Bromophenol blue; 0.04 g/mL Dithiothreitol in double distilled H_2O.
11. TSA Buffer: 100 mM HEPES pH 7.4 (*see* **Note 1**).
12. SYPRO™ Orange protein gel stain (5000× concentrate in DMSO).
13. Test compounds: 10 or 1 mM in DMSO.
14. Phosphopeptide probe (Ac-pYLPQTV-NH_2): 1 mM in DMSO.
15. Phosphopeptide with a fluorophore tag (FAM-GpYLPQTV-NH_2): 1 mM in DMSO (*see* **Note 2**).
16. STAT3 protein stock solution: 20 μM (*see* **Note 3**).

3 Methods

3.1 TSA—Thermal Shift Assay

This assay evaluates the suitability of the target protein to conventional TSA (*see* **Note 4**).

3.1.1 Fluorescence-Based Thermal Denaturation Assay Protocol

1. Prepare a "Serial Dilution Plate" of STAT3 protein using a clear, U-bottom, 96-well plate.
   (a) Use the STAT3 protein stock solution and pipette 200 μL into the first well of the 96-well plate (A1, which will have 200 μL of 20 μM of STAT3).
   (b) Add 100 μL of TSA buffer in the wells B1, C1, D1, and E1 (*see* **Note 5**).
   (c) Transfer 100 μL from A1 to B1, mix, and then transfer 100 μL from B1 to C1, etc. up until E1.
   (d) The final volume in wells A1–D1 will be 100 μL and well E1 will contain 200 μL. [One column is enough for one TSA assay with three triplicates].
   (e) Add 200 μL of TSA buffer to well F1.
2. Prepare the "Assay Plate" using a white, thin-walled, skirted, 96-well PCR plate.
   (a) Transfer 25 μL of A1 from the "Serial Dilution Plate" to each of the wells A1, A2, and A3 of the "Assay Plate." Repeat this process to transfer 25 μL of B1, C1, D1, E1, and F1 to a corresponding series of triplicate wells in the "Assay plate."
   (b) Add 20 μL of TSA Buffer to each of the wells above.
   (c) Prepare a 50× SYPRO™ Orange stock, by adding 5 μL of 5000× SYPRO™ Orange to 495 μL of TSA buffer in an 1.5 mL Eppendorf tube. Use the 50× SYPRO™ Orange stock and add 5 μL to each well in the "Assay Plate" above.
   (d) In the "Assay Plate," include an additional six wells for experimental controls:
      (i) To three wells, include 25 μL of STAT3 protein stock solution and 25 μL of TSA buffer.
      (ii) To three wells, include 50 μL of TSA Buffer.
3. Incubate the "Assay Plate" at room temperature for 5 min (*see* **Note 6**).
4. Load the plate into the qPCR machine. The thermal melt program should include heating the samples from 20 to 85 °C in steps of 0.5 °C with 30 s equilibration time between each temperature step. The fluorescence intensity is recorded following each temperature equilibration step (*see* **Notes 7** and**8**).

5. The fluorescence data is plotted as a function of temperature, and the negative first derivative is calculated. The local maximum on the plot gives the T_m of the protein (*see* **Note 9**).

3.1.2 Inhibitor Library Screen

This assay evaluates a library of test compounds for potential engagement with STAT3 (or another SH2 domain-containing protein).

1. Prepare an "Inhibitor Compound Plate" using a clear, U-bottom, 96-well plate as detailed above.
   (a) For each test compound/inhibitor, use a 2 mM stock (in DMSO) and transfer 50 μL into three consecutive wells (i.e., for Inhibitor 1, transfer 100 μL to wells A1, A2, A3; for Inhibitor 2, transfer 100 μL to wells A4, A5, A6, etc.).
   (b) Reserve at least six wells on the plate for controls:
      (i) To three wells, transfer 100 μL of 2 mM of phosphopeptide.
      (ii) To three wells, transfer 100 μL of DMSO.
2. Prepare the "Assay Plate." Follow the same set-up/use the same number of wells as the "Inhibitor Plate" above. This assay is designed for at least ten wells, and please scale all volumes proportionally for the number of assays required.
   (a) Prepare the STAT3 protein solution, by combining 60 μL of STAT3 (20 μM) + 450 μL TSA buffer. Mix the solution, and transfer 42.5 μL to each well.
   (b) Using a multichannel pipette transfer 2.5 μL of each inhibitor in the "Inhibitor Compound Plate" to the "Assay Plate."
   (c) Incubate the plate for 10 min at room temperature.
   (d) Add 5 μL of 25× SYPRO™ Orange to each well and mix.
   (e) Incubate the plate for 10 min at room temperature.
3. Run the plate as in Subheading 3.1.1, **step 4**.
4. Determine the T_m for each protein–inhibitor solution. For any inhibitors that show a change in T_m, run a dose–response curve. This refers to repeating the experiments in **steps 1–3** above, with varying concentrations of ligand. This will help validate the candidate ligand as a hit molecule in TSA.
5. Prepare the "Inhibitor Control Plate." Follow the same set-up/use the same number of wells as the "Inhibitor Plate" above.

(a) Add 42.5 μL of buffer to each well.

(b) Using a multichannel pipette transfer 2.5 μL of each inhibitor in the "Inhibitor Compound Plate" to the "Inhibitor Control Plate."

(c) Add 5 μL of 25× SYPRO™ Orange to each well and mix.

(d) Incubate the plate for 10 min at room temperature.

6. Run the plate as in Subheading 3.1.1, **step 4**. Compounds that exhibit intrinsic fluorescence are not amenable for the assay.

3.1.3 Inhibitor Library Screen for Nonconventional TSA Profiles

This assay evaluates a library of test compounds for potential engagement with STAT3 (or another SH2 domain-containing protein).

1. Prepare an "Inhibitor Compound Plate" using a clear, U-bottom, 96-well plate.

   (a) For each test compound/inhibitor, use a 1 mM stock (in DMSO) and transfer 100 μL into three consecutive wells. (i.e., for Inhibitor 1 transfer 100 μL to wells A1, A2, A3; for Inhibitor 2 transfer 100 μL to wells A4, A5, A6, etc.)

   (b) Reserve at least six wells on the plate for controls:

      (i) To three wells, transfer 100 μL of 1 mM of phosphopeptide.

      (ii) To three wells, transfer 100 μL of DMSO.

2. Prepare the "Assay Plate." Follow the same set-up/use the same number of wells as the "Inhibitor Plate" above.

   (a) Prepare a stock of Protein–FAM–Phosphopeptide solution: 10 μL STAT3 + 2 μL FAM-Phosphopeptide + 938 μL TSA Buffer. Allow the solution to stand for 10 min after mixing. This volume will be sufficient for at least 18 wells, and scale accordingly (200 nM STAT3 and 2 μM phosphopeptide). Gently invert the mixture and let the mixture stand for 10 min. Transfer 47.5 μL into each well.

   (b) Using a multichannel pipette transfer 2.5 μL of each inhibitor in the "Inhibitor Compound Plate" to the "Assay Plate."

   (c) Incubate the plate for 10 min at room temperature.

3. Run the plate as in Subheading 3.1.1, **step 4**.

4. Determine the integral of the first derivative plot. A reduction greater than 10% indicates potential binding and the inhibitor can be advanced to downstream assays.

3.2 CETSA—Cellular Thermal Shift Assay

The method consists of two parts—generation of the melting curve and determination of isothermal dose–response. The procedure for both two steps is essentially identical, except for the fact that to generate the melting curve, the concentration of the compound stays constant and a range of temperatures is tested, while for the dose–response experiment, a range of compound concentrations is tested and the temperature stays constant (around T_m).

3.2.1 Cell Culture

1. Expand MV4-11 cells in full RPMI cell culture medium to an optimal density (1×10^6 cells/mL) using sterile cell culture techniques (*see* **Note 10**).
2. Seed 3×10^6 MV4-11 cells per condition in 10 mL of full RPMI media into separate T25 flask (*see* **Note 11**).
3. Add DMSO stock solution of compound to individual flasks to get the desired final concentration of the compound. Add the same volume of DMSO to the remaining flask serving as the vehicle or solvent control. Mix the cell suspension well by gently swirling or pipetting up and down (*see* **Note 12**).
4. Incubate cells with the compound for a defined period in the CO_2 incubator at 37 °C.
5. Collect the cell suspension and transfer the cells to marked 15 mL conical tubes.
6. Centrifuge the conical tubes at 300 *g* for 5 min at room temperature to pellet the cells, and then remove the culture medium.
7. During the waiting time interval of the previous step, switch on and preheat the PCR machine by starting the respective program.
8. Resuspend the cell pellets with 5 mL of PBS and centrifuge at 300 *g* for 5 min at room temperature. Carefully remove and discard the supernatant.
9. Add 400 μL of PBS supplemented with 1× protease inhibitors to each respective tube and carefully resuspend the cell pellet.
10. Distribute each cell suspension, i.e., with DMSO control or with the test compound, into ten different PCR tubes with 30 μL of cell suspension in each tube (~3 million cells per tube). Mark each tube or strip with a designated temperature value. Briefly spin down the tubes. Avoid pelleting the cells as well as formation of the air bubbles in the tubes. The tubes are kept at room temperature.

3.2.2 Heating

1. Use the gradient PCR thermocycler to heat the PCR-tube strips at their designated temperature for 3 min.

2. Let the blocks reach the desired temperature before placing the tubes in. Consistent timing between tubes in both the heating and cooling steps, as well as between heat stages is crucial.
3. Immediately after heating, incubate the tubes at room temperature for 3 min. After this 3 min incubation, immediately snap-freeze the samples.
4. In the meantime, label 1.5 mL microfuge tubes for the samples and precool them on ice or in the cold-room.

3.2.3 Cell Lysis

1. Transfer the lysates to the 1.5 mL microfuge tubes.
2. Freeze-thaw the lysates three times using liquid nitrogen. Thaw samples on ice. The tubes should be intensely vortexed after each thawing.
3. Briefly vortex the tubes and centrifuge the cell lysate containing tubes at 16,000 *g* for 20 min at 4 °C.
4. For each tube, transfer 30 μL of each supernatant without touching the pellet to a new tube. Keep the samples on ice. The soluble fraction is now ready for analysis by Western blotting.

3.2.4 SDS-PAGE and Western Blot

1. Start the SDS-PAGE by mixing 15 μL of each respective clarified cell lysate with 4 μL of reducing loading buffer in new tubes; vortex briefly, briefly spin down the samples in a microcentrifuge and heat all the tubes at 95 °C for 10 min.
2. Load 19 μL of each sample into the wells on the same gel. For easier interpretation, it is convenient to load the samples from the same temperature endpoint next to each other.
3. Proceed with the standard Western blot protocol [12] using primary antibody against STAT3. For the loading control of the Western blot, it is advisable to probe for a target that displays minimal precipitation at the chosen temperatures [13].

3.2.5 Quantification

1. Quantify the different protein bands by using the ImageJ quantification tools [14].
2. To evaluate the data, use a data processing software program (e.g., GraphPad Prism). Data should be first normalized by setting the highest and lowest value in each set to 100% and 0%, respectively (*see* **Note 13**).

4 Notes

1. The buffer for TSA assays is relatively versatile and any of Good's buffers [15] are compatible, as well as Tris–HCl and/or glycerol. DMSO concentrations above 10% should be

avoided, as well ionic concentrations higher than 500 mM. Even small amounts of detergent are incompatible with SYPRO™ Orange assays in Subheadings 3.1.1 and 3.1.2, although the thermal denaturation assay in Subheading 3.1.3 is more tolerant to variations in buffer composition.

2. This fluorescently labeled phospho-tyrosine peptide is only required for Subheading 3.1.3, and the sequence should be customized based on the SH2 domain under investigation.
3. The protein stock solution can be prepared in the TSA buffer. Depending on the concentration of the original protein solution, it is not always necessary to dialyze or buffer exchange the protein into TSA buffer prior to starting the assays. For example, if the protein concentration is 200 μM in 20 mM sodium phosphate, pH 7.0, 150 mM NaCl, 5% glycerol a straight dilution into TSA buffer will likely be sufficient. This is because all of these components will be heavily diluted and will likely be negligible in the TSA assay, and this assay is mostly examining relative shifts in T_m. However, this should be properly documented in the uncommon cases of large variations arising between independent assays.
4. This assay is proposed for a total well volume of 50 μL. However, this volume can be reduced to lower levels, such as 25 μL or 10 μL, and all volumes would have to be proportionally scaled. However, any errors in pipetting can be substantially magnified with smaller volumes, and extra care should be taken to mix and collect samples.
5. This set-up allows technical/biological replicates to be maintained side-by-side, which can aid in data processing. If this is not preferred, the well placement of the serial dilutions can be inversed.
6. Ensure there are no air bubbles in the plate as this can disrupt the fluorescence reading. Similarly, ensure that all of the solution is mixed and pooled at the bottom of the well. Centrifuging the plate for 20 s can facilitate this. This centrifugation step can also be done for any step.
7. SYPRO™ Orange has an excitation and emission at 472 and 570 nm, respectively. Therefore, the filters/channels on the qPCR that most closely overlap with these wavelengths should be used. On the Bio-Rad qPCR, this can be accomplished through the FRET channel.
8. The ramp rate for the temperature can vary for different experiments, and some proteins may demonstrate a kinetic T_m, where the melting temperature varies with the ramp rate. Therefore, maintain the same rate for screening the entire library is important, and 0.5 °C represents a reasonable compromise.

9. Each qPCR provides its own data analysis software for determining the first derivative, and this is generally automatically built into the program. If the first derivative plot yields a local maximum (i.e., a two-state unfolding transition), the protein may be amenable to conventional TSA. From the series of experiments performed in Subheading 3.1, select the protein concentration that maximizes the signal-to-noise but does not oversaturate the detector and use this concentration in Subheading 3.2.
10. Any other cell line expressing STAT3 can be used. Previous validation for a function of STAT3 would be needed to facilitate later biological read outs.
11. If working with adherent cells, this step should be performed the day before to allow the cells to attach.
12. Avoid DMSO concentrations above 1% (v/v). Note that some cell lines might be more sensitive, so it is advisable to do a pilot experiment to determine maximal DMSO tolerance.
13. For statistical purposes, this experiment should be performed at least three times, on different days.

References

1. Niesen FH, Berglund H, Vedadi M (2007) The use of differential scanning fluorimetry to detect ligand interactions that promote protein stability. Nat Protoc 2:2212–2221
2. Kranz JK, Schalk-Hihi C (2011) Protein thermal shifts to identify low molecular weight fragments. Methods Enzymol 493:277–298
3. Gao K, Oerlemans R, Groves MR (2020) Theory and applications of differential scanning fluorimetry in early-stage drug discovery. Biophys Rev 12(1):85–104
4. Redhead M, Satchell R, McCarthy C et al (2017) Thermal shift as an entropy-driven effect. Biochemistry 56:6187–6199
5. de Araujo ED, Orlova A, Neubauer HA et al (2019) Structural implications of STAT3 and STAT5 SH2 domain mutations. Cancers (Basel) 11:1757
6. Orlova A, Wagner C, De Araujo ED et al (2019) Direct targeting options for STAT3 and STAT5 in cancer. Cancers (Basel) 11:1930
7. de Araujo ED, Manaswiyoungkul P, Erdogan F et al (2019) A functional in vitro assay for screening inhibitors of STAT5B phosphorylation. J Pharm Biomed Anal 162:60–65
8. De Araujo ED, Kanelis V (2014) Successful development and use of a thermodynamic stability screen for optimizing the yield of nucleotide binding domains. Protein Expr Purif 103:38–47
9. Kosack L, Wingelhofer B, Popa A et al (2019) The ERBB-STAT3 axis drives Tasmanian devil facial tumor disease. Cancer Cell 35:125–139.e9
10. Molina DM, Jafari R, Ignatushchenko M et al (2013) Monitoring drug target engagement in cells and tissues using the cellular thermal shift assay. Science (80-) 341:84–87
11. Jafari R, Almqvist H, Axelsson H et al (2014) The cellular thermal shift assay for evaluating drug target interactions in cells. Nat Protoc 9:2100–2122
12. Mahmood T, Yang PC (2012) Western blot: technique, theory, and trouble shooting. N Am J Med Sci 4:429–434
13. Delport A, Hewer R (2022) A superior loading control for the cellular thermal shift assay. Sci Rep 12:6672
14. Schneider CA, Rasband WS, Eliceiri KW (2012) NIH Image to ImageJ: 25 years of image analysis. Nat Methods 9:671–675
15. Good NE, Winget GD, Winter W et al (1966) Hydrogen ion buffers for biological research. Biochemistry 5:467–477

Chapter 12

Dissecting Selectivity Determinants of Small-Molecule Inhibitors of SH2 Domains Via Fluorescence Polarization Assays

Angela Berg, Julian Gräb, Barbara Klüver, and Thorsten Berg

Abstract

Fluorescence polarization (FP) assays can be used to identify small-molecule inhibitors that bind to SH2 domain-containing proteins. We have developed FP assays by which to identify inhibitors of the SH2 domains of the two closely-related transcription factors STAT5a and STAT5b. Point mutation of selected amino acids in the putative binding site of the protein is a valuable tool by which to gain insight into the molecular mechanism of binding. In this chapter, we describe the cloning and application of point mutant proteins in order to transfer the binding preference of selected SH2 domain-binding STAT5b inhibitors to STAT5a, with results that highlight the importance of considering a role for residues outside the SH2 domain in contributing to the binding interactions of SH2 domain inhibitors.

Key words STAT5a, STAT5b, Point mutant proteins, Fluorescence polarization, Catechol bisphosphate, Fosfosal, SH2 domains

1 Introduction

FP assays are a valuable tool for assessing small-ligand binding to SH2 domain-containing proteins [1, 2], with competitive FP assays allowing the IC_{50} values of inhibitors of peptide binding to be calculated. We have developed FP assay methodologies for identifying inhibitors of binding to the SH2 domains of the two closely-related transcription factors STAT5b [3] and STAT5a [4], which are constitutively activated in multiple human tumor types [5]. Both STAT5a and STAT5b can be activated by the erythropoietin receptor (EpoR) [6] and the assays for STAT5a and STAT5b use the same EpoR-derived peptide as the fluorescent-labeled tracer. Although the STAT5a and STAT5b proteins can compensate for each other in some situations, they also have nonoverlapping roles [7], which have not yet been fully elucidated. The SH2 domains of the two STAT5 proteins are 93% identical on the

Teresa Carlomagno and Maja Köhn (eds.), *SH2 Domains: Functional Modules and Evolving Tools in Biology*, Methods in Molecular Biology, vol. 2705, https://doi.org/10.1007/978-1-0716-3393-9_12,

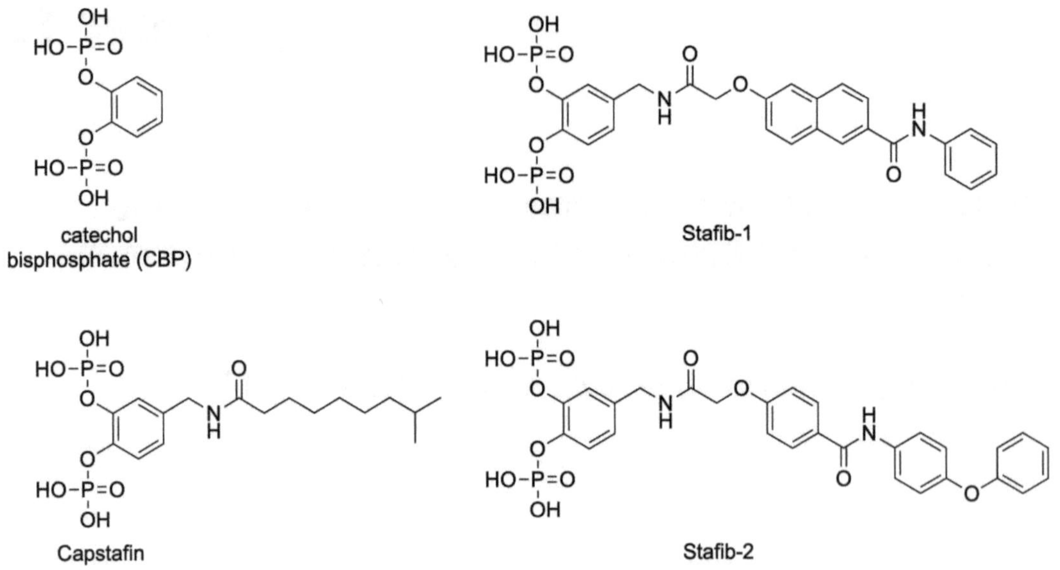

Fig. 1 Structures of catechol bisphosphate and its more active derivatives Stafib-1, Stafib-2, and Capstafin

amino acid level, yet despite this high degree of homology, catechol bisphosphate (CBP) [8] and its more active derivatives Stafib-1 [8], Stafib-2 [9], and Capstafin [10] are able to distinguish between the two proteins, with all four inhibitors showing a strong preference for STAT5b (Fig. 1).

Interaction with the conserved residue Arg618 in the SH2 domain has been shown to be important for CBP binding to STAT5b [8]. In an attempt to verify that the molecular origin of the preference of CBP and its derivatives for STAT5b also lies within the SH2 domain, we first produced a point mutant STAT5b protein in which all six of the SH2 domain residues which differ from those in the STAT5a SH2 domain were exchanged for the corresponding STAT5a residues [11]. We expected that this technique would reduce inhibitor binding to the levels observed with STAT5a. Interestingly, this was not the case, with CBP and derivatives instead being even more active against the sixfold mutant than against wild-type STAT5b in competitive inhibition FP assays [11]. We then made two single point mutant proteins in which residue 566, one of 3 divergent amino acids in the SH2-adjacent linker domain, was exchanged between STAT5a and STAT5b. STAT5b has an arginine residue at this position, the guanidinium moiety of which may be able to engage in electrostatic interactions with one of the negatively charged groups of catechol bisphosphate and its derivatives. This would not be the case for the tryptophan residue of STAT5a. In this way, we demonstrated that the actual determinant of STAT5b specificity for catechol bisphosphates is the presence of Arg566 in a flexible loop of the linker domain [11].

In this chapter, we describe the steps that are taken to clone point mutant STAT5a and STAT5b proteins via site-directed mutagenesis of the expression plasmids. We then describe how to express the proteins in bacteria, followed by protein purification via chromatography columns using the C-terminal His-tag. Finally, we describe the methodology by which to analyze wild-type and mutant proteins in FP assays.

2 Materials

2.1 Site-Directed Mutagenesis

1. Ampicillin: dissolve in ultrapure water at 100 mg/mL to obtain a 1000× stock solution. Sterile filter and store at −20 °C in appropriate aliquots. Working concentration is 100 μg/mL.
2. Autoclaved Luria broth (LB) medium is supplemented with ampicillin at working concentration when required, directly before use.
3. Autoclaved LB agar (3 g agar per 200 mL LB) is supplemented with antibiotic briefly before pouring plates, after the solution has cooled to 55 °C.
4. The PCR reaction is run using Pfu Ultra II polymerase and 10× Pfu Ultra II buffer (both from Agilent) and 10 mM dNTPs (e.g., New England Biolabs).
5. The enzyme DpnI (we used the product from New England Biolabs) is required to digest the template DNA after the PCR.
6. A 1% agarose-TBE gel is used to verify the presence of plasmid bands after PCR. This requires 6× loading dye and a standard DNA ladder (e.g., SmartLadder from Eurogentec).

2.2 Protein Expression and Purification

1. Chloramphenicol: dissolve in ethanol at 34 mg/mL to obtain a 1000× stock solution. Store at −20 °C without aliquoting. Working concentration is 34 μg/mL.
2. Isopropyl-β-D-thiogalactopyranoside (IPTG): dissolve in ultrapure water at 1 M to obtain a 1000× stock solution. Sterile filter and store at −20 °C in 1 mL aliquots. Working concentration is 1 mM.
3. Phenylmethanesulfonyl fluoride (PMSF): dissolve in ethanol at 100 mM to obtain a 100× stock solution. Store at −20 °C without aliquoting. Working concentration is 1 mM.
4. $NiSO_4$ charge buffer: dissolve $NiSO_4$ in ultrapure water to give a 50 mM working solution. Sterile filter and store at room temperature.
5. Binding buffer: prepare an 8× stock with 4 M NaCl, 160 mM Tris and 80 mM imidazole in ultrapure water, adjusted to pH 7.9 with HCl. Sterile filter and store at 4 °C. Working

binding buffer contains 0.5 M NaCl, 20 mM Tris and 10 mM imidazole, pH 7.9.

6. Wash buffer: prepare an 8× stock with 4 M NaCl, 160 mM Tris and 480 mM imidazole in ultrapure water, adjusted to pH 7.9 with HCl. Sterile filter and store at 4 °C. Working wash buffer contains 0.5 M NaCl, 20 mM Tris and 60 mM imidazole, pH 7.9.
7. Elution buffer: prepare a 4× stock with 2 M NaCl, 80 mM Tris and 2 M imidazole in ultrapure water, adjusted to pH 7.9 with HCl. Sterile filter and store at 4 °C. Working elution buffer contains 0.5 M NaCl, 20 mM Tris and 0.5 M imidazole, pH 7.9.
8. Dialysis tubing is chosen according to the size of the protein, with a molecular weight cut-off at or below half of the molecular weight of the protein. Prepare dialysis tubing according to the manufacturer's instructions.
9. Dithiothreitol (DTT): dissolve in ultrapure water at 1 M concentration to obtain a 1000× stock solution. Sterile filter and store in 1 mL aliquots at −20 °C. Working concentration is 1 mM.
10. Dialysis buffer: 100 mM NaCl, 50 mM HEPES, 1 mM EDTA, 10% glycerol (v/v), 0.1% NP-40 substitute (v/v), pH 7.5 in ultrapure water. Sterile filter and store at 4 °C. Add 1 mL of 1 M DTT just before use (final concentration 1 mM DTT).

2.3 Fluorescence Polarization Assays

1. The tracer used in STAT5a and STAT5b FP assays has the sequence 5-carboxyfluorescein-GY(PO_3H_2)LVLDKW. Dissolve at 2 mM in dry DMSO. Prepare 20 μM stock solution by diluting with dry DMSO. Store at −20 °C in 10 μL aliquots.
2. FP assay buffer: prepare a 2× stock in ultrapure water with 100 mM NaCl, 20 mM Tris, 2 mM EDTA, and 0.2% NP-40 substitute (v/v), pH 8.0. Sterile filter and store at 4 °C. Prepare the working solution just before use by diluting with ultrapure water and adding 1 mM DTT. Working assay buffer contains 50 mM NaCl, 10 mM Tris, 1 mM EDTA, 0.1% NP-40 substitute, 1 mM DTT, pH 8.0.
3. Compound preparation for inhibition curves: weigh out accurately 0.5–1.0 mg compound and dissolve in dry DSMO at 10 mM (or 5 mM/2.5 mM for more active compounds). Pipette up and down to dissolve and incubate at room temperature if necessary. The dissolved compound can be stored at −20 °C in appropriate aliquots.

3 Methods

3.1 Site-Directed Mutagenesis

1. Select a plasmid template to be used as the starting point for site-directed mutagenesis. Here, we used two previously-described plasmids that contain either amino acids (aa) 137-707 of human STAT5a or aa 137-703 of human STAT5b in a modified pMAL-TEV2 vector incorporating sequence from the plasmid pQE70. This vector carries an N-terminal MBP-tag, a C-terminal 6× His-tag, and an ampicillin resistance gene [3, 8]. Choose which residues are to be mutated. An alignment of the protein sequences of STAT5a and STAT5b indicates that the SH2 domains of the two proteins differ at 6 positions: aa 636 (Pro in STAT5a, Gln in STAT5b), 639 (Asn in STAT5a, Met in STAT5b), 640 (Leu in STAT5a, Phe in STAT5b), 644 (Lys in STAT5a, Met in STAT5b), 664 (Ser in STAT5a, Asn in STAT5b), and 679 (Phe in STAT5a, Tyr in STAT5b). We therefore chose to exchange the amino acids at all six divergent positions in STAT5b for the corresponding residues in STAT5a, resulting in a STAT5b protein containing the STAT5a SH2 domain. Two additional proteins with a single crossover mutation at the divergent position 566 (Trp in STAT5a, Arg in STAT5b) in the linker domain were also cloned.
2. To introduce a single mutation, design pairs of PCR primers incorporating the desired mutation at or near the center of an approximately 40 base sequence. We use the free software PrimerX, at Bioinformatics.org, to assist with primer design. To mutate multiple amino acids within one protein, use successive rounds of mutagenesis that introduce the mutations sequentially (*see* **Note 1**).
3. The PCR reaction used is an adapted version of the Agilent QuikChange site-directed mutagenesis protocol, with a total reaction volume of 50 μL. Pipette together 5 μL 10× Pfu Ultra II buffer, 3 μL DMSO, 1 μL template plasmid DNA (approximately 100 ng), 1.25 μL of each primer (10 μM stock in water), 1 μL 10 mM dNTPs, and 36.5 μL autoclaved ultrapure water, then add 1 μL Pfu Ultra II polymerase (Agilent) and pipette gently to mix. After a 95 °C hot start step for 1 min, perform 25 cycles of: 95 °C for 50 s, 60 °C for 50 s, and 68 °C for 4 min. Add a final 5 min extension step at 68 °C after cycling is complete. These parameters work well for the majority of primers and templates. After the final extension step, add 1 μL DpnI to the reaction, pipette gently to mix and incubate at 37 °C for 1 h to digest the template DNA.

4. PCR products are run on a diagnostic 1% agarose-TBE gel to verify the presence of plasmid bands. Combine 5 μL PCR product with 5 μL ultrapure water and 2 μL 6× loading dye and include a lane with 5 μL standard DNA ladder (*see* **Note 2**).
5. Transform the PCR product into competent bacteria. Incubate 50 μL of competent DH5α cells with 3 μL of PCR product for 30 min on ice. Heat-shock in a 42 °C water bath for 45 s and then immediately return to ice for 2 min. Add 500 μL of LB medium without antibiotic and incubate at 37 °C for 1 h, with shaking. Centrifuge briefly and discard excess medium, then resuspend the bacteria in the remaining volume and plate onto a single LB-agar plate containing ampicillin. Incubate overnight at 37 °C.
6. Ensure that there are no overlapping colonies by preparing streak plates from several isolated colonies per transformation plate, using a sterile loop and LB-agar plates containing ampicillin. Incubate at 37 °C overnight.
7. Inoculate one 2 mL overnight culture per streak plate, using ampicillin-supplemented LB medium. Incubate at 37 °C overnight with 180 rpm shaking and prepare minipreps using a commercial miniprep kit.
8. Sequence the clones using primers which cover the full length of the open reading frame. Choose a clone in which the desired mutation has been incorporated (without further mutations). It is generally not necessary to sequence the remaining part of the vector.

3.2 Protein Expression and Purification

1. Transform the plasmid into a suitable bacterial host strain for protein expression. Incubate 9 μL of competent Rosetta BL21DE3 cells (Novagen) with 10 ng plasmid in 1 μL ultrapure water for 5 min on ice. Heat-shock in a 42 °C water bath for 30 s and then immediately return to ice for 2 min. Add 50 μL of LB medium without antibiotic and incubate at 37 °C for 1 h, with shaking. Plate the entire sample onto a single LB-agar plate containing ampicillin and chloramphenicol (*see* **Note 3**). Incubate at 37 °C overnight. Pick a single isolated colony and prepare a streak plate using a sterile loop and an LB-agar plate containing ampicillin and chloramphenicol.
2. Prepare starter cultures for protein expression using LB medium supplemented with ampicillin and chloramphenicol. Inoculate 40 mL starter culture in a 100 mL flat-bottomed flask with eight colonies from the streak plate. Incubate at 37 °C overnight with 180 rpm shaking.
3. Inoculate 1 L LB medium supplemented with ampicillin and chloramphenicol with 40 mL starter culture. Since the STAT5a and STAT5b expression plasmids do not give high protein

yields, it is advisable to prepare several 1 L cultures per protein. Use 3 L flasks with baffles. Add three drops of molecular biology grade antifoam emulsion per flask. Incubate the culture at 37 °C with 180 rpm shaking until an OD_{600} of 0.4 is reached, as measured by spectrophotometer. At this point, the culture is transferred to a cooled incubator at 18 °C, with 180 rpm shaking, and grown further to an OD_{600} between 0.7 and 0.8.

4. Induce protein expression by adding 1 mL of 1 M IPTG (final concentration 1 mM). Continue shaking at 18 °C for 5 h before harvesting.
5. Transfer the bacterial culture to autoclaved buckets with lids and centrifuge at 4500 rpm (maximum 4400 g) for 30 min at 4 °C. Discard the supernatant. Weigh each bucket before use and again together with the pellet, in order to calculate the pellet weight. Freeze the pellets at −80 °C to assist in cell lysis. Pellets are stable at −80 °C for several months.
6. Thaw the pellets on ice and resuspend in 10 mL binding buffer per 1 g pellet weight by pipetting the liquid repeatedly. All steps from this point on take place either on ice or at 4 °C. Lyse the cells either by homogenization, using three passes through a cooled homogenizer (we used the Emulsiflex-C5 homogenizer from Avestin) at 10,000–12,000 psi, or sonication on ice using a Sonopuls HD70 sonicator (Bandelin) for 7× 2 min at standard intensity, with a 70/30 pulse/pause interval and at least 2 min between sonication steps. For sonication, separate the volume into multiple 50 mL tubes containing not more than 40 mL each. Add a final concentration of 1 mM PMSF after lysis.
7. Centrifuge the lysate in 50 mL tubes at 10,500 rpm (maximum 12,800 g) for 30 min at 4 °C to pellet cell debris. Filter the cleared supernatant through a 0.45 μm filter syringe-tip filter to remove any remaining insoluble components.
8. Prepare Poly-Prep chromatography columns (Bio-Rad) by adding 1.25 mL of 50% His·Bind resin (Merck Millipore) to give a bed volume of 0.625 mL. One column is sufficient for 40–60 mL of cleared lysate; prepare as many columns as necessary for the total lysate volume. After the liquid has run through, rinse each column with 1.9 mL (3 bed volumes) of autoclaved ultrapure water and charge with 3.2 mL (5 bed volumes) of $NiSO_4$ charge buffer. Finally, apply 1.9 mL of binding buffer.
9. Apply the cleared lysate to the columns at 4 °C. Keep 20 μL each of the cleared lysate, the column run-through fraction and each subsequent fraction for analysis via SDS-PAGE. Add 6.25 mL of cold binding buffer containing 1 mM PMSF, to

wash remaining unbound protein from the column, followed by 3.8 mL of wash buffer containing 1 mM PMSF. Wait after each step until the liquid has all run into the column before adding the next buffer, and collect each fraction in a separate 15 mL tube. Finally, elute the bound protein from the column by adding 1.9 mL elution buffer containing 1 mM PMSF.

10. Dialyze the elution fraction against at least 100-fold excess of binding buffer to remove excess imidazole. Do not add PMSF to buffers from this point on. Use prepared dialysis tubing and change the buffer once, with each incubation lasting at least 4 h at 4 °C, with gentle stirring. Dilute the dialyzed protein five-fold with binding buffer before applying to the second His·Bind resin column.
11. Prepare a single chromatography column with 0.625 mL His·Bind resin bed volume as described in **step 8**. Apply the diluted protein to the column at 4 °C, followed by 6.25 mL of cold binding buffer and 3.8 mL of wash buffer, collecting each fraction in a separate 15 mL tube. Finally, elute the bound protein from the column by adding 0.625 mL elution buffer. Repeat the elution step twice, to give a total of three elution fractions.
12. Run a 10% polyacrylamide SDS gel to assess the protein content of each fraction, and combine clean elution fractions with sufficient protein for dialysis.
13. Dialyze the combined elution fractions against at least 100-fold excess of dialysis buffer containing 1 mM DTT. Use prepared dialysis tubing and change the buffer twice, with each incubation lasting at least 4 h at 4 °C, with gentle stirring.
14. Determine the concentration of the protein using a reducing agent-compatible BCA Assay (Pierce). Aliquot the protein into suitable aliquot volumes (usually 30 or 50 μL), snap freeze in liquid nitrogen and store at −80 °C. Proteins are stable at −80 °C for several years.

3.3 Fluorescence Polarization Assays

3.3.1 Binding Assays

1. Before a competitive inhibition assay can be carried out, it is necessary to establish the dissociation constant (K_d) for peptide binding, which corresponds to the protein concentration at which half-maximal binding occurs. Introducing a mutation to the protein is likely to change the K_d value. It is therefore necessary to make a new binding curve for each expressed protein.
2. Prepare a stock of working FP assay buffer containing 2% DMSO to use during preparation of the binding curve (*see* **Note 4**). Leave at room temperature, and perform each of the following steps at room temperature.

3. Thaw an aliquot of protein on ice and dilute with FP assay buffer to give 120 μL of 3840 nM protein in a 1.5 mL tube. Keep the protein stock on ice (*see* **Note 5**). Prepare a series of 9 twofold dilutions at room temperature, transferring 60 μL protein solution to 60 μL fresh FP assay buffer at each step (1920, 960, 480, 240, 120, 60, 30, 15, 7.5 nM; these concentrations represent 1.5× the final concentration). This is sufficient to generate a single binding curve. Note that STAT5a tends to express less well than STAT5b, and for this reason, it may be preferable to take 1920 nM (1280 nM final) as the highest concentration for STAT5a-binding curves if protein stocks are limited. Incubate the dilution series for 1 h at room temperature. In the meantime, transfer 50 μL of each dilution step into a new 1.5 mL tube and prepare a blank (75 μL of FP assay buffer) and a negative control (which will receive peptide only; add 50 μL of FP assay buffer at this stage).
4. Prepare a 30 nM stock of the 5-carboxyfluorescein (CF)-labeled peptide by adding 1.5 μL of 20 μM stock solution to 1 mL FP assay buffer. Mix by pipetting or vortexing briefly at low speed. If the 30 nM stock is to be stored for more than a few minutes, wrap the tube in foil to protect the solution from excessive light exposure. Add 25 μL of the 30 nM peptide solution to the negative control and each of the protein dilutions and mix either by pipetting up and down or by brief vortexing at low speed. This represents the zero timepoint; the degree of FP is read after 15 min and 1 h, and at further hourly intervals as required (*see* **Note 6**).
5. Transfer three aliquots of 20 μL from each sample into three adjacent wells in a flat-bottomed black polystyrene 384-well plate. The average reading of the three wells is taken to produce a single curve; this minimizes errors during the FP reading.
6. To read, excite the fluorophore with linearly polarized light of 485 nm wavelength using an FP-equipped plate reader such as the Tecan Infinite F500. Read the fluorescence emission at 535 nm parallel and perpendicular to the plane of excitation, with the buffer-containing wells defined as the blanks to allow the instrument to subtract background fluorescence before calculating the degree of fluorescence polarization. Cover the plate with a black lid and keep at room temperature between measurements.
7. Subtract the average negative control reading (peptide only) from each averaged protein dilution reading and plot the binding curve in Origin (OriginLab) or similar graph analysis software. Fit a suitable 4-parameter curve through the data points (in Origin, a Hill1 curve) and determine the K_d value of the protein–peptide interaction. K_d values are presented as the

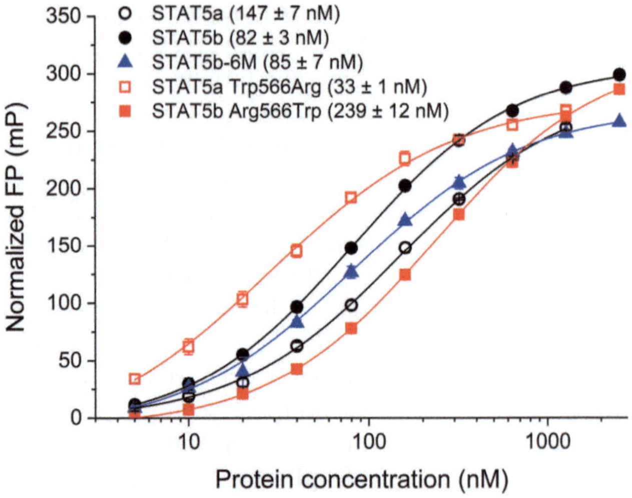

Fig. 2 Binding curves for the interaction of the peptide 5-CF-GY(PO_3H_2)LVLDKW with the wild-type and point mutant proteins. The K_d values for each curve are given in brackets. (The underlying data have previously been published [11])

mean of 3 independent experiments, plus/minus standard deviation. Verify that the assay window is large enough, and the binding curve is stable with respect to time (*see* **Note 6**). Figure 2 shows the binding curves obtained for wild-type STAT5a and STAT5b, and each of the mutant proteins.

3.3.2 Competitive Inhibition Assays

1. The protein concentration to be used for competitive inhibition assays should correspond to, or only slightly exceed, the K_d value of the binding curve for that particular protein–peptide combination (*see* **Note 7**). The protein concentrations to be used for the inhibition curves are thus: STAT5a 147 nM, STAT5b 82 nM, STAT5b-6M 85 nM, STAT5a Trp566Arg 33 nM, and STAT5b Arg566Trp 239 nM.
2. Prepare a 1.02× stock of FP assay buffer. This will later be diluted to 1× by addition of compound in DMSO, which contributes 2% DMSO to the final composition. Leave at room temperature, and perform each of the following steps at room temperature.
3. Prepare a dilution series of the compound in dry DMSO, beginning with 10 mM stock solution and transferring 10 μL each time to a tube containing fresh DMSO and continuing until there are 12–14 dilutions in total. These will be diluted 50-fold to give the final concentrations (10 mM corresponds to 200 μM final, 5 mM to 100 μM, and so on). For compound–protein combinations with lower IC_{50} values, reduce the stock concentration to 5 or 2.5 mM (to give an inhibition curve up to 100 or 50 μM).

4. Transfer 1.6 μL of each dilution step into a new 1.5 mL tube for the assay. Prepare three extra tubes with 1.6 μL DMSO each for a blank (FP assay buffer alone), a negative control (buffer plus peptide) and a positive control (protein plus peptide).
5. Prepare a 2 mL solution of protein in FP assay buffer at 1.05 times the desired final concentration. This will be diluted to 1× by addition of compound and peptide. Mix by pipetting or brief vortexing. Pipette 76 μL of the protein dilution into the positive control and each of the compound tubes. Add 76 μL FP assay buffer to the negative control and 78.4 μL to the blank. Mix by brief vortexing at low speed, ensuring that the solution remains in the bottom of the tube afterward. Incubate for 1 h at room temperature.
6. Prepare a 333 nM peptide stock by adding 2 μL of 20 μM peptide stock solution to 120 μL of 1.02× FP assay buffer and mixing by pipetting or brief vortexing. After a 1 h incubation of compound and protein, add 2.4 μL of the 333 nM peptide to the negative control, the positive control and each of the compound dilutions.
7. Transfer 3 × 20 μL of each sample to a 384-well plate and read the fluorescence polarization as for the binding curve. Read after 15 min and 1 h, and at further 1 h intervals as required.
8. Subtract the average negative control reading (peptide only) from the averaged positive control and each averaged compound dilution reading. Calculate the percentage remaining protein–peptide binding at each concentration of inhibitor. For this, use the equation of the binding curve of the protein–peptide pair (*see* **Note 8**). Enter the FP value (y) in the presence of inhibitor into the binding curve equation and solve for x to give the effective remaining protein available for peptide binding. Convert this value into a percentage of the maximum possible (represented by the positive control). Plot the resulting inhibition curve in Origin or similar software, with a logarithmic x-axis, and fit a suitable 4-parameter curve through the data points (in Origin, a Logistic curve). To calculate the half-maximal inhibitory concentration (IC_{50}) of the inhibitor with the protein used, solve the inhibition curve equation for x at $y = 50$. The IC_{50} is expressed as the mean of 3 independent experiments, plus/minus standard deviation. Figure 3 shows the inhibition curves obtained for each of the proteins with CBP (*see* **Note 9**).

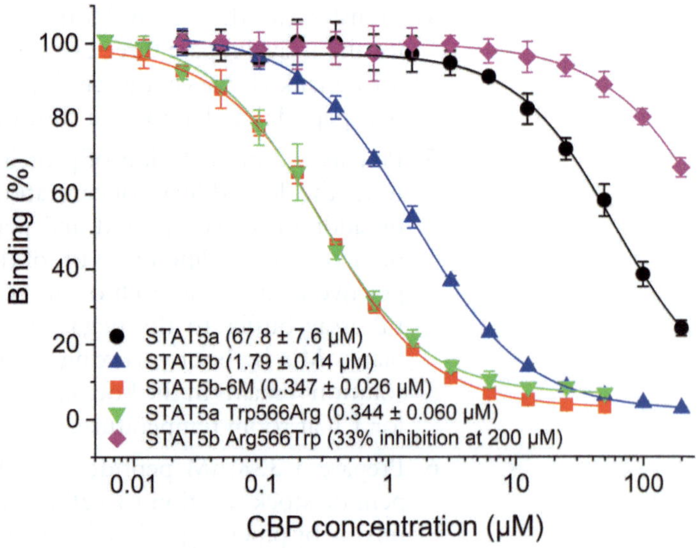

Fig. 3 Inhibition of the wild-type and mutant proteins by the STAT5b inhibitor CBP. Note that the dilution series for CBP with the STAT5b-6M and STAT5a Trp566Arg proteins begin and end two dilution points lower than with the wild-type proteins. Because CBP is more active against these two mutants, the reduced concentrations are necessary to ensure that the beginning of the curve has a sufficient plateau for a good curve fit. The IC_{50} values for inhibition of each protein by CBP are given in brackets. (The underlying data have previously been published [11])

4 Notes

1. When designing primers for subsequent rounds of mutagenesis when multiple point mutations are to be introduced into a protein, it is important to remember to incorporate the mutations which have already been made into each successive primer sequence. Attempting to introduce too many mutations during a single round of mutagenesis can impair binding of the primer to the template DNA.
2. The PCR product at this point is not linearized, which prevents an accurate assessment of the size of the plasmid as compared to the ladder. However, any plasmid bands present represent a PCR product, since the template has been DpnI-digested. If it is necessary to verify the size of the plasmid, digest 5 μL of the PCR product first with an appropriate restriction enzyme.
3. The Rosetta expression strain supplies tRNAs for codons which are rarely used in *E. coli*, and these are encoded by a plasmid with a chloramphenicol resistance gene.

4. The FP assay buffer used in the binding curve contains 2% DMSO. This is necessary for consistency with the competitive inhibition assay, in which the compound used introduces 2% DMSO.
5. Remaining unused protein can usually be stored at 4 °C for a few days after thawing before losing activity. Activity can be verified after storage by preparing a blank, negative and positive control and measuring the assay window. If the window is not significantly smaller than with fresh protein, the stored protein can still be used.
6. Some point mutations affect the stability of a protein, which may detract from the quality of the competitive inhibition assay. To verify that this is not the case, the binding curve with the peptide should be measured over several hours (e.g., up to and including the 4 h timepoint). The K_d of the curve should remain stable over this time period. Alternatively, if a point mutation is within the binding pocket of the peptide used in the FP assay, it is possible for peptide binding to be significantly reduced or even lost such that the binding curve fails to produce a sufficient assay window at any timepoint. In either of these cases, the mutant protein is not suitable for use.
7. This is important to ensure that the assay window (the difference between the FP readings for unbound peptide and peptide in the presence of protein) is of a reasonable size, ideally at least 100 mP, but to avoid approaching the saturation of binding observed at higher protein concentrations.
8. It is important to use the binding curve equation from the corresponding timepoint to avoid introducing errors in case of minor loss of activity of the binding curve over time.
9. Similar results were obtained for the other CBP-derived inhibitors [11]. These results indicate that the presence of arginine at position 566 is the determining factor for binding specificity, despite the fact that the CBP-based inhibitors compete with the labeled peptide for binding to the SH2 domain. This highlights the importance of considering that inhibitors of the SH2 domain may also interact with amino acid residues in other protein domains in close spatial proximity to the inhibitor binding site when choosing residues to mutate. For inhibitors of STAT proteins in particular, the creation of a binding pocket with the flexible loop of the linker domain surrounding the STAT5a and STAT5b residue 566 needs to be taken into consideration.

Acknowledgments

This work was generously supported by the Deutsche Forschungsgemeinschaft (BE 4572/4-1) and the European Union and the Free State of Saxony, European Regional Development Fund.

References

1. Owicki JC (2000) Fluorescence polarization and anisotropy in high throughput screening: perspectives and primer. J Biomol Screen 5: 297–306
2. Lea WA, Simeonov A (2011) Fluorescence polarization assays in small molecule screening. Expert Opin Drug Discov 6:17–32
3. Müller J, Schust J, Berg T (2008) A high-throughput assay for signal transducer and activator of transcription 5b based on fluorescence polarization. Anal Biochem 375:249–254
4. Berg A, Berg T (2017) A small-molecule screen identifies the antitrypanosomal agent suramin and analogues NF023 and NF449 as inhibitors of STAT5a/b. Bioorg Med Chem Lett 27: 3349–3352
5. Miklossy G, Hilliard TS, Turkson J (2013) Therapeutic modulators of STAT signalling for human diseases. Nat Rev Drug Discov 12: 611–629
6. Quelle FW, Wang D, Nosaka T et al (1996) Erythropoietin induces activation of Stat5 through association with specific tyrosines on the receptor that are not required for a mitogenic response. Mol Cell Biol 16:1622–1631
7. Hennighausen L, Robinson GW (2008) Interpretation of cytokine signaling through the transcription factors STAT5A and STAT5B. Genes Dev 22:711–721
8. Elumalai N, Berg A, Natarajan K et al (2015) Nanomolar inhibitors of the transcription factor STAT5b with high selectivity over STAT5a. Angew Chem Int Ed 54:4758–4763
9. Elumalai N, Berg A, Rubner S et al (2017) Rational development of Stafib-2: a selective, nanomolar inhibitor of the transcription factor STAT5b. Sci Rep 7:819
10. Elumalai N, Berg A, Rubner S, Berg T (2015) Phosphorylation of capsaicinoid derivatives provides highly potent and selective inhibitors of the transcription factor STAT5b. ACS Chem Biol 10:2884–2890
11. Gräb J, Berg A, Blechschmidt L et al (2019) The STAT5b linker domain mediates the selectivity of catechol bisphosphates for STAT5b over STAT5a. ACS Chem Biol 14:796–805

Chapter 13

Lipid Binding of SH2 Domains

Wonhwa Cho, Kyli Berkley, and Ashutosh Sharma

Abstract

The Src homology 2 (SH2) domain is a modular protein interaction domain that specifically recognizes the phosphotyrosine (pY) motif of a target molecule. We recently reported that a large majority of human SH2 domains tightly bind membrane lipids, and many show high lipid specificity. Most of them can bind a lipid and the pY motif coincidently because their lipid-binding sites are topologically distinct from pY-binding pockets. Lipid binding of SH2 domain-containing kinases and phosphatases is functionally important because it exerts exquisite spatiotemporal control on protein–protein interaction and cell signaling activities mediated by these proteins. Here, we describe two assays, surface plasmon resonance analysis and fluorescence quenching analysis, which allow quantitative determination of the affinity and specificity of SH2–lipid interaction and high-throughput screening for SH2 domain–lipid-binding inhibitors.

Key words SH2 domains, Lipids, Plasma membrane, Surface plasmon resonance, Fluorescence quenching

1 Introduction

Cell regulation, including cell signaling, involves myriad of protein–protein interactions (PPI) [1–4]. PPI involves either stable interactions among folded domains that establish macromolecular complexes or transient interactions among those proteins mediating cellular signaling pathways and regulatory processes. The latter typically involves small (≈ 100 amino acids) modular protein interaction domains (PIDs), such as Src-homology 2 (SH2), Src-homology 3 (SH3), and *P*SD95, *D*lg1, *Z*O-1 (PDZ) domains [5–7], and short linear motifs [8]. Physiological significance of PID-mediated PPI has been well supported by a large number of PIDs encoded by human genome [5–7] and many human diseases, including cancer, caused by dysfunctional PID-mediated PPI [8, 9]. PIDs can recognize either unmodified linear motifs (e.g., PDZ and SH3 domains) or posttranslationally modified motifs (e.g., SH2 domain).

Teresa Carlomagno and Maja Köhn (eds.), *SH2 Domains: Functional Modules and Evolving Tools in Biology*, Methods in Molecular Biology, vol. 2705, https://doi.org/10.1007/978-1-0716-3393-9_13,

The SH2 domain is the first identified PID that has long served as a model system for studying cellular PPI [6, 10, 11]. It specifically recognizes the phosphotyrosine (pY) motif of target molecule during diverse pY signaling pathways [10]. The pY signaling system in humans consists of ≈ 90 protein tyrosine kinases (PTKs) serving as signal writers, 107 protein tyrosine phosphatases (PTP) as signal erasers, and 111 SH2 domain-containing proteins as major signal readers, creating a complex network of signaling molecules [6]. While PTKs and PTPs govern the amplitude and the duration of pY signaling, SH2 domain-containing proteins control the specificity of pY signaling as they link various PTKs to downstream molecules and nucleate signalosomes [10]. Structural analyses of SH2 domains and their complexes have revealed that they specifically recognize pY and a few residues immediately C-terminal to pY using a conserved pY-binding pocket and a variable secondary binding site, respectively [6]. However, quantitative analysis has shown that SH2 domains have variable affinity and a significant degree of promiscuity for the pY motifs [12–14]. This may be necessary for reversibility and redundancy of cell signaling pathways but raises a fundamental question as to how high-fidelity cellular function and regulation can be achieved through SH2 domain-mediated PPI. Although protein compartmentalization [15, 16], signalosome formation [17, 18], and pY-independent secondary protein interactions [19] have been suggested to augment the specificity of SH2 domain-mediated PPI, a universal mechanism that confers high spatiotemporal specificity on SH2 domain-mediated PPI remains elusive.

Recently, we and others have shown that lipids can directly interact with PIDs, most notably PDZ domains [20–24] and SH2 domains [25–27]. Our genome-wide characterization of human SH2 domains showed that ≈ 90% of tested SH2 domains tightly bind plasma membrane (PM) lipids and that many of them have high phosphoinositide (PtdIns*Ps*) specificity [26]. Most of them bind PtdIns*Ps* using the sites that are separate from pY-binding pockets and can thus coincidently bind a PtdIns*P* and a pY motif [26]. The morphology and the molecular location of their lipid-binding sites are highly variable, allowing for flexible lipid-mediated control of pY signaling pathways [26]. Functional studies of diverse SH2 domain-containing proteins, including PTKs and PTPs, showed that lipid binding of their SH2 domain is crucial for their cell signaling activities [25–27]. These results show how lipids enable exquisite spatiotemporal control of PPI and cell signaling activities through their interaction with the SH2 domain.

Lipid-binding SH2 domains are identified in PTKs, PTPs, and adaptors/scaffolds [26], suggesting that lipids may control not only the spatiotemporal specificity but also the amplitude and the duration of pY signaling. In particular, a large number of SH2 domain-containing PTKs have been implicated in human diseases, including cancer, diabetes, and immune disorders, and thus have been targets for active drug development [28]. Since SH2–lipid-binding controls not only the enzymatic activity of these proteins but also their scaffolding function [25–27], targeting lipid binding of SH2 domains would be a promising and viable approach to developing specific and potent kinase inhibitors [29]. Despite this potential, quantitative analysis of membrane lipid binding of SH2 domains is technically challenging for various reasons. First, it is often difficult to express and purify isolated SH2 domains in sufficient quantity for biophysical analysis due to their low stability and solubility. Secondly, lipid–protein interaction is more difficult to quantitatively measure and analyze than other interactions, such as protein–protein or protein–ligand interactions. Here, we describe generally applicable methods for expressing and purifying recombinant SH2 domains, quantitatively analyzing their interaction with membrane lipids, and screening for and characterizing SH2 domain–lipid-binding inhibitors.

1.1 General Experimental Strategies

1.1.1 SH2 Domain Expression

Bacterial expression of isolated SH2 domains often suffers from a low expression yield because exposed hydrophobic residues can lower solubility of the proteins [26]. Thus, optimization of the SH2 domain constructs and protein expression conditions is essential for production of sufficient SH2 domain proteins for biochemical and biophysical analyses. We have found that expressing SH2 domains as a fusion protein with enhanced green fluorescence protein (EGFP) improves the expression yield for most SH2 domains as the exceptionally stable EGFP fold may serve as a chaperone [26]. Also, since EGFP does not have affinity for membrane lipids [30], the EGFP tag with a proper linker does not affect the membrane binding of a diverse group of the fusion partner proteins tested so far [26, 30]. For most SH2 domains, N-terminal and C-terminal EGFP tags have essentially the same effect [26]. For some SH2 domains, however, N-terminal and C-terminal EGFP tags may differentially affect the structure and function of the proteins. For those proteins, the location of the EGFP tag and the sequence and length of the linker between EGFP and the host protein should be adjusted for optimal protein expression. EGFP-tagged SH2 domains are typically expressed with a His_6 tag and purified by affinity chromatography using a Ni-chelate column. The length of the His tag may be extended up to 10 to increase the affinity of the fusion protein to the resin. For some proteins, most notably electrostatically neutral proteins, the His_6 tag may significantly enhance their binding to anionic membranes. In such a case, the His_6 tag should be enzymatically removed after affinity purification or the assay performed at slightly elevated pH (e.g., pH = 7.8) to ensure that all His side chains are deprotonated. Alternatively, the protein can be expressed as a glutathione-S-transferase (GST)-tagged protein, purified by glutathione-based affinity chromatography, and the GST tag removed before the assay.

1.1.2 Quantitative Lipid–SH2 Domain-Binding Assays

Membrane binding of soluble proteins has been measured by various biochemical and biophysical methods [31, 32]. Sedimentation assays using lipid vesicles [33] have been most commonly used to assess membrane binding of proteins. However, difficulties associated with accurately quantifying membrane-bound versus free proteins and variable pelleting efficiency associated with different lipid vesicles have limited their utility. The lipid overlay assay has also been commonly used due to its convenience but it suffers from many drawbacks, including low sensitivity, poor reliability, and an inability to yield quantitative information [34]. Also, lipids are presented in a poorly defined, nonphysiological state in this assay. The surface plasmon resonance (SPR) analysis allows robust quantitative analysis of membrane–protein interactions and has thus been a mainstay in biophysical analysis of membrane–protein interaction [35, 36]. It monitors in real-time the change in the reflective index of the medium next to the sensor chip surface in terms of resonance units (RU) as the analyte traveling through a microfluidic channel binds the surface of a sensor chip. Main advantages of the SPR analysis include high sensitivity, no requirement for protein labeling, and an ability to provide real-time kinetic information. However, it also has drawbacks, including the necessity of expensive commercial instrumentation and sensor chips, rigorous controls to eliminate nonspecific binding, uncertainty about the physical nature of lipids coated on the sensor chip, and binding measurements under nonequilibrium conditions. Also, except for high-end commercial models, most SPR instruments are not suited for high-throughput lipid–protein-binding analysis and inhibitor/activator screening. As an alternative, we developed a high-throughput membrane-binding assay that is based on fluorescence quenching of fluorescence proteins, such as EGFP, fused to a membrane-binding protein [30, 37]. The assay utilizes a dark quencher-containing lipid, such as N-dimethylaminoazobenzenesulfonyl-phosphatidylethanolamine (dabsyl-PE), incorporated in lipid vesicles, which quenches the fluorescence intensity of EGFP as an EGFP-tagged protein binds the membrane. This simple and rapid assay can be optimized for sensitive, accurate, and reproducible quantitative determination of lipid affinity and specificity of diverse proteins as well as for high-throughput screening of small molecules that can modulate their membrane binding [30, 37]. However, the method may suffer from a low dynamic range for data analysis for those EGFP-tagged proteins showing low quenching efficiency upon membrane binding [30, 37].

1.1.3 High-Throughput Screening for Lipid–SH2 Domain Inhibitors

The fluorescence quenching-based lipid-binding assay has been optimized for high-throughput screening of small molecule libraries for their inhibitory activity against the SH2 domain–lipid binding [29]. An inhibitor of EGFP–SH2–lipid binding would relieve the fluorescence quenching of EGFP caused by membrane binding of an EGFP-SH2 domain and thus enhance the fluorescence emission intensity of EGFP. The key to success in high-throughput screening by this assay is to optimize the conditions to initially drive membrane binding of the EGFP-SH2 domain while maintaining binding loose enough to be reversed by an accessible dose (e.g., 5 μM) of inhibitors. The *Z'* factor is a statistical benchmark to assess the suitability of an assay for high-throughput screening and is a measure of the reproducibility in the difference in the dynamic range of the assay across many wells. An assay with ideal reproducibility displays a *Z'* factor of 1 and a *Z'* factor greater than 0.5 is considered acceptable for a good high-throughput assay. The *Z'* factor > 0.7 was achieved for high-throughput screening of EGFP–SH2–lipid-binding inhibitors [29].

2 Materials

2.1 SH2 Domain Expression and Purification

1. Growth media: Luria Broth containing 50 μg/mL kanamycin or 100 μg/mL ampicillin.
2. Isopropyl β-D-1-thiogalactopyranoside stock solution: 1 M in water.
3. Ni-NTA-affinity agarose resin (ProteIndex HiBond Ni-NTA Agarose Settled Resin).
4. 30-mL disposable plastic gravity flow column with 20–50 μM frit.
5. Amicon Ultra 0.5 mL Centrifugal Filter, 10–30 K sized.
6. Lysis buffer: 50 mM Tris–HCl buffer (pH 7.9) containing 300 mM NaCl, 10 mM imidazole, 10% glycerol, 01% Triton X-100, 1 mM 4-(2-aminoethyl)benzenesulfonyl fluoride hydrochloride.
7. Wash buffer 1: 50 mM Tris–HCl, 300 mM NaCl, 50 mM imidazole, pH 7.9.
8. Wash buffer 2: 20 mM Tris–HCl, 160 mM NaCl, 100 mM imidazole, pH 7.9.
9. Elution buffer: 20 mM Tris–HCl, 160 mM NaCl, 200 mM imidazole, pH 7.9.
10. Assay buffer: 20 mM Tris–HCl, 160 mM NaCl, pH 7.4.

2.2 Preparation of Lipid Stock Solutions and Lipid Vesicles

1. Regular lipids: Dissolve 1-palmitoyl-2-oleoyl-*sn*-glycero-3-phosphocholine (POPC), 1-palmitoyl-2-oleoyl-*sn*-glycero-3-phosphoethanolamine (POPE), 1-palmitoyl-2-oleoyl-*sn*-glycero-3-phosphoserine (POPS), soy phosphatidylinositol (PI), sphingomyelin, or cholesterol in the highest grade oxygen-free chloroform to yield 10 mg/mL stock solutions. Store them in Teflon-sealed vials at −20 °C.

2. PtdIns*P* stock solutions: Dissolve 10 mg phosphatidylinositol 3-phosphate (PI(3)P), phosphatidylinositol 4-phosphate (PI (4)P), or phosphatidylinositol 5-phosphate (PI(5)P) in chloroform/methanol/water (4:4:2 v/v/v). Dissolve 10 mg phosphatidylinositol 3,4-bisphosphate ($PI(3,4)P_2$), phosphatidylinositol 3,5-bisphosphate ($PI(3,5)P_2$), or phosphatidylinositol 4,5-bisphosphate ($PI(4,5)P_2$, Cat. No.: 850165) in chloroform/methanol/water (5:3:2 v/v/v). Dissolve 10 mg phosphatidylinositol-3,4,5-trisphosphate ($PI(3,4,5)P_3$) in chloroform/methanol/water (5:4:1, v/v/v). Store them in Teflon-sealed vials at −20 °C.
3. Dabsyl-PE is synthesized from POPE [11]. Dissolve POPE (50 mg) in chloroform (2 mL) and add the solution to a solution of dabsyl chloride (22.6 mg) and triethylamine (0.2 mL) in chloroform (5 mL). Stir the mixture for 6 h at room temperature in the dark and remove the solvent in vacuo. Dissolve the residue in dichloromethane/methanol (9:1) and purify the compound by silica column chromatography using the same solvent mixture as eluent. Evaporate the solvent in vacuo to afford dabsyl-PE as an orange solid. Prepare a stock solution of dabsyl-PE in chloroform (10 mg/mL) and store it in a Teflon-sealed vial at −20 °C.
4. A microextruder with a 100-nM polycarbonate filter.

2.3 Surface Plasmon Resonance Analysis

1. BIACORE X-100 or other SPR instrument (Cytiva Life Sciences).
2. BIACORE L1 chip (Cytiva Life Sciences).
3. Running buffer: 20 mM Tris–HCl, 160 mM NaCl, pH 7.4.

2.4 Fluorescence Quenching Assay

1. Fluorescence plate reader (e.g., BioTek Synergy Neo Multi-Mode Reader).
2. Polystyrene 96-well plate.
3. Assay buffer: 20 mM Tris–HCl, 160 mM NaCl, pH 7.4.

3 Methods

3.1 Preparation of SH2 Domains: Protein Expression and Purification

1. Inoculate 500 mL of Luria broth containing either 50 μg/mL kanamycin or 100 μg/mL ampicillin with 20 mL of *Escherichia coli* BL21 (DE3) pLysS cells harboring an EGFP- and His_6-tagged SH2 domain.
2. Shake cells at 37 °C for 3–4 h until the absorbance of the media at 600 nM reaches 0.5–0.7 (*see* **Note 1**).

3. Add isopropyl β-D-1-thiogalactopyranoside to a final concentration of 1.0 mM to cells to induce overexpression of the protein. Grow cells for additional 12–18 h at 15–17 °C.
4. Harvest cells by centrifugation (5000 rpm, 10 min) and resuspend cell pellets in the Lysis buffer (*see* **Note 2**).
5. Lyse cells by sonication and collect the supernatant after centrifugation at 4 °C (16,000 rpm, 30 min).
6. Add 1 mL of Ni-NTA-affinity resin into the cell lysate and gently shake the mixture for 1–2 h at 4 °C.
7. Apply the mixture to a column and wash the column several times with the Wash buffer 1.
8. Wash the column 2–3 times with Wash buffer 2.
9. Elute the protein with the Elution buffer (*see* **Note 3**). Concentrate the collected fractions in an Amicon Ultra 0.5 mL Centrifugal Filter, 10–30 K sized and exchange the buffer solution to the Analysis buffer.
10. Check the purity of the protein sample with sodium dodecyl sulfate-polyacrylamide gel electrophoresis (*see* **Note 4**).
11. Determine the protein concentration by the Bradford assay.

3.2 Preparation of Large Unilamellar Vesicles (LUVs)

1. Prepare a lipid mixture (e.g., POPC/POPE/POPS/PI/cholesterol/PtdIns(4,5)P_2 (20:30:30:9:10:1 in mole ratio) as an inner plasma membrane (IPM) mimetic) and evaporate the solvent under a gentle stream of N_2.
2. Add 1 mL of Analysis buffer to bring the total concentration to 500 μM.
3. Gently shake the lipid mixture for 30 min.
4. Prepare LUVs by passing the lipid mixture through a micro-extruder with a 100-nM polycarbonate filter >20 times.

3.3 Quantification of Membrane Binding of SH2 Domains by SPR Analysis (See Note 5)

3.3.1 Rapid Screening of Membrane-Binding Activity of SH2 Domains Using IPM-Mimetic LUVs (See Note 6)

1. Coat the control and sample surfaces of the L1 chip (with the flow rate of 5 μL/min) in the BIACORE X100 instrument (Cytiva Life Sciences) with POPC and IPM-mimetic LUVs, respectively. To ensure the consistent sensor surface coverage by lipids, adjust the amount of the injected lipid vesicle solutions until virtually identical resonance units (RU) have been achieved for all lipid vesicles (*see* **Note 7**).
2. Inject a fixed concentration (e.g., 100 nM) (*see* **Note 8**) of EGFP-SH2 domain samples at 20 μL/min and monitor the association phase to obtain a near-saturation RU value (R_{eq}) (*see* **Note 9**).
3. Compare the R_{eq} value with that of a standard IPM-binding domain, such as the pleckstrin homology (PH) domain of phospholipase C-δ (PLCδ -PH) (*see* **Note 10**).

3.3.2 Estimation of Phosphoinositide Specificity of SH2 Domains (See **Note 11**)

1. Coat the control and sample channels of the L1 chip (with the flow rate of 5 μL/min) in the BIACORE X100 instrument with POPC/POPS (80:20) and POPC/POPS/PtdInsP (77: 30:3) LUVs, respectively.
2. Inject a fixed concentration (e.g., 100 nM) of EGFP-SH2 domain samples at 5–10 μL/min until a R_{eq} value is reached (*see* **Note 12**).

3.3.3 Determination of K_d values for Membrane Affinity and Lipid Specificity

1. Coat the control and sample channels of the L1 chip (with the flow rate of 5 μL/min) in the BIACORE X100 instrument with POPC LUVs and LUVs of a specific lipid composition (e.g., POPC/POPS/PI(4,5)P_2 (77:30:3), respectively.
2. Inject varying concentrations (e.g., 0–500 nM) of EGFP-SH2 domain samples at 5 μL/min until a R_{eq} is reached at each protein concentration (*see* **Note 13**).
3. Plot using graphing software (e.g., KaleidaGraph (Synergy Software)) the R_{eq} values against the protein concentrations (P_o), and determine the K_d value for binding by nonlinear least squares analysis of the binding isotherm using the equation, $R_{eq} = R_{max}/(1 + K_d/P_o)$ where R_{max} indicates the maximal R_{eq} value (*see* **Note 14**).

3.4 Quantification of Membrane Binding of SH2 Domains by Fluorescence Quenching Assay

3.4.1 Rapid Screening of Membrane-Binding Activity of SH2 Domains Using IPM-Mimetic LUVs

1. Add a fixed concentration of EGFP-SH2 domains (e.g., 100 nM) to each well of a nontreated black polystyrene 96-well plate.
2. Add a fixed concentration of IPM-mimetic LUVs containing dabsyl-PE (e.g., POPC/POPE/POPS/PI/cholesterol/dabsyl-PE/PtdIns(4,5)P_2 (20:30:20:9:10:10:1)) to these wells and adjust the final volume to 200 μL with the Assay buffer (*see* **Note 15**).
3. Fill three rows with the same protein and lipid mixtures for triplex determinations.
4. Select one row for background correction for nonspecific quenching using the same proteins and POPC/dabsyl-PE (90:10) vesicles.
5. Incubate the plate at 25 °C with gentle shaking for 5 min.
6. Monitor a decrease in EGFP fluorescence emission intensity at 509 nM (ΔF) with excitation set at 488 nM using a fluorescence plate reader.
7. Subtract the background fluorescence values from binding data.
8. Compare the average ΔF values of EGFP-SH2 domains with that of a standard IPM-binding protein, such as EGFP-PLCδ-PH

*3.4.2 Determination of Phosphoinositide Specificity of SH2 Domains (See **Note 16**)*

1. Add a fixed concentration of protein (e.g., 100 nM) to each well of a nontreated black polystyrene 96-well plate.
2. To each row, add vesicles with varying lipid composition (e.g., POPC/dabsyl-PE/PtdIns*P* (95 − *x*:5:*x*; *x* = 0–10 mole%)). Keep the total lipid concentration constant (e.g., 10 μM) and adjust the final volume of the mixture in the assay buffer to 200 μL.
3. Select one row for background correction for nonspecific binding and quenching with the same protein and POPC/dabsyl-PE (95:5) vesicles.
4. Incubate the plate at 25 °C with gentle shaking for 5 min.
5. Monitor the decrease in EGFP fluorescence intensity at 509 nm with excitation set at 488 nM using a fluorescence plate reader.
6. Subtract the background fluorescence values from binding data.
7. Analyze resulting membrane-binding data of a protein using the equation: $\Delta F/\Delta F_{max} = 1/(1 + K/[\text{PtdIns}P]_{\%})$. ΔF and ΔF_{max} indicate the fluorescence quenching and the maximal quenching, respectively, and $[\text{PtdIns}P]_{\%}$ and K are mole% of a PtdIns*P* and the $[\text{PtdIns}P]_{\%}$ value causing half-maximal binding, respectively. Comparison of $[\text{PtdIns}P]_{\%}$ for seven PtdIns*Ps* will allow accurate determination of PtdIns*P* specificity of a particular SH2 domain.

*3.4.3 Membrane Affinity Determination (See **Note 17**)*

1. Add a fixed concentration of an EGFP-SH2 domain (e.g., 100 nM) to each well of a nontreated black polystyrene 96-well plate.
2. Add an increasing concentration of LUVs with fixed composition (e.g., POPC/POPS/PI(4,5)P_2/dabsyl-PE (72:20:3:5)) to these wells and adjust the final volume to 200 μL with the Assay buffer. Vary the final lipid concentrations to the value that yields a near saturation signal (e.g., 50 μM).
3. Fill triplicate rows with the same protein and lipid mixtures for triplex determination.
4. Select one row for background correction for nonspecific binding and quenching with the same protein and POPC/dabsyl-PE (95:5) vesicles.
5. Incubate the plate at 25 °C with gentle shaking for 5 min.
6. Monitor a decrease in EGFP fluorescence emission intensity at 509 nM (ΔF) with excitation set at 488 nM using a fluorescence plate reader.
7. Subtract the background fluorescence values from binding data.

8. Calculate the K_d for a particular protein and lipid vesicle combination by nonlinear least-squares analysis using the equation: $\Delta F = \Delta F_{max}/(1 + K_d/[L])$, where $[L]$ and K_d indicate the free lipid concentration of the vesicles and the $[L]$ value giving rise to half-maximal binding, respectively. Since L_o (total lipid concentration) >> P_o (total SH2 domain concentration), we assume that $L_o \approx [L]$.

3.4.4 High-Throughput Screening of SH2 Domain–Lipid-Binding Inhibitors

1. Add fixed concentrations of a protein and a lipid vesicle in 200 μL of the assay buffer to each well of the plate. Typically, the protein concentration is 100 nM and the lipid vesicle concentration is adjusted to give ≈80% of maximal FP quenching.
2. Add a fixed concentration of different inhibitors to each well. Minimize the volume of DMSO (i.e., ≤2% v/v) to avoid protein denaturation. The inhibitor concentration is typically set at 20 μM for initial screening and is gradually lowered for the following rounds of screening to increase the detection threshold.
3. For background correction for each row, run the assay without lipid vesicles in the mixture in the next row.
4. Incubate the plate at 25 °C with gentle shaking for 10 min.
5. Monitor the increase in EGFP fluorescence intensity ($\Delta\Delta F$) at 509 nm with excitation set at 488 nm.
6. Subtract the background fluorescence values from binding data.
7. From $\Delta\Delta F$ *values*, identify potential lead compounds.

3.4.5 Determination of Inhibition Parameters (See ***Note 18****)*

1. Add fixed concentrations of a protein and a lipid vesicle in 200 μL of the assay buffer to each well of the plate. Typically, the protein concentration is 100 nM and the lipid vesicle concentration is adjusted to give ≈80% of maximal FP quenching.
2. Add an increasing concentration of an inhibitor to each well of a given row. Keep the total volume of DMSO (i.e., ≤2% v/v) constant for all wells.
3. Incubate the plate at 25 °C with gentle shaking for 10 min.
4. Monitor the increase in EGFP fluorescence emission intensity ($\Delta\Delta F$) of each well at 509 nm with excitation set at 488 nm.
5. For background correction for each row, run the assay without lipid vesicles in the mixture in the next row.
6. Subtract the background fluorescence values from inhibition data.

7. Analyze inhibition of membrane binding of a protein by an inhibitor using an equation: $\Delta\Delta F = \Delta\Delta F_{max}/(1 + IC_{50}/[I])$ [38]. $\Delta\Delta F$ and $\Delta\Delta F_{max}$ indicate the fluorescence intensity increase of EGFP by a given concentration of inhibitor and maximal $\Delta\Delta F$, respectively. [I] and IC_{50} are the free inhibitor concentration and the inhibition constant.

4 Notes

1. For less stable SH2 domains, the culture can be grown at 30 °C (or a lower temperature) for 5–6 h until the absorbance at 600 nm of the media reaches 0.5–0.7.
2. 20 mL of Lysis buffer should be used for a protein pellet harvested from 500 mL of media. If a more concentrated sample is needed, two or more pellets can be combined.
3. 200 mM imidazole is sufficient for eluting most SH2 domain proteins, but the final imidazole concentration can be varied from 150 to 250 mM.
4. For SPR analysis, protein samples must be pure to suppress false-positive signals due to nonspecific binding of protein impurities to the sensor chip.
5. All SPR analysis protocols are described for the basic BIA-CORE X100 model. They can be adapted to and optimized for any more advanced models. SPR measurements can be performed with SH2 domains with and without an EGFP tag. Since SPR signals depend on the molecular mass of the analyte proteins, however, it is advised to consistently use EGFP-SH2 domains.
6. Since most SH2 domain-containing proteins function near or at the IPM, IPM-mimetic LUVs are used for rapid screening of their membrane affinity. One can also use more a generic lipid mixture, such as POPC/POPS (70:30) for initial screening.
7. It is extremely difficult to check the physical status of lipid vesicles coated onto the L1 sensor chip. Earlier studies suggested that the size and lipid composition could affect whether coated vesicles remain intact on the chip or are transformed into a different structure, such as a supported lipid monolayer or bilayer [36]. It is therefore advised to use the vesicles with the same size and lipid composition to obtain consistent results for different SH2 domains. When vesicles with different lipid compositions must be used, one can check if the vesicles consistently remain intact on the chip by performing control experiments with vesicles encapsulating a fluorophore, such as 6-carboxyfluorescein [36].

8. The concentration indicates that of the stock solution. The diffusion and dilution in the microfluidic flow channel is expected to be minimal during SPR measurements. Proteins samples are kept on ice and warmed to the analysis temperature immediately before injection. Samples may need to be centrifuged periodically to remove precipitates.
9. Although the SPR analysis allows real-time monitoring of both association and dissociation phases, it is typically difficult to determine the rate constant for and extract the meaningful mechanistic information from the dissociation phase of lipid–protein interaction. The dissociation phase is often too slow for robust kinetic analysis and highly sensitive to various experimental variables, including nonspecific adsorption of proteins to the sensor chip surface and the quality of the sensor chip. The R_{eq} values for the association phase are in general well correlated with the extent of surface binding of analyte proteins with comparable molecular mass and are less subject to experimental artifacts as long as pure protein samples are used for measurements. However, there are exceptions as some proteins may show lower R_{eq} values than expected from their membrane-binding activity due to their unique mode of membrane interaction. It is therefore necessary to determine K_d values for absolute quantitative comparison of membrane affinity of SH2 domains (*see* Subheading 3.3.3).
10. Any prototypal lipid-binding domain (e.g., PH domains) can be used as a standard.
11. Lipid selectivity determined by SPR analysis is typically reported as the relative R_{eq} values for different lipids at a given protein concentration [25, 26]. Although simple and intuitive, this type of analysis can sometimes yield misleading and erroneous results because some proteins show widely different R_{eq} values when bound to different lipid surfaces. Thus, a more reliable parameter that represents the fraction of the membrane-bound protein molecules at a given protein concentration would be a normalized value of R_{eq}/R_{max}, where R_{max} indicates the maximal R_{eq} value when a given lipid surface is fully saturated with the protein molecules. R_{max} can be estimated for different lipid species either by a single-point measurement employing the highest protein concentration experimentally feasible (e.g., 1–5 μM) or by the curve fitting of the R_{eq} versus the protein concentration plot (*see* Subheading 3.3.3). Although this approach allows semiquantitative estimation of the PtdInsP selectivity, accurate determination of PtdInsP selectivity requires determination of K_d values for all seven PtdInsPs (*see* Subheading 3.3.3).

12. For more robust estimation of R_{eq} values, the flow rate should be set at a lower value than used for rapid screening.
13. For accurate K_d determination, the slowest possible flow rate should be employed, which allows accurate R_{eq} determination at each protein concentration.
14. Since the concentration of lipids coated on the sensor chip cannot be accurately determined, K_d is defined as P_o yielding half-maximal binding with a fixed lipid concentration [26]. Notice that the equation is described in terms of not the free ([P]) but the total protein concentration (P_o) since it is practically impossible to determine [P] values in SPR measurements.
15. Typically, 5–10 mol% dabsyl-PE is used for the assay, but the dabsyl-PE content can be raised to 15 mole% if the quenching efficiency of a particular EGFP-SH2 domain is not high enough to give a robust dynamic range.
16. PtdInsP specificity of SH2 domains can be determined by a rapid single point measurement using lipid vesicles with a fixed PtdInsP concentration (e.g., POPC/dabsyl-PE/PtdIns*P* (92: 5:*3*)) or more accurately by varying the PtdInsP concentration in the vesicles (e.g., POPC/dabsyl-PE/PtdIns*P* (95 − x:5:x; $x = 0$–10 mole%)). Since the fluorescence quenching assay can be performed rapidly, the latter approach is recommended as long as sufficient protein and lipid samples are available.
17. Notice that here K_d is defined in terms of total lipid concentration of vesicles. K_d can be also defined in terms of mole% of PtdIns*P* in the vesicles if binding is measured with varying concentrations of PtdIns*P* with a fixed total lipid concentration (e.g., POPC/dabsyl-PE/PtdIns*P* (95 − x:5:x; $x = 0$–10 mole %)).
18. Unlike competitive inhibitors of enzymes, inhibitors of lipid–protein interaction cause varying degrees of maximal inhibition at saturating concentrations depending on the mode and efficiency of their inhibitory action. It is because the membrane-binding sites of soluble proteins typically comprise a pocket that specifically binds a cognate lipid and the flanking surface that makes nonspecific, mostly electrostatic, contact with the membrane [18]. Since the latter type of interaction also contributes to the overall membrane affinity of these proteins [18] and would not be blocked by inhibitors targeting specific lipid pockets, some of these proteins may be able to interact with the membrane to some degree even in the presence of a saturating concentration of an inhibitor targeting the lipid-binding pocket. We thus define two parameters to assess their inhibitory efficacy: maximal inhibition (I_{max}) and the concentration required for half-maximal inhibition (IC_{50}) [29]. I_{max} and IC_{50}

values of inhibitors were determined using the equation, $I/I_{max} = \Delta\Delta F/\Delta\Delta F_{max} = (1 + IC_{50}/[I])$, where I and $[I]$ *indicate %inhibition and free inhibitor concentration, respectively.*

Acknowledgments

The work is supported by a National Institutes of Health grant R35GM122530.

References

1. Gavin AC, Aloy P, Grandi P et al (2006) Proteome survey reveals modularity of the yeast cell machinery. Nature 440:631–636
2. Guruharsha KG, Rual JF, Zhai B et al (2011) A protein complex network of Drosophila melanogaster. Cell 147:690–703
3. Havugimana PC, Hart GT, Nepusz T et al (2012) A census of human soluble protein complexes. Cell 150:1068–1081
4. Kiel C, Beltrao P, Serrano L (2008) Analyzing protein interaction networks using structural information. Annu Rev Biochem 77:415–441
5. Pawson T, Nash P (2003) Assembly of cell regulatory systems through protein interaction domains. Science 300:445–452
6. Pawson T (2004) Specificity in signal transduction: from phosphotyrosine-SH2 domain interactions to complex cellular systems. Cell 116:191–203
7. Bhattacharyya RP, Remenyi A, Yeh BJ et al (2006) Domains, motifs, and scaffolds: the role of modular interactions in the evolution and wiring of cell signaling circuits. Annu Rev Biochem 75:655–680
8. Van Roey K, Uyar B, Weatheritt RJ et al (2014) Short linear motifs: ubiquitous and functionally diverse protein interaction modules directing cell regulation. Chem Rev 114:6733–6778
9. Corbi-Verge C, Kim PM (2016) Motif mediated protein-protein interactions as drug targets. Cell Commun Signal 14:8
10. Lim WA, Pawson T (2010) Phosphotyrosine signaling: evolving a new cellular communication system. Cell 142:661–667
11. Liu BA, Nash PD (2012) Evolution of SH2 domains and phosphotyrosine signalling networks. Philos Trans R Soc Lond Ser B Biol Sci 367:2556–2573
12. O'Rourke L, Ladbury JE (2003) Specificity is complex and time consuming: mutual exclusivity in tyrosine kinase-mediated signaling. Acc Chem Res 36:410–416
13. Machida K, Mayer BJ (2005) The SH2 domain: versatile signaling module and pharmaceutical target. Biochim Biophys Acta 1747: 1–25
14. Ladbury JE, Arold S (2000) Searching for specificity in SH domains. Chem Biol 7:R3–R8
15. Scott JD, Pawson T (2009) Cell signaling in space and time: where proteins come together and when they're apart. Science 326:1220–1224
16. Good MC, Zalatan JG, Lim WA (2011) Scaffold proteins: hubs for controlling the flow of cellular information. Science 332:680–686
17. Bray D (1998) Signaling complexes: biophysical constraints on intracellular communication. Annu Rev Biophys Biomol Struct 27:59–75
18. Cho W (2006) Building signaling complexes at the membrane. Sci STKE 2006:pe7
19. Bae JH, Lew ED, Yuzawa S et al (2009) The selectivity of receptor tyrosine kinase signaling is controlled by a secondary SH2 domain binding site. Cell 138:514–524
20. Zimmermann P (2006) The prevalence and significance of PDZ domain-phosphoinositide interactions. Biochim Biophys Acta 1761:947–956
21. Feng W, Zhang M (2009) Organization and dynamics of PDZ-domain-related supramodules in the postsynaptic density. Nat Rev Neurosci 10:87–99
22. Chen Y, Sheng R, Kallberg M et al (2012) Genome-wide functional annotation of dual-specificity protein- and lipid-binding modules that regulate protein interactions. Mol Cell 46: 226–237
23. Sheng R, Chen Y, Yung Gee H et al (2012) Cholesterol modulates cell signaling and protein networking by specifically interacting with PDZ domain-containing scaffold proteins. Nat Commun 3:1249
24. Sheng R, Kim H, Lee H et al (2014) Cholesterol selectively activates canonical Wnt

signalling over non-canonical Wnt signalling. Nat Commun 5:4393
25. Sheng R, Jung DJ, Silkov A et al (2016) Lipids regulate Lck protein activity through their interactions with the Lck Src homology 2 domain. J Biol Chem 291:17639–17650
26. Park MJ, Sheng R, Silkov A et al (2016) SH2 domains serve as lipid-binding modules for pTyr-signaling proteins. Mol Cell 62:7–20
27. Kim E, Kim DH, Singaram I et al (2018) Cellular phosphatase activity of C1-Ten/Tensin2 is controlled by Phosphatidylinositol-3,4,5-triphosphate binding through the C1-Ten/Tensin2 SH2 domain. Cell Signal 51:130–138
28. Kraskouskaya D, Duodu E, Arpin CC et al (2013) Progress towards the development of SH2 domain inhibitors. Chem Soc Rev 42: 3337–3370
29. Singaram I, Sharma A, Pant S et al (2023) Targeting lipid-protein interaction to treat Syk-mediated acute myeloid leukemia. Nat Chem Biol 19:239–250
30. Kim H, Afsari HS, Cho W (2013) High-throughput fluorescence assay for membrane-protein interaction. J Lipid Res 54:3531–3538
31. Cho W, Bittova L, Stahelin RV (2001) Membrane binding assays for peripheral proteins. Anal Biochem 296:153–161
32. Narayan K, Lemmon MA (2006) Determining selectivity of phosphoinositide-binding domains. Methods 39:122–133
33. Rebecchi M, Peterson A, McLaughlin S (1992) Phosphoinositide-specific phospholipase C-delta 1 binds with high affinity to phospholipid vesicles containing phosphatidylinositol 4,5-bisphosphate. Biochemistry 31:12742–12747
34. Dowler S, Kular G, Alessi DR (2002) Protein lipid overlay assay. Sci STKE 2002:pl6
35. Stahelin RV (2013) Surface plasmon resonance: a useful technique for cell biologists to characterize biomolecular interactions. Mol Biol Cell 24:883–886
36. Stahelin RV, Cho W (2001) Differential roles of ionic, aliphatic, and aromatic residues in membrane-protein interactions: a surface plasmon resonance study on phospholipases A2. Biochemistry 40:4672–4678
37. Cho W, Kim H, Hu Y (2016) High-throughput fluorometric assay for membrane-protein interaction. Methods Mol Biol 1376: 163–174
38. Dua R, Cho W (1994) Inhibition of human secretory class II phospholipase A2 by heparin. Eur J Biochem 221:481–490

Chapter 14

Exploring the Binding Interaction Between Phosphotyrosine Peptides and SH2 Domains by Proximal Crosslinking

Rui Wang, Yishu Bao, and Jiang Xia

Abstract

Proximal crosslinking refers to the site-specific conjugation reaction between a synthetic ligand with a bioorthogonal reactive group incorporated at a particular site and a protein of interest (POI). The binding interaction positions a reactive group of a native amino acid of the POI to the proximity of the reactive group in the ligand. The covalent conjugation increases the molecular weight of the POI, shows an upshift in the polyacrylamide gel, and gives a fluorescent band if the ligand is fluorescently labeled. Here, we summarize a method to covalently conjugate phosphotyrosine peptides and SH2 domains that contain cysteine residues. This method yields covalent peptide blockers for a set of SH2 proteins and elucidates the binding interaction between phosphotyrosine peptides and SH2 domains.

Key words Proximal crosslinking, Src homology 2, Phosphotyrosine peptides, Site-specific conjugation

1 Introduction

Src homology 2 (SH2) domains are a big family in a wide variety of signal molecules, including protein kinases (SRC, LCK) [1], protein phospholipases (PLCγ) [2], transcription factors (STAT) [3], adaptor proteins (NCK, GRB2) [4], and others. The classic structures of SH2 domains typically contain two pockets on their surface: a phosphotyrosine (pTyr) binding pocket formed by the N-terminal half containing the α-helix A (αA) and strands βA − βD and a hydrophobic specificity pocket formed by the C-terminal half of the SH2 domain to accommodate a hydrophobic residue located C-terminal to the pTyr in the peptide ligand (Fig. 1) [5]. The phosphotyrosine residue is essential for SH2 binding because it provides nearly half of the total binding energy

Rui Wang and Yishu Bao contributed equally with all other contributors.

Teresa Carlomagno and Maja Köhn (eds.), *SH2 Domains: Functional Modules and Evolving Tools in Biology*, Methods in Molecular Biology, vol. 2705, https://doi.org/10.1007/978-1-0716-3393-9_14,

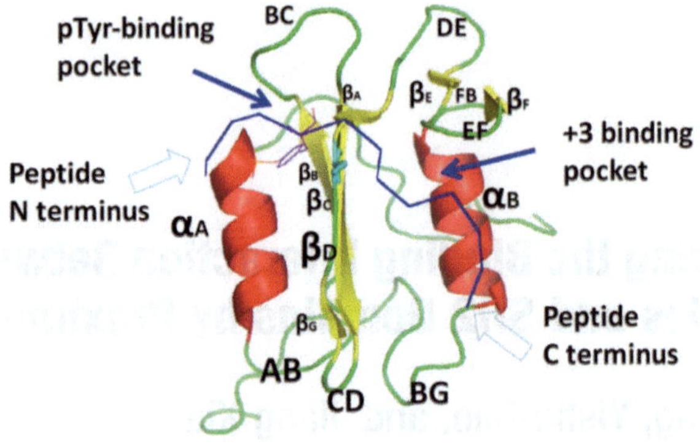

Fig. 1 The canonical binding mode of SH2 domains (PDB ID 2PLD)

[6]. Generally, the SH2 domain is flexibly tethered to a larger protein through local intracellular spaces.

Specific recognition of pTyr-binding ligands is a critical feature of SH2 domain recognition; phosphorylation shows approximately a binding affinity 1000 times higher than that of unphosphorylated peptides. Through binding to pTyr sequences, SH2 domains play a key role in the signal transduction emanating from a tyrosine kinase to control cell proliferation, differentiation, migration, or survival [7, 8]. Probing the binding affinities of SH2 domains with phosphopeptide libraries in which a central pTyr is flanked by degenerate positions has revealed that different SH2 domains preferentially recognize distinct amino acids at the pTyr +1 to +3 positions. Therefore, the residues C-terminal to pTyr determine the binding preference, albeit ambiguously. The binding interaction between pTyr-containing peptides and SH2 domains is dominated by the pTyr site, which contributes to more than half of the binding energy. Because of this feature, designing a specific blocker for a particular SH2 domain based on a peptide that contains pTyr and natural amino acids is difficult; one needs another feature besides pTyr to reach the SH2 specificity. Here, we leverage a unique cysteine residue in an SH2 domain, the C-terminal SH2 domain of PLCγ1, and harness the principle of proximity-induced reaction to design a specific, covalently reactive peptide blocker for this particular SH2 domain. The PLCγ1-c SH2 domain contains a cysteine (Cys58) at the bridging β strand βD, close to the pY + 1 position of the peptide ligand. Proximal crosslinking, which results in covalent conjugation of selected residues (e.g., cysteine, lysine, arginine, histidine, etc.) on a protein of interest (POI), represents one effective approach in the field of covalent conjugation due to its spontaneity, site specificity, and biocompatibility [9, 10]. This proximal crosslinking distinguishes a particular cysteine from other cysteines in the same protein or a particular cysteine-carrying protein among many similar proteins in the proteome.

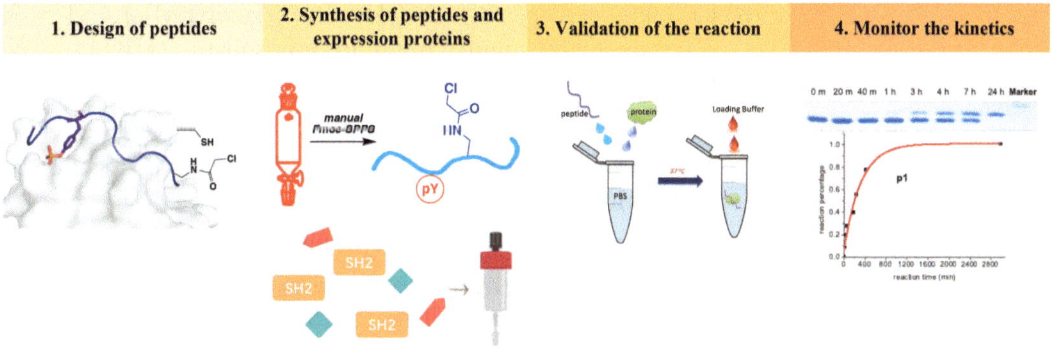

Fig. 2 Procedures for exploring the binding interaction between phosphotyrosine peptides and SH2 domains

The α-chloroacetyl group rapidly conjugates with the thiol group on cysteine residues that are in its proximity but very slowly with nucleophiles at distal positions in the same protein [11, 12]. Therefore, we utilized reactive peptides to explore the inherent flexibility of the SH2 domain–pTyr peptide binding model [13]. This strategy is based on a principle, namely proximity-induced reactivity (also known as proximity effect or proximity-enabled reactivity), in which the binding interaction brings an electrophile installed on a synthetic peptide to the proximity of a cysteine residue of the protein. Therefore, the spatial closeness leads to a covalent reaction between the SH2 domain and pTyr peptide. The proximity-induced reaction also ensures a selectivity on the target protein.

Achieving a successful proximal crosslinking reaction between phosphotyrosine peptides and SH2 domains includes the following steps (Fig. 2).

1. Design of proximal reactive peptides based on the structure of the target protein.
2. Synthesis of a set of reactive phosphotyrosine peptides with different reactive group positions and expression of SH2 domain proteins.
3. Validation of the site-specific covalent reaction in the test tube.
4. Final step is to monitor the crosslinking kinetics.

Details are described in the following sections [13].

2 Materials

2.1 Peptide Synthesis, Purification, and Characterization

Unless otherwise specified, all reagents and solvents are purchased from commercial sources and are used without further purification.

1. Rink Amide-ChemMatrix® resins.
2. Unnatural amino acid: Fmoc-Dap(Mtt)-OH, Fmoc-Lys(Mtt)-OH, Fmoc-Tyr(PO(OBzl)OH)-OH.

3. Coupling reagents: 2-(1H-benzotriazol-1-yl)-1, 1, 3, 3-tetramethyluronium hexafluorophosphate (HBTU), *N*-hydroxybenzotriazole (HOBt).
4. *N, N*-diisopropylethylamine (DIPEA).
5. Ninhydrin test solution: 0.3 g ninhydrin and 0.6 mL acetic acid in 20 mL ethanol.
6. Fmoc-deprotected solution: 20% piperidine in DMF (v/v).
7. Cleavage cocktail: trifluoroacetic acid (TFA)/deionized water (DI H_2O)/triisopropylsilane (TIPS) (95:2.5:2.5).
8. Di-isopropylcarbodiimide (DIC).
9. Dichloromethane (DCM).
10. *N, N*-dimethylformamide (DMF).
11. Chloroacetic acid.
12. 5(6)-Carboxyfluorescein (5(6)-FAM).
13. 4-Dimthylaminopyridine (DMAP).
14. Mixer: EYELA® CM-1000 (Shanghai Ailang Instrument Co., Ltd., China).
15. Preparative C18 RP-HPLC column: here a Vydac (218TP510) RP-HPLC column was used (C18, 5 μM, 10 mM ID × 250 mM, Alltech Associates, Inc., USA).
16. LC-2010A HPLC systems (Shimadzu, Kyoto, Japan).
17. Mobile phase: solvent A is 0.1% (v/v) TFA in water, solvent B is 0.1% (v/v) TFA in acetonitrile.
18. MALDI (*see* **step 20**) target plate: here an MTP 384 target plate polished steel BC was used (Bruker Daltonik GmbH, Germany).
19. Matrix solution: 2 mg α-Cyano-4-hydroxycinnamic acid (HCCA) in 1 mL 1:1 ACN/DI H_2O (v/v).
20. Matrix-assisted laser desorption/ionization-time of flight mass spectrometry (MALDI-TOF MS, Bruker Daltonics, Germany).
21. Ninhydrin Reagent.

2.2 Recombinant Proteins Expression and Purification

Prepare all media and buffers in deionized water.

1. *E. coli* expression strain BL21 (DE3).
2. Luria-Bertani (LB) media: 10 g/L tryptone, 5 g/L yeast extract, 5 g/L NaCl, autoclaved.
3. Antibiotics: 100 μg/mL ampicillin or 50 μg/mL kanamycin in LB media. All the antibiotics are stored at −20 °C before use.
4. 1 M isopropyl-β-D-thiogalactopyranoside (IPTG) stock solution: 238 mg IPTG in 1 mL water, sterile-filtered (*see* **Note 1**).

5. Binding buffer: 20 mM Tris, pH 7.5, 500 mM NaCl, 3 mM dithiothreitol (DTT), and 0.1 mM phenylmethanesulfonyl fluoride (PMSF).
6. Elution buffer: 50 mM Tris, pH 8.0, 500 mM NaCl, 500 mM imidazole.
7. General buffer: 50 mM Tris, pH 8.0, 100 mM NaCl, 1 mM EDTA.
8. Human rhinovirus (HRV) 3C protease.
9. Temperature-controlled centrifuge.
10. Fast protein liquid chromatography (FPLC) system (ÄKTA pure 25).
11. Ni-NTA agarose (Qiagen).
12. SDS-PAGE equipment and gels.
13. DNA sequences and a plasmid carrying the DNA sequence of the SH2 domain in a pET28a vector are required. For example, the DNA can be inserted between NdeI and HindIII in pET28a.
14. Selection plates: Mix 4 g LB powder, 3 g agar powder, and 200 mL water. Autoclave, wait until the solution cool down to 60 °C, mix in antibiotics, pool the solution on agar plates, and let the plates cool down to room temperature.

2.3 Tricine–SDS-PAGE Gel

Prepare all media and buffers in deionized water.

1. AB-3 stock solution: acrylamide-bisacrylamide (AB)-3 stock solution (49.5% T, 3% C mixture), which dissolves 48 g of acrylamide and 1.5 g of bisacrylamide in 100 mL of water. Store at 4 °C.
2. AB-6 stock solution: acrylamide-bisacrylamide (AB)-6 stock solution (49.5% T, 6% C mixture), which dissolves 48 g of acrylamide and 3.0 g of bisacrylamide in 100 mL of water. Store at 4 °C.
3. Gel buffer (3×): 3.0 M Tris, 1.0 M HCl, 0.3% SDS, pH 8.45. Store at 4 °C.
4. 16% separating gel: AB-6 2.5 mL, gel buffer (3×) 2.5 mL, glycerol 0.75 g, 10% ammonium persulfate (APS) 25 μL, Tetramethylethylenediamine (TEMED) 2.5 μL, added water to 7.5 mL.
5. 10% spacer gel: AB-3 300 μL, gel buffer (3×) 0.5 mL, glycerol 0.15 g, 10% APS 7.5 μL, TEMED 0.75 μL, added water to 1.5 mL.
6. 4% stacking gel: AB-3 250 μL, gel buffer (3×) 0.75 mL, 10% APS 22.5 μL, TEMED 2.25 μL, added water to 3.0 mL.

7. Cathode buffer (10×): 1.0 M Tris–HCl, 1.0 M Tricine, 1% SDS (pH ~8.25).
8. Anode buffer (10×): 1 M Tris, 0.225 M HCl, pH 8.9.
9. Sample buffer (20 mL): 5 mL 0.5 M Tris pH 6.8, 4 mL 20% SDS, 1 mL 2-mercaptoethanol, 4 mL 50% glycerol, 0.004 g bromophenol blue, 6 mL water.
10. Ammonium persulfate: 10% solution in water.

2.4 Peptide–Protein Covalent Conjugation Reactions

Prepare all media and buffers in deionized water.

1. SDS-PAGE loading buffer (5×): 0.25 M Tris–HCl, 10% SDS, 0.5 M dithiothreitol, 0.1% bromophenol blue, 50% glycerol (pH 6.8).
2. Glycine buffer (1×): 125 mM Glycine, 0.1% Triton X-100, pH 2.0.
3. Coomassie brilliant blue dye: 1.06% Coomassie Brilliant Blue, 50% methanol, 10% acetic acid in H_2O.
4. SDS-PAGE equipment and gels.
5. Imager (here, Typhoon TRIO+ Variable Mode Imager (GE Healthcare, USA)).

3 Methods

3.1 Design and Synthesis of SH2 Domains-Specific Reactive Phosphotyrosine (pY) Peptides

3.1.1 Design of SH2 Domain-Specific Reactive pY Peptides

1. According to the features of the crystal structure of the protein–ligand complex (e.g., PDB ID: 2PLD), a set of α-chloroacetyl-containing pY peptides are designed. For example, Src kinase SH2 domains preferentially recognize their binding motif designated as pYEEI (derived from viral hamster polyma middle T antigen protein), and Grb2 SH2 domains prefer the pYVNV motif.
2. The first pTyr peptide template is derived from a peptide ligand from the X–Y linker of PLCγ1-c surrounding pY783 (PGFpY-VEAN; called **LA** here) [14].
3. The second pTyr peptide template is a doubly phosphorylated peptide from Syk tyrosine kinase (DTEVpYESPpYADP; called **LB** here) [15].

3.1.2 Manual Fmoc Solid-Phase Peptide Synthesis (See **Note 2**, Fig. 3)

1. Resin-handling procedure: place the dry Rink-resin in a peptide synthesis reactor. Then, add solvents to cover resin with 3 times bed volume and shake the reaction vessel gently to remove air bubbles and form a suspension of resin for 30 min (*see* **Note 3**).
2. Coupling method: mixtures of corresponding Fmoc-protected amino acids (5 equiv.), HOBt (5 equiv.), HBTU (5 equiv.) and DIPEA (10 equiv.) in a minimum volume of DMF are added to

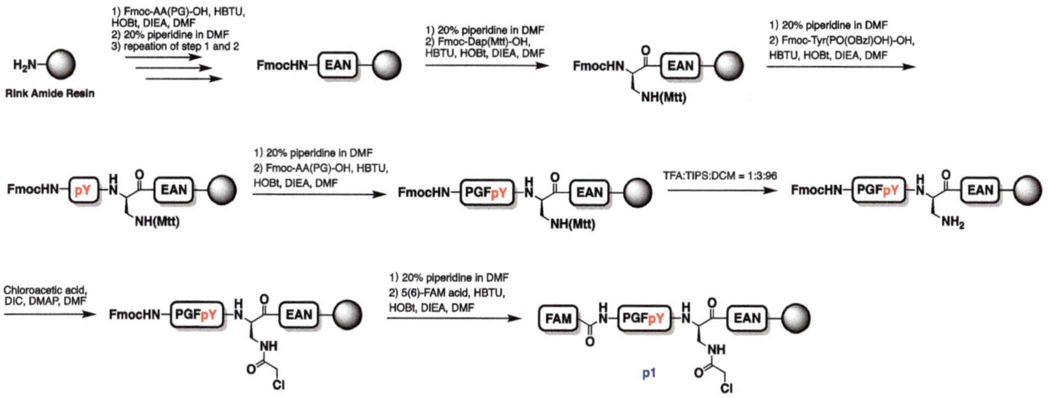

Fig. 3 General procedures for manual Fmoc solid-phase peptide synthesis (take **p1** as an example)

resin (*see* **Note 4**). Allow the mixture to agitate at room temperature for 35 min. Completeness of the coupling reactions is checked by Ninhydrin Reagent (here, N7285, Merck). If resin still gives a positive color test, repeat the coupling reaction with fresh reagents. After accomplishment of the coupling reaction, the resin is washed thoroughly with DMF (3 × 5 mL) and DCM (3 × 5 mL).

3. Fmoc group removal: addition of 20% piperidine/DMF for 15 min twice to the resin to remove the Fmoc group from the N-terminus of the resin-bound peptide chain. Then wash the resin with DMF (5 times).
4. To incorporate the unnatural amino acid X_1 (Fmoc-Dap(Mtt)-OH) or X_2 (Fmoc-Lys(Mtt)-OH): the same as the above coupling methods, only need to replace the natural amino acid with unnatural amino acid X and extend reaction time to 45 min.
5. To incorporate Fmoc-Tyr(PO(OBzl)OH)-OH: the same as above coupling methods, only need to replace the natural amino acid with unnatural amino acid X and extend the reaction time to 45 min.
6. 4-Methyltrityl (Mtt) groups deprotection: addition of TFA/DCM/TIPS (1/96/3) in resin at r. t. shaking for 5 min and repeating 6–8 times (*see* **Note 5**).
7. To incorporate chloroacetyl, the side chain amine is subsequently coupled with a ten-fold excess of chloroacetic acid, DIC, and DMAP in DMF for 3 h. A covalent bond will form between a surface Cys residue on the SH2 domain and the α-chloroacetyl group incorporated into the peptide ligand when they are brought into proximity by binding interaction (Fig. 4).

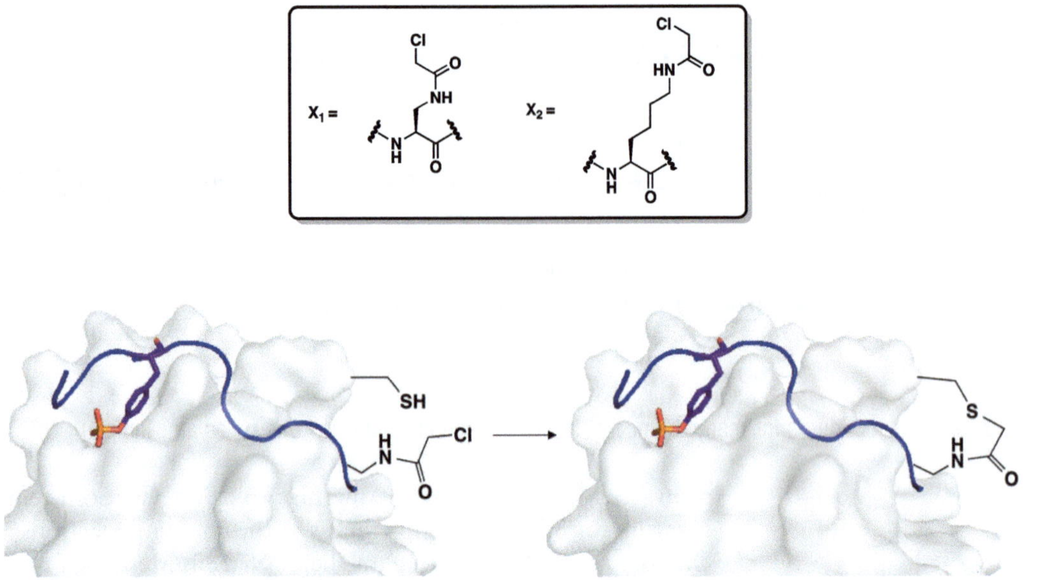

Fig. 4 Diagram depicting the proximal crosslinking strategy used in this study to covalently tether an SH2 domain (shown in surface representation) to a bound pY peptide (blue) upon binding interaction. The chemical structure of the unnatural amino acid, **X**, (2S)-2-amino-3-[(α-chloroacetyl)amino]-propionic acid, is shown in the inset

8. The phosphotyrosine peptide ligands **p1** to **p7**, by site-specific incorporation of the chloroacetyl-containing residues X_1 or X_2 at pY + 1, pY + 2, or pY + 3 positions of the above two peptides will be achieved (Fig. 5).
9. Labeling with the fluorescent group in the N terminus: the same as the above coupling methods, only need to replace the natural amino acid with 5(6)-FAM acid and let it react for 6 h (*see* **Note 6**).
10. After completion of coupling-deprotection cycles, wash the resin with DMF and DCM (5 times) and dry under a vacuum before global deprotection by the TFA cleavage cocktail. Treat dried resins with the freshly prepared cleavage cocktail for 2–3 h.
11. Filter TFA mixtures and precipitate the filtrates by adding cold ether. Dry in argon before HPLC analysis is performed.

3.1.3 Purification of Peptides

1. Perform RP-HPLC analyses using a C18 column at a flow rate of 3.0 mL/min. All HPLC chromatograms are recorded at 215 nm.
2. Purify and isolate all peptides using a linear gradient from 5% to 95% of acetonitrile for 40 min in 0.1% aqueous TFA solution.

Peptide	Name	Sequence
p1	LA+1	P - G - F - pY - X_1 - E - A - N
p2	LA+3	P - G - F - pY - V - E - X_1 - N
p3	LA+3$_{X2}$	P - G - F - pY - V - E - X_2 - N
p4	LB+2	D - T - E - V - pY - E - X_1 - P - pY - A - D - P
p5	LB+2$_{X2}$	D - T - E - V - pY - E - X_2 - P - pY - A - D - P
p6	LB+3	D - T - E - V - pY - E - S - X_1 - pY - A - D - P
p7	LB+3$_{X2}$	D - T - E - V - pY - E - S - X_2 - pY - A - D - P
p8	AA-3	X_1 - A - A - pY - A - A - A - A
p9	AA-2	A - X_1 - A - pY - A - A - A - A
p10	AA-1	A - A - X_1 - pY - A - A - A - A
p11	AA+1	A - A - A - pY - X_1 - A - A - A
p12	AA+2	A - A - A - pY - A - X_1 - A - A
p13	AA+3	A - A - A - pY - A - A - X_1 - A
p14	AA+4	A - A - A - pY - A - A - A - X_1

Fig. 5 The sequence of the reactive peptides designed in this study. $\mathbf{X_1}$: Dap(α-chloroacetyl), $\mathbf{X_2}$: Lys (α-chloroacetyl). Besides a number, each peptide is given a name to indicate its origin. For example, **LA + 1** means the peptide is derived from **LA**, with an original sequence of PGFpYVEAN, and the modification is $\mathbf{X_1}$ at pY + 1 position. For replacement by amino acid $\mathbf{X_2}$, $\mathbf{X_2}$ is added as subscript. The peptides were synthesized with FAM at the N-terminus to facilitate detection

3.1.4 MALDI-TOF MS Analysis of Peptides Isolated from RP-HPLC Analysis

1. Mix peptide solutions with the HCCA matrix solution (1:1) in target plate.
2. After the sample is dry, the mass analyses of the peptides are performed in an MALDI-TOF MS.

3.2 Recombinant Protein Expression and Purification

3.2.1 Protein Expression

1. Transformation of the recombinant plasmid into *Escherichia coli* strain BL21 (DE3).
2. Thaw a tube of 100 μL BL21 (DE3) competent *E. coli* cells on ice until the last ice crystals disappear.
3. Add 1–5 μL of the plasmid DNA solution (containing 1 pg–100 ng plasmid DNA) to the cell mixture. Carefully flick the tube 4–5 times to mix cells and DNA. Do not vortex.
4. Place the mixture on ice for 30 min. Do not mix.
5. Heat shock at exactly 42 °C for exactly 60 s. Do not mix.
6. Place on ice for 5 min. Do not mix.
7. Pipette 900 μL LB at room temperature without antibiotics into the mixture.
8. Place at 37 °C for 60 min. Shake vigorously (200 rpm) or rotate.
9. Warm selection plates to 37 °C.

10. Centrifuge the tubes at 5000 *g* for 1 min, and resuspend the pellet gently using 100 μL LB.
11. Spread 50–100 μL from each tube onto a selection plate and incubate overnight at 37 °C. Alternatively, incubate at 30 °C overnight to allow colonies to develop.
12. Pick single colonies and incubate in 1 mL LB media supplemented with antibiotics and let the culture grow overnight.
13. Inoculate the overnight culture to LB media supplemented with antibiotics at 1:100 ratio.
14. Shake at 37 °C until the optical density at 600 nm (OD 600) reaches 0.6–0.8.
15. Add IPTG to a final concentration of 1 mM to induce protein expression, and shake the culture overnight at 16 °C.
16. Harvest the cells by centrifugation at 5000 *g* for 10 min. Store the pellets for protein purification.

3.2.2 Protein Purification

All SH2 proteins are purified using Nickel-NTA affinity chromatography through their His tags. Proteins can be further purified using size-exclusion chromatography (**step 6**), and when needed, recombinant protein tags can be cleaved by protease and further purified by size-exclusion chromatography (**step 7**) (*see* **Note 7**).

1. Resuspend the harvested cells with binding buffer, lyse the cells by sonication.
2. Centrifuge the cell lysate at 13,000 *g* for 30 min at 4 °C.
3. Purify recombinant proteins from lysates by Ni-NTA agarose. Protein is eluted by an imidazole gradient from 10 to 500 mM, using binding buffer and elution buffer.
4. Analyze the collected fractions by SDS-PAGE gel.
5. Collect pure protein fractions and exchange with the general buffer.
6. Further, purify the eluted protein using Superdex 200 column equilibrated with general buffer.
7. (Optional, tag removal) Exchange eluted protein with general buffer. Remove the tag by HRV 3C protease incubation at 4 °C overnight. Purify the protein preparation without the tag by Superdex 75 column equilibrated with general buffer.
8. Flash freeze the protein under liquid nitrogen and store at − 80 °C.

3.3 Proximity-Induced Crosslinking Reaction

1. Incubate the pTyr peptides with the purified PLCγ1-c SH2 protein in the dark at pH 7.4 and 37 °C, at a final concentration of 30 μM protein in PBS.

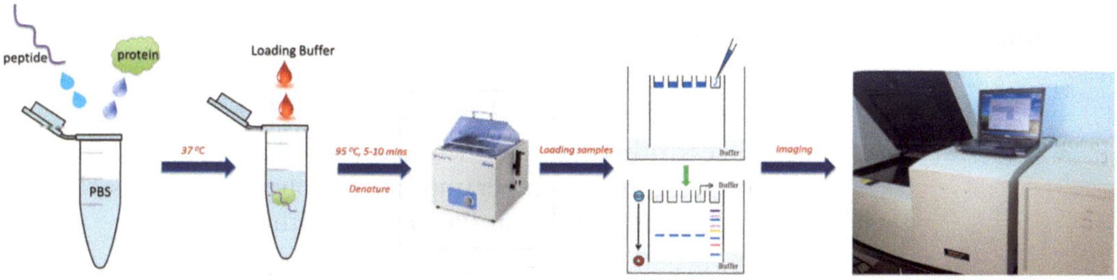

Fig. 6 The general procedure of analyzing the covalent conjugation reactions

2. Set a ratio of the concentrations of proteins and peptides by varying the concentration of the peptide.
3. Add 5× loading dye and denature under heating of 95 °C for 5 min to stop the reaction.
4. Resolve the denatured samples on Tricine-SDS-PAGE electrophoresis in 1× Glycine buffer.
5. Image the gel in an Imager at 488 nm laser (here, Typhoon TRIO+ Variable Mode used).
6. Stain the gels with Coomassie Blue dye. Image under an imager, and quantify the band intensities using gel quantifying software (here, Bio-Rad) (Fig. 6).
7. Calculate the conjugation yield based on the band intensities of protein and protein conjugate.

3.4 Monitoring the Crosslinking Kinetics

1. Prepare the crosslinking reactions at an [SH2 domain]: [pTyr peptide] ratio of 1:5. Here, 30 μM SH2 domain proteins were incubated with 150 μM pTyr peptides at a final concentration at 37 °C in the dark.
2. Take the reaction aliquots at different time points (0, 15, 30, 60, 120, 240, and 900 min). The negative control is SH2 proteins alone without peptides.
3. After crosslinking incubation, take 20 μL of the reaction mixtures and add 5 μL 5× loading dye to denature under heating of 95 °C for 5 min to stop the reaction.
4. Evaluate the results by examining the percentage of pTyr peptide-crosslinked SH2 domain over the total amount of SH2 domain by Tricine-SDS-PAGE after stained of Coomassie Blue.

4 Notes

1. IPTG is stored as 1 M stocks.
2. Unless otherwise specified, all solid-phase reactions are performed at room temperature and in polypropylene vessels fitted with the fritted disk at the bottom of the vessels.
3. Normally, polystyrene-based resins swell with DCM, while polyethylene glycol-polystyrene (PEG-PS) resins swell with DMF.
4. If necessary, add extra DMF to ensure complete coverage of the resin bed.
5. Sometimes, the result of deprotection of Mtt groups is not ideal due to the long-time storage of reagents. We can increase the amount of TFA to 5%.
6. Fluorescent dyes, including FITC and TAMRA, can also be used for labeling.
7. Most of the proteins eluted from affinity chromatography should be at least 90% purity according to SDS-PAGE. For some proteins like the scaffold protein Homer, besides affinity purification, size-exclusion chromatography may be utilized to acquire high purity samples.

References

1. Kästle M, Merten C, Hartig R et al (2020) Tyrosine 192 within the SH2 domain of the Src-protein tyrosine kinase p56^{Lck} regulates T-cell activation independently of Lck/CD45 interactions. Cell Commun Signal 18(1):183
2. Bae YS, Lee HY, Jung YS et al (2016) Phospholipase Cγ in Toll-like receptor-mediated inflammation and innate immunity. Adv Biol Regul 63:92–97
3. de Araujo ED, Orlova A, Neubauer HA et al (2019) Structural implications of STAT3 and STAT5 SH2 domain mutations. Cancers (Basel) 11(11):1757
4. Gupta RW, Mayer BJ (1998) Dominant-negative mutants of the SH2/SH3 adapters Nck and Grb2 inhibit MAP kinase activation and mesoderm-specific gene induction by eFGF in Xenopus. Oncogene 17(17): 2155–2165
5. Kaneko T, Huang H, Zhao B et al (2010) Loops govern SH2 domain specificity by controlling access to binding pockets. Sci Signal 3(120):ra34
6. Bradshaw JM, Mitaxov V, Waksman G (1999) Investigation of phosphotyrosine recognition by the SH2 domain of the Src kinase. J Mol Biol 293(4):971–985
7. Pawson T (2004) Specificity in signal transduction: from phosphotyrosine-SH2 domain interactions to complex cellular systems. Cell 116(2):191–203
8. Machida K, Mayer BJ (2005) The SH2 domain: versatile signaling module and pharmaceutical target. Biochim Biophys Acta 1747(1):1–25
9. Gallagher SS, Sable JE, Sheetz MP et al (2009) An in vivo covalent TMP-tag based on proximity-induced reactivity. ACS Chem Biol 4(7):547–556
10. Kurien BT, Scofield RH (2015) Multiple immunoblots by passive diffusion of proteins from a single SDS-PAGE gel. Methods Mol Biol 1312:77–86
11. Bai R, Pei XF, Boyé O et al (1996) Identification of cysteine 354 of beta-tubulin as part of the binding site for the A ring of colchicine. J Biol Chem 271(21):12639–12645
12. Nonaka H, Fujishima SH, Uchinomiya SH et al (2010) Selective covalent labeling of tag-fused GPCR proteins on live cell surface with a

synthetic probe for their functional analysis. J Am Chem Soc 132(27):9301–9309

13. Wang R, Leung PYM et al (2018) Reverse binding mode of phosphotyrosine peptides with SH2 protein. Biochemistry 57:5257–5269

14. Groesch TD, Zhou F, Mattila S et al (2006) Structural basis for the requirement of two phosphotyrosine residues in signaling mediated by Syk tyrosine kinase. J Mol Biol 356(5): 1222–1236

15. Hajicek N, Charpentier TH, Rush JR et al (2013) Autoinhibition and phosphorylation-induced activation of phospholipase C-gamma isozymes. Biochemistry 52(28):4810–4819

Chapter 15

Synthesis and Biochemical Evaluation of Monocarboxylic GRB2 SH2 Domain Inhibitors

Tao Xiao, Min Zhang, and Haitao Ji

Abstract

This protocol discloses the synthesis of monocarboxylic inhibitors with a macrocyclic peptide scaffold to bind with the GRB2 SH2 domain and disrupt the protein–protein interactions (PPIs) between GRB2 and phosphotyrosine-containing proteins.

Key words GRB2, SH2 domain, Inhibitors, Macrocyclic peptides, Synthesis, Protein–protein interactions, Diastereomeric selectivity

1 Introduction

Growth factor receptor-bound protein 2 (GRB2) is a key adaptor protein for many receptor or nonreceptor tyrosine kinase pathways. After upstream signal activation, phosphotyrosine (pTyr)-containing sequences pYXNX (pY, X, and N represent phosphotyrosine, any residue, and asparagine, respectively) adopt a type I β-turn conformation to bind to the central Src homology 2 (SH2) domain of GRB2. The *N*- and *C*-terminal SH3 domains of GRB2 then bring the nucleotide exchange factor SOS to the cell membrane to activate the Ras–mitogen-activated protein kinase (MAPK) signaling cascade. Aberrant activation of GRB2-dependent signaling pathways significantly contributes to cancer development and progression [1–4].

Highly appreciable medicinal chemistry efforts have been carried out to discover inhibitors that bind to the GRB2 SH2 domain and disrupt the protein–protein interactions (PPIs) between GRB2 and pYXNX-containing proteins [5–9]. However, despite excellent in vitro low nanomolar biochemical activities, reported GRB2 inhibitors do not display satisfactory cellular and in vivo activities for further clinical development [10–15]. The phosphate or phosphate mimetic in these compounds carrying −2 charges at physiological

Teresa Carlomagno and Maja Köhn (eds.), *SH2 Domains: Functional Modules and Evolving Tools in Biology*, Methods in Molecular Biology, vol. 2705, https://doi.org/10.1007/978-1-0716-3393-9_15,

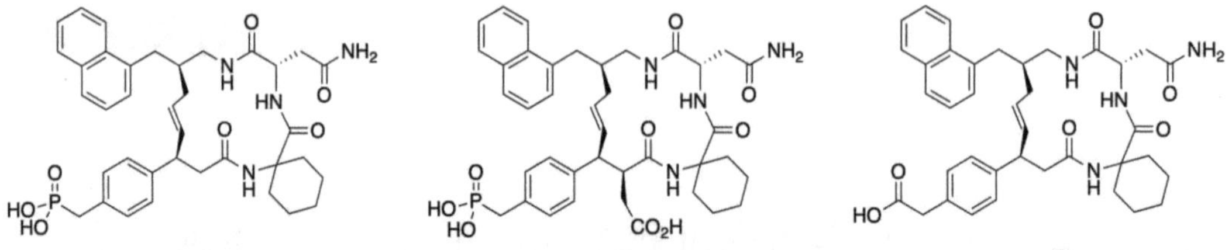

Fig. 1 Macrocyclic GRB2 inhibitors. (Adapted with permission from ref. 19. Copyright 2021 Elsevier)

pH were thought to cause poor cell permeability [5, 6, 16–18]. Among all of the reported GRB2 inhibitors, *i*/*i* + 3 macrocyclized tetrapeptides developed by Burke and coworkers are very interesting. Compounds **1** [10] and **2** [11] in Fig. 1 are the representative GRB2 inhibitors of this series.

We substituted the phosphono group in **1** with a carboxylic acid group and designed compound **3** in Fig. 1 to examine the binding affinity difference between **1**, **3**, and the $EGFR_{1068}$-containing phosphotyrosyl peptide [19]. Out of all EGFR sequences, the $EGFR_{1068}$-containing phosphopeptide ($pEGFR_{1068}$) is known to have the highest binding affinity with GRB2. The carboxylic acid methyl ester of **3**, compound **4,** was also synthesized for comparison. Herein, we discuss the protocol to synthesize **3** and its methyl ester **4,** the fluorescence anisotropy competitive inhibition assays to examine their activities for the disruption of the $pEGFR_{1068}$/GRB2 PPI, and the parallel artificial membrane permeability assays (PAMPAs) for the passive diffusion through the membrane.

2 Materials

2.1 Chemical Reagents

1. Evans' chiral auxiliaries: (*S*)-(+)-4-phenyl-2-oxazolidinone, (*S*)-4-benzyl-2-oxazolidinone, (*S*)-(−)-4-isopropyl-2-oxazolidinone, (*S*)-(−)-4-benzyl-5,5-dimethyl-2-oxazolidinone, and (*S*)-4-*tert*-butyl-2-oxazolidinone.
2. Reagents for organic synthesis: sodium hydride (60% dispersion in mineral oil), ethyl 2-(diethoxyphosphoryl) acetate, 1-naphthaldehyde, palladium on carbon (10 wt.% on carbon), sodium hydroxide (pellets/certified ACS), hydrochloric acid (ACS reagent, 37%), allyl iodide, thionyl chloride (reagent grade, 97%), sodium bis(trimethylsilyl)amide (2.0 M in THF), lithium aluminum hydride, triethylamine, *p*-toluenesulfonyl chloride, di-*tert*-butyl iminodicarboxylate, cesium carbonate, Boc-L-asparagine, *N*, *N*-diisopropylethylamine, 1-[Bis(dimethylamino)methylene]-1H-1,2,3-triazolo[4,5-b] pyridinium 3-oxide hexafluoro phosphate (HATU), trifluoroacetic acid, 1-aminocyclohexanecarboxylic acid, acryloyl

chloride, palladium (II) acetate, tris (*o*-tolyl)-phosphine, cuprous thiophenoxide, lithium hydroxide, hydrogen peroxide, *N,N'*-diisopropylcarbodiimide (DIPCDI), HOBt, Grubbs II catalyst, triethylsilane, methyl iodide, potassium carbonate, sodium sulfite, sodium sulfate, ammonium chloride, sodium bicarbonate, and sodium chloride. All these reagents were used without further purification.

3. Solvents for organic synthesis: hexanes, tetrahydrofuran, ethyl acetate, ethanol, dichloromethane, 1,4-dioxane, dimethyl sulfoxide, dimethylformamide, diethyl ether, chloroform, methanol, and acetone.
4. Solvents and reagents for high-performance liquid chromatography (HPLC) analysis: trifluoroacetic acid (HPLC grade, 99.5+%), acetonitrile (HPLC grade), and water (HPLC grade).

2.2 Chromatography

1. Thin-layer chromatography (TLC) SiliaPlate® 250 μM 20 × 20 cm plates with F_{254} indicator.
2. Ultraviolet (UV) light at 254 nM and/or the phosphomolybdic acid (PMA) staining solution consisting of 10 wt.% phosphomolybdic acid in ethanol.
3. Silica gel column chromatography: SiliaFlash® F60 40–63 μM silica gel (230–400 mesh).
4. Preparative HPLC columns. Here, a semipreparative Phenomenex C18 column (Luna 5 μL C18(2) 100 Å, 10 × 250 mM) was used.

2.3 Characterization of the Synthesized Compounds

1. Nuclear magnetic resonance (NMR) spectrometer. Here, spectra were recorded using a Bruker AVANCE III HD 500 (500 MHz) and chemical shifts were reported as values in parts per million (ppm), with the reference resonance peaks set at 7.26 ppm ($CDCl_3$) and 2.50 ppm [$(CD_2H)_2SO$] for ^1H NMR spectra, and 77.16 ppm ($CDCl_3$) and 39.52 ppm (DMSO-d_6) for ^{13}C NMR spectra.
2. Solvents: $CDCl_3$ and DMSO-d_6.
3. High-performance liquid chromatography-mass spectrometry (HPLC–MS) system with an electrospray ionization (ESI) source. Here, an Agilent 6120 single quadrupole mass spectrometer with a 1220 infinity LC system (HPLC–MS) and an Agilent 1260 Infinity II HPLC system with a quaternary pump, a vial sampler, and a diode array detector (DAD) was used.
4. HPLC columns. Here, a Kromasil 300–5–C_{18} column (4.6 × 250 mM) and a Phenomenex C18 column (Luna 5 μL C18(2) 100 Å, 4.6 × 250 mM) were used.
5. High-resolution mass spectrometer. Here, an Agilent G6230BA TOF LCMS Mass Spectrometer with a TOF mass detector was used.

2.4 Reagents, Peptides, and Proteins for Biochemical Assays

1. Human GRB2 (residues 1–217) protein: Here, cDNA was cloned into a pEHISTEV vector carrying a *N*-terminal His_6–TEV tag and expressed in *Escherichia coli* BL21 DE3.
2. Human $EGFR_{1068}$ phosphopeptide (residues 1088–1100), Ac-PVPEpYINQSVPKRK-NH_2, and *N*-terminally fluorescein (FITC)-labeled human $EGFR_{1068}$ phosphopeptide (residues 1088–1100), FITC-Ahx-PVPEpYINQSVPKRK-NH_2, purified to the purity of >95%.
3. LB medium, kanamycin, isopropyl β-D-1-thiogalactopyranoside (IPTG), tris(hydroxymethyl)aminomethane (Tris), sodium chloride, dithiothreitol (DTT), 4-(2-hydroxyethyl)-1-piperazineethanesulfonic acid (Hepes), Triton X-100, γ-globulin.

2.5 Protein Purification

1. Ni-NTA affinity chromatography.
2. size-exclusion chromatography with a HiLoad 26/600 Superdex 200 pg column.
3. ÄKTA Pure FPLC system.
4. The Tris buffer for the GRB2 purification: 20 mM Tris–HCl (pH 8.5), 100 mM NaCl, and 2 mM DTT.

2.6 Fluorescence Anisotropy Assays

1. 96-well Microfluor 2 black plates.
2. Plate reader. Here, the Synergy 2 plate reader (Biotek) was used.
3. The fluorescence anisotropy assay buffer: 25 mM Hepes (pH 7.4), 100 mM NaCl, 0.01% Triton X-100, and 100 μg/mL γ-globulin.

2.7 Parallel Artificial Membrane Permeability Assays (PAMPAs)

1. 96-well filtration plates as the artificial membrane support and the receiver plate. Here, MAIPNTR10 from Millipore Sigma were used.
2. The artificial membrane solution consisting of 1% egg lecithin in *n*-dodecane.
3. Donor plates. Here, MATRNPS50 from Millipore Sigma were used.
4. Phosphate buffer (0.1 M, pH 7.4) for the PAMPA assay: 3.1 g of $NaH_2PO_4 \cdot H_2O$ and 10.9 g of Na_2HPO_4 (anhydrous) to distilled H_2O to make a volume of 1 L.
5. HPLC instrument: an Agilent 1260 Infinity II HPLC system equipped with a quaternary pump, a vial sampler, and a DAD detector was used for the quantitative analysis.
6. HPLC column: the Phenomenex C18 column (Luna 5 μL C18 (2) 100 Å, 4.6 × 250 mM).
7. HPLC mobile phases: mobile phase A: H_2O (0.1% trifluoroacetic acid, TFA), and mobile phase B: MeCN.

3 Methods

3.1 Synthesis of Compounds

3.1.1 Synthesis of Key Intermediate 15 (Fig. 2)

1. Ethyl 3-(naphthalen-1-yl)acrylate (**5**). Add dropwise ethyl 2-(diethoxyphosphoryl) acetate (21.53 g, 96.03 mmol) to a mixture of NaH (60% dispersion in mineral oil, 4.1 g, 102 mmol) in THF at 0 °C, and stir the mixture for 0.5 h. Then, add dropwise 1-naphthaldehyde (10.00 g, 64.0 mmol) to the mixture at 0 °C. Stir the mixture at room temperature for 14 h, and then pour the mixture into cooled water, extract with EtOAc, wash with brine, dry over sodium sulfate, concentrate in vacuo, and purify with silica gel column chromatography. Here, 13.6 g oil (yield 94%) were obtained.
2. 3-(Naphthalen-1-yl)propanoic acid (**6**). Stir a mixture of **5** (168 g, 742 mmol) and Pd/C (10 wt.% on carbon, 5 g) in EtOH under one hydrogen atmosphere for 14 h at room temperature. Filter the mixture, and add a solution of NaOH (46.1 g, 1150 mmol) in water to the filtrate. Stir the mixture at 60 °C for 1 h, and remove EtOH in vacuo. Add 1 M HCl aqueous solution to pH = 2, and collect the solid by filtration, wash with water, and dry to give 124 g white solid (yield 83%). ^{1}H NMR (500 MHz, DMSO-d_6) δ 12.20 (s, 1H), 8.09–8.04 (m, 1H), 7.95–7.90 (m, 1H), 7.78 (dt, J = 8.0, 1.2 Hz, 1H), 7.54 (dddd, J = 21.7, 8.1, 6.8, 1.4 Hz, 2H), 7.43 (dd, J = 8.0,

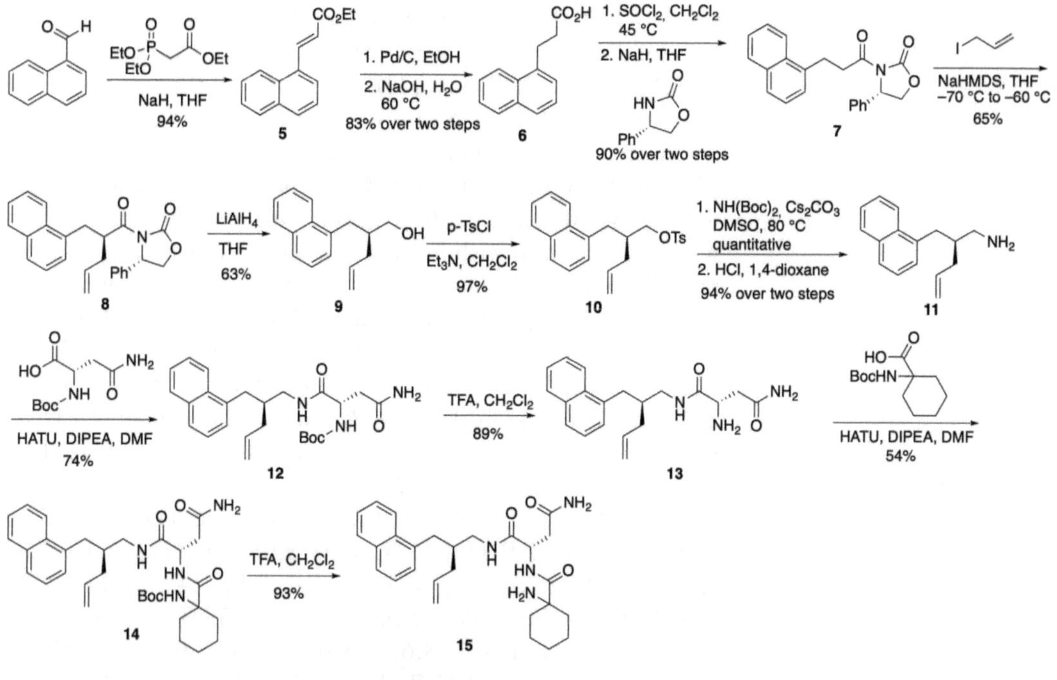

Fig. 2 The synthetic route for key intermediate **15**. (Adapted with permission from ref. 19. Copyright 2021 Elsevier)

7.0 Hz, 1H), 7.38 (dd, J = 7.0, 1.4 Hz, 1H), 3.30 (t, J = 7.7 Hz, 2H), 2.64 (dd, J = 8.2, 7.2 Hz, 2H). ^{13}C NMR (126 MHz, DMSO-d_6) δ 174.3, 137.3, 133.9, 131.7, 129.1, 127.2, 126.6, 126.2, 126.1, 126.1, 123.9, 35.1, 27.9. MS (ESI) m/z = 223.2 [M + Na]$^+$.

3. (S)-3-(3-(Naphthalen-1-yl)propanoyl)-4-phenyloxazolidin-2-one (7). Stir the solution of **6** (20.0 g, 99.9 mmol) and $SOCl_2$ (23.8 g, 200 mmol) in CH_2Cl_2 at 45 °C for 4 h, concentrate in vacuo, and dissolve in CH_2Cl_2. Add (S)-4-phenyloxazolidin-2-one (16.3 g, 99.9 mmol) to another mixture of NaH (60% dispersion in mineral oil, 11.9 g, 300 mmol) in THF at 0 °C. Stir the mixture for 1 h at 0 °C, and add dropwise the pre-acyl chloride/CH_2Cl_2 solution at −50 °C. Stir the mixture at room temperature for 14 h, pour into cooled water, extract with EtOAc, wash with brine, dry over sodium sulfate, concentrate in vacuo, and purify with silica gel column chromatography to give 31.0 g white solid (yield 90% over two steps). ^{1}H NMR (500 MHz, DMSO-d_6) δ 8.10–8.04 (m, 1H), 7.96–7.89 (m, 1H), 7.79 (dt, J = 8.3, 1.0 Hz, 1H), 7.59–7.49 (m, 2H), 7.44–7.29 (m, 7H), 5.49 (dd, J = 8.6, 3.6 Hz, 1H), 4.74 (t, J = 8.7 Hz, 1H), 4.17 (dd, J = 8.8, 3.6 Hz, 1H), 3.32–3.18 (m, 4H). ^{13}C NMR (126 MHz, DMSO-d_6) δ 171.3, 153.8, 139.9, 136.7, 133.4, 131.2, 128.8 (2C), 128.6, 127.9, 126.7, 126.1, 126.0, 125.8 (2C), 125.6, 125.6, 123.5, 70.3, 57.0, 36.0, 26.7. MS (ESI) m/z = 368.2 [M + Na]$^+$.

4. (S)-3-((S)-2-(Naphthalen-1-ylmethyl)pent-4-enoyl)-4-phenyloxazolidin-2-one (**8**, *see* **Note 1**). Add dropwise 2.0 M NaHMDS in anhydrous THF (7.6 mL, 15.06 mmol) to the solution of 7 (4 g, 11.6 mmol) in anhydrous THF at −70 °C. Then, stir the solution for 0.5 h, and add dropwise allyl iodide (3.11 g, 18.5 mmol) at −60 °C (*see* **Note 2**). Stir the solution for another 0.5 h at −60 °C, pour into cooled water, extract with EtOAc, wash with brine, dry over sodium sulfate, concentrate in vacuo, and purify with silica gel column chromatography to give 2.9 g white solid (yield 65%). ^{1}H NMR (500 MHz, DMSO-d_6) δ 8.15–8.07 (m, 1H), 7.93 (dd, J = 8.1, 1.5 Hz, 1H), 7.81 (d, J = 8.1 Hz, 1H), 7.54 (dddd, J = 21.5, 7.9, 6.8, 1.3 Hz, 2H), 7.43 (dd, J = 8.2, 7.0 Hz, 1H), 7.39–7.21 (m, 6H), 5.70–5.55 (m, 1H), 5.36 (dd, J = 8.5, 3.6 Hz, 1H), 4.95 (dq, J = 14.4, 1.7 Hz, 2H), 4.48 (t, J = 8.7 Hz, 1H), 4.41–4.30 (m, 1H), 4.08 (dd, J = 8.8, 3.6 Hz, 1H), 3.38 (dd, J = 14.0, 7.9 Hz, 1H), 3.20–3.08 (m, 1H), 2.46–2.34 (m, 1H), 2.22 (dddt, J = 13.7, 6.8, 5.5, 1.4 Hz, 1H). ^{13}C NMR (126 MHz, DMSO-d_6) δ 173.9, 153.3, 139.6, 135.0, 134.7, 133.4, 131.5, 128.6, 128.6 (2C), 128.0, 127.2, 127.0, 126.1, 125.8 (2C), 125.7, 125.4, 123.7, 117.5, 69.8, 57.2, 35.7, 33.3. MS (ESI) m/z = 408.2 [M + Na]$^+$.

5. (*S*)-2-(Naphthalen-1-ylmethyl)pent-4-en-1-ol (**9**). Add $LiAlH_4$ (3.95 g, 103.76 mmol) to a solution of **8** (20 g, 51.9 mmol) in anhydrous THF at −20 °C (*see* **Note 3**). Then, stir the solution for 0.5 h at 0 °C, pour into cooled water, extract with EtOAc, wash with brine, dry over sodium sulfate, concentrate in vacuo, and purify with silica gel column chromatography to give 7.4 g oil (yield 63%). ^{1}H NMR (500 MHz, DMSO-d_6) δ 8.12–8.03 (m, 1H), 7.96–7.86 (m, 1H), 7.77 (dt, *J* = 8.2, 1.0 Hz, 1H), 7.58–7.46 (m, 2H), 7.42 (dd, *J* = 8.2, 7.0 Hz, 1H), 7.34 (dd, *J* = 7.0, 1.3 Hz, 1H), 5.83 (ddt, *J* = 17.3, 10.2, 7.1 Hz, 1H), 5.08–4.96 (m, 2H), 4.59 (t, *J* = 5.1 Hz, 1H), 3.38 (dt, *J* = 10.5, 5.3 Hz, 1H), 3.30 (dt, *J* = 10.6, 5.4 Hz, 1H), 3.14 (dd, *J* = 13.7, 7.0 Hz, 1H), 2.87 (dd, *J* = 13.7, 7.1 Hz, 1H), 2.16 (dtt, *J* = 14.1, 7.1, 1.3 Hz, 1H), 2.04 (dddd, *J* = 14.1, 7.2, 4.3, 1.4 Hz, 1H), 1.90–1.80 (m, 1H). ^{13}C NMR (126 MHz, DMSO-d_6) δ 137.0, 136.9, 133.6, 131.6, 128.6, 127.3, 126.5, 125.8, 125.5, 125.3, 123.9, 116.4, 62.7, 41.5, 35.2, 33.8. MS (ESI) m/z = 249.2 $[M + Na]^+$.

6. (*S*)-2-(Naphthalen-1-ylmethyl)pent-4-en-1-yl 4-methylbenzenesulfonate (**10**). Add *p*-toluenesulfonyl chloride (8.09 g, 42.4 mmol) to the solution of **9** (8 g, 35.4 mmol) and Et_3N (7.14 g, 70.7 mmol) in CH_2Cl_2 at room temperature. Then, stir the solution for 14 h, pour into water, extract with CH_2Cl_2, wash with brine, dry over sodium sulfate, concentrate in vacuo, and purify with silica gel column chromatography to give 13.1 g oil (yield 97%). ^{1}H NMR (500 MHz, DMSO-d_6) δ 7.99–7.86 (m, 2H), 7.80–7.74 (m, 1H), 7.74–7.67 (m, 2H), 7.55–7.47 (m, 2H), 7.44–7.38 (m, 2H), 7.33 (dd, *J* = 8.2, 7.0 Hz, 1H), 7.16 (dd, *J* = 7.0, 1.2 Hz, 1H), 5.75–5.65 (m, 1H), 5.04–4.91 (m, 2H), 3.89 (qd, *J* = 9.7, 4.4 Hz, 2H), 3.08–2.85 (m, 2H), 2.40 (s, 3H), 2.19–1.99 (m, 3H). ^{13}C NMR (126 MHz, DMSO-d_6) δ 144.9, 135.4, 135.1, 133.5, 132.0, 131.3, 130.1 (2C), 128.7, 127.6 (2C), 127.2, 126.9, 126.1, 125.6, 125.3, 123.5, 117.5, 71.6, 54.9, 34.7, 33.1, 21.1. MS (ESI) m/z = 403.2 $[M + Na]^+$.

7. (*S*)-2-(Naphthalen-1-ylmethyl)pent-4-en-1-amine (**11**, *see* **Note 4**). Add **10** (8 g, 21.0 mmol) to a mixture of di-*tert*-butyl iminodicarboxylate (5.03 g, 23.2 mmol) and Cs_2CO_3 (7.53 g, 23.1 mmol) in DMSO. Then, stir the mixture for 2 h at 80 °C, pour into water, extract with EtOAc, wash with brine, dry over sodium sulfate, concentrate in vacuo, and purify with silica gel column chromatography to give 11 g oil. Add HCl/1,4-dioxane to the solution of this oil (102 g, 240 mmol) in 1,4-dioxane at 0 °C. Then, stir the solution for 14 h at room temperature, and remove the solvent in vacuo. Mix the residue with Na_2CO_3/water, extract with CH_2Cl_2, wash with brine,

dry over sodium sulfate, and concentrate in vacuo to give 51 g oil (yield 94%). ^{1}H NMR (500 MHz, DMSO-d_6) δ 8.17–8.03 (m, 1H), 7.96–7.87 (m, 1H), 7.78 (dt, *J* = 8.3, 1.1 Hz, 1H), 7.59–7.46 (m, 2H), 7.46–7.39 (m, 1H), 7.39–7.30 (m, 1H), 5.84 (ddt, *J* = 17.2, 10.2, 7.2 Hz, 1H), 5.16–4.95 (m, 2H), 3.11 (dd, *J* = 13.8, 7.2 Hz, 1H), 2.97–2.85 (m, 1H), 2.67–2.51 (m, 2H), 2.25–2.09 (m, 1H), 2.08–1.96 (m, 1H), 1.86 (p, *J* = 6.5 Hz, 1H). ^{13}C NMR (126 MHz, DMSO-d_6) δ 136.8 (2C), 133.6, 131.6, 128.6, 127.3, 126.5, 125.9, 125.5, 125.4, 124.0, 116.7, 66.3, 48.6, 43.7, 35.5, 34.5. MS (ESI) m/z = 226.2 [M + H]$^+$.

8. *tert*-Butyl ((*S*)-4-amino-1-(((*S*)-2-(naphthalen-1-ylmethyl)pent-4-en-1-yl)amino)-1,4-dioxobutan-2-yl)carbamate (**12**). Add HATU (69.9 g, 184 mmol) to a mixture of **11** (31.9 g, 142 mmol), Boc-Asn-OH (32.9 g, 142 mmol) and DIPEA (36.7 g, 284 mmol) in DMF was room temperature. Then, stir the mixture for 14 h at room temperature, and remove the solvent in vacuo. Mix the residue with water, extract with CH_2Cl_2, wash with brine, dry over sodium sulfate, concentrate in vacuo, and purify with silica gel column chromatography to give 46 g white solid (yield 74%). ^{1}H NMR (500 MHz, DMSO-d_6) δ 8.02 (d, *J* = 8.4 Hz, 1H), 7.91 (dd, *J* = 8.0, 1.4 Hz, 1H), 7.85–7.74 (m, 2H), 7.52 (dddd, *J* = 22.5, 8.0, 6.7, 1.4 Hz, 2H), 7.42 (dd, *J* = 8.1, 7.0 Hz, 1H), 7.37 (dd, *J* = 7.1, 1.4 Hz, 1H), 7.32–7.22 (m, 1H), 6.87 (t, *J* = 5.7 Hz, 2H), 5.15–4.95 (m, 2H), 4.24 (td, *J* = 8.1, 5.6 Hz, 1H), 3.25–3.09 (m, 1H), 3.08–2.81 (m, 3H), 2.41 (qd, *J* = 15.0, 6.7 Hz, 2H), 2.17–1.85 (m, 3H), 1.35 (s, 9H). ^{13}C NMR (126 MHz, DMSO-d_6) δ 171.5 (2C), 155.1, 136.5, 136.5, 133.5, 131.6, 128.6, 127.3, 126.6, 125.9, 125.5, 125.3, 123.9, 116.8, 78.1, 51.5, 41.5, 37.2, 35.6, 34.6, 28.2. MS (ESI) m/z = 462.3 [M + Na]$^+$.
9. (*S*)-2-Amino-N^1-((*S*)-2-(naphthalen-1-ylmethyl)pent-4-en-1-yl)succinamide (**13**). Add TFA (90 mL) to the solution of **12** (46 g, 105 mmol) in CH_2Cl_2 at room temperature. Then, stir the solution for 14 h at room temperature, and remove the solvent in vacuo. Mix the residue with Na_2CO_3/water, extract with EtOAc, wash with brine, dry over sodium sulfate, and concentrate in vacuo to give 31.5 g white solid (yield 89%). ^{1}H NMR (500 MHz, DMSO-d_6) δ 8.07–7.99 (m, 1H), 7.96 (t, *J* = 6.0 Hz, 1H), 7.91 (dd, *J* = 8.0, 1.5 Hz, 1H), 7.78 (d, *J* = 8.0 Hz, 1H), 7.52 (dddd, *J* = 20.6, 8.0, 6.8, 1.3 Hz, 2H), 7.46–7.32 (m, 3H), 6.82 (s, 1H), 5.84 (ddt, *J* = 17.1, 10.2, 6.8 Hz, 1H), 5.14–4.98 (m, 2H), 3.48 (dd, *J* = 8.8, 4.2 Hz, 1H), 3.17 (dt, *J* = 12.5, 6.0 Hz, 1H), 3.01 (ddd, *J* = 13.1, 11.2, 5.9 Hz, 2H), 2.91 (dd, *J* = 13.9, 6.7 Hz, 1H), 2.41 (dd, *J* = 15.0, 4.2 Hz, 1H), 2.19 (dd, *J* = 15.0, 8.9 Hz, 1H),

2.14–1.94 (m, 3H), 1.86 (s, 2H). ^{13}C NMR (126 MHz, DMSO-d_6) δ 174.4, 172.7, 136.5, 136.5, 133.5, 131.6, 128.6, 127.3, 126.6, 125.9, 125.5, 125.4, 123.9, 116.9, 52.1, 41.5, 40.5, 38.7, 35.7, 34.7. MS (ESI) m/z = 340.2 [M + H]$^+$.

10. *tert*-Butyl (1-(((*S*)-4-amino-1-(((*S*)-2-(naphthalen-1-ylmethyl) pent-4-en-1-yl)amino)-1,4-dioxobutan-2-yl)carbamoyl)cyclohexyl)carbamate (**14**). Add HATU (44.7 g, 118 mmol) to a mixture of **13** (30.7 g, 90.5 mmol), 1-aminocyclohexanecarboxylic acid (24.2 g, 99.5 mmol), and DIPEA (23.4 g, 181 mmol) in DMF at room temperature. Then, stir the mixture for 14 h at room temperature, and remove the solvent in vacuo. Mix the residue with water, extract with CH_2Cl_2, wash with brine, dry over sodium sulfate, concentrate in vacuo, purify with silica gel column chromatography, and recrystallize to give 27.3 g white solid (yield 54%). ^{1}H NMR (500 MHz, DMSO-d_6) δ 8.05 (dd, *J* = 20.5, 8.1 Hz, 2H), 7.90 (dd, *J* = 8.0, 1.6 Hz, 1H), 7.77 (d, *J* = 8.1 Hz, 1H), 7.66–7.45 (m, 3H), 7.45–7.37 (m, 2H), 7.35 (dd, *J* = 7.0, 1.3 Hz, 1H), 7.23 (s, 1H), 6.97–6.75 (m, 1H), 5.81 (ddt, *J* = 17.3, 10.2, 6.9 Hz, 1H), 5.19–4.90 (m, 2H), 4.40 (q, *J* = 6.1 Hz, 1H), 3.28 (dd, *J* = 13.4, 6.8 Hz, 1H), 3.05 (dd, *J* = 13.9, 6.5 Hz, 1H), 2.89 (dd, *J* = 13.8, 7.0 Hz, 1H), 2.81 (dt, *J* = 10.6, 4.8 Hz, 1H), 2.75–2.61 (m, 1H), 2.43 (dd, *J* = 15.5, 5.1 Hz, 1H), 2.06 (ddt, *J* = 16.6, 12.5, 6.3 Hz, 2H), 1.94 (td, *J* = 15.0, 14.0, 8.0 Hz, 2H), 1.81–1.65 (m, 2H), 1.56 (d, *J* = 9.6 Hz, 2H), 1.51–1.37 (m, 3H), 1.36 (s, 1H), 1.23 (s, 9H), 1.11 (s, 1H). ^{13}C NMR (126 MHz, DMSO-d_6) δ 174.3, 172.6, 170.7, 155.3, 136.5, 136.4, 133.5, 131.6, 128.5, 127.3, 126.6, 125.9, 125.4, 125.3, 124.0, 116.9, 78.7, 58.6, 50.2, 41.9, 34.6, 32.9, 30.3, 28.1 (2C), 24.9, 20.9 (2C). MS (ESI) m/z = 587.4 [M + Na]$^+$.

11. (*S*)-2-(1-Aminocyclohexane-1-carboxamido)-N^1-((*S*)-2-(naphthalen-1-ylmethyl)pent-4-en-1-yl)succinamide (**15**). Add TFA (55 mL) to the solution of **14** (27.3 g, 48.3 mmol) in CH_2Cl_2 at room temperature. Then, stir the solution for 14 h at room temperature, and remove the solvent in vacuo. Mix the residue with Na_2CO_3/water, extract with EtOAc, wash with brine, dry over sodium sulfate, and concentrate in vacuo to give 20.8 g white solid (yield 93%). ^{1}H NMR (500 MHz, DMSO-d_6) δ 8.46 (s, 1H), 8.02 (dd, *J* = 8.3, 1.2 Hz, 1H), 7.91 (dd, *J* = 8.1, 1.4 Hz, 1H), 7.78 (dd, *J* = 7.1, 4.7 Hz, 2H), 7.53 (dddd, *J* = 25.7, 7.9, 6.8, 1.3 Hz, 2H), 7.42 (dd, *J* = 8.2, 7.0 Hz, 1H), 7.39–7.34 (m, 2H), 6.88 (s, 1H), 5.81 (ddt, *J* = 17.1, 10.2, 6.9 Hz, 1H), 5.09–5.01 (m, 2H), 4.47 (t, *J* = 6.2 Hz, 1H), 3.22–3.15 (m, 1H), 3.07–2.83 (m, 3H), 2.54 (dd, *J* = 15.2, 6.5 Hz, 1H),

2.43 (dd, J = 15.2, 6.0 Hz, 1H), 2.10–1.85 (m, 5H), 1.78–1.62 (m, 2H), 1.57–1.39 (m, 5H), 1.34–1.26 (m, 2H), 1.19–1.08 (m, 1H). ^{13}C NMR (126 MHz, DMSO-d_6) δ 178.0, 172.2, 171.5, 137.0, 134.0, 132.1, 129.0, 127.8, 127.1, 126.5, 126.0, 125.8, 124.5, 117.4, 57.1, 50.2, 36.0, 35.1, 35.0, 34.9, 25.7, 21.2, 21.1. MS (ESI) m/z = 465.3 $[M + H]^+$.

3.1.2 Synthesis of Compounds 3 and 4 (Fig. 3)

1. Acrylic anhydride (**16**). Add acryloyl chloride (2.7 g, 30 mmol) dropwise over 5 min to an ice-cooled solution of acrylic acid (2.0 g, 28 mmol) and Et_3N (2.8 g, 28 mmol) in THF (50 mL), and stir the solution at room temperature for 16 h. Collect the ammonium chloride precipitate in a fritted glass filter, and remove the solvent from the filtrate by rotary evaporation. Then, dissolve the residue in CH_2Cl_2 (25 mL), wash twice with dilute aqueous $NaHCO_3$ (50 mL each) and once with saturated aqueous NaCl (50 mL), and dry over Na_2SO_4. After

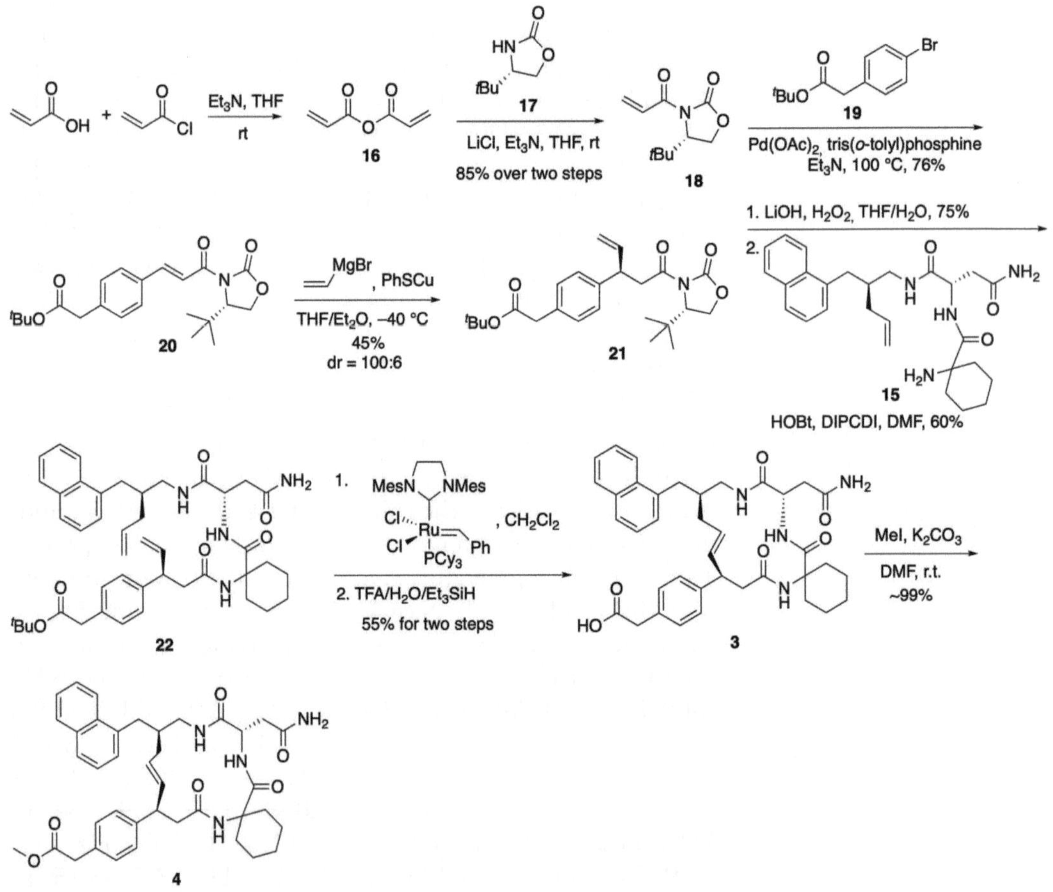

Fig. 3 The synthetic route for compounds **3** and **4**. (Adapted with permission from ref. 19. Copyright 2021 Elsevier)

filtration, remove the solvent by rotary evaporation. Use the product without further purification.

2. (*S*)-3-Acryloyl-4-(*tert*-butyl)oxazolidin-2-one (**18**). Use acrylic anhydride obtained from the previous step immediately and dissolve it in THF. Then, prepare compound **17** by following the literature method [20]. ^{1}H NMR (500 MHz, $CDCl_3$) δ 6.02 (s, 1H), 4.37 (t, J = 9.0 Hz, 1H), 4.19 (dd, J = 9.0, 5.8 Hz, 1H), 3.59 (ddd, J = 9.0, 5.8, 1.1 Hz, 1H), 0.91 (s, 9H). Add Et_3N (0.38 g, 3.75 mmol) to a suspension of **17** (0.43 g, 3 mmol) and lithium chloride (0.16 g, 3.75 mmol) in 30 mL THF. Then, add the acrylic anhydride solution, and stir the resulting mixture for 12 h. Remove the solvent in vacuo, and add 1 M HCl (100 mL). Extract the mixture with dichloromethane, wash the combined organic layers with saturated $NaHCO_3$ aqueous solution and brine, and dry over Na_2SO_4. Concentrate in vacuo and purify by silica gel column chromatography (hexanes/ethyl acetate = 4/1) to give acrylamide **18** as white solid (0.50 g, yield 85%). ^{1}H NMR (500 MHz, $CDCl_3$) δ 7.52 (dd, J= 17.0, 10.4 Hz, 1H), 6.53 (dd, J= 17.0, 1.7 Hz, 1H), 5.90 (dd, J = 10.4, 1.8 Hz, 1H), 4.51 (dd, J = 7.9, 1.3 Hz, 1H), 4.35–4.22 (m, 2H), 0.95 (s, 9H).

3. *tert*-Butyl (*S*,*E*)-2-(4-(3-(4-(*tert*-butyl)-2-oxooxazolidin-3-yl)-3-oxoprop-1-en-1-yl)phenyl)acetate (**20**). Add 80 mL anhydrous Et_3N to a mixture of **19** (4.3 g, 15.7 mmol), **18** (3.7 g, 18.9 mmol), $Pd(OAc)_2$ (353 mg, 1.57 mmol), and tris (*o*-tolyl)-phosphine (958 mg, 3.15 mmol), and reflux the resulting solution under argon overnight. A significant amount of precipitate is formed and silvers the flask wall. Evaporate Et_3N, and dissolve the residue in 150 mL CH_2Cl_2. Remove the solid palladium compound by filtration. Then, wash the filtrate with H_2O and brine, dry over Na_2SO_4, and evaporate to dryness to give yellow solid. Use column chromatography to purify the crude product on silica gel (hexanes/EtOAc = 4/1 ~ 1/1, v/v) to give the desired product as pale yellow solid (4.6 g, yield 76%). ^{1}H NMR (500 MHz, $CDCl_3$) δ 7.92 (d, J = 15.5 Hz, 1H), 7.83 (d, J = 15.5 Hz, 1H), 7.57 (d, J = 8.0 Hz, 2H), 7.30 (d, J = 8.0 Hz, 2H), 4.58 (dd, J= 7.4, 1.8 Hz, 1H), 4.37–4.24 (m, 2H), 3.54 (s, 2H), 1.43 (s, 9H), 0.97 (s, 9H). ^{13}C NMR (126 MHz, $CDCl_3$) δ 170.3, 165.5, 154.7, 146.1, 137.5, 133.3, 129.7, 128.7, 116.8, 81.1, 65.2, 61.0, 42.6, 36.0, 28.0, 25.6. MS (ESI) m/z = 388.3 $[M + H]^+$.

4. *tert*-Butyl 2-(4-((*R*)-5-((*S*)-4-(*tert*-butyl)-2-oxooxazolidin-3-yl)-5-oxopent-1-en-3-yl)phenyl)acetate (**21**, *see* **Note 5**). Add vinyl magnesium bromide in THF (1.0 M, 4.6 mL, 4.6 mmol) dropwise to a slurry of cuprous thiophenoxide (267 mg, 1.5 mmol) in 100 mL of anhydrous Et_2O under argon at −40 °C. Then, allow the mixture to warm to 0 °C when a color

change from brown to black-green which indicated the formation of complex PhSCu (RMgX)n. Stir the mixture at room temperature for an additional 1 h, then add dropwise a solution of **20** (500 mg, 1.3 mmol) in 100 mL anhydrous THF. Stir the mixture at −40 °C with TLC monitoring. At approximately 3 h, all starting material will be consumed. Pour the mixture into an ice-cooled NH_4Cl solution, and remove the solid cuprous salts by filtration. Then, remove THF by evaporation, extract the residue with EtOAc, wash with H_2O and saturated brine, and dry over Na_2SO_4. Concentrate in vacuo and purify by silica gel chromatography (hexanes:EtOAc from 4:1 to 2:1) to afford (*R*)-**21** as white solid (241 mg, yield 45%). ^{1}H NMR (500 MHz, $CDCl_3$) δ 7.20 (s, 4H), 6.03 (ddd, *J* = 17.2, 10.3, 7.0 Hz, 1H), 5.13–5.04 (m, 2H), 4.30 (dd, *J* = 7.7, 1.5 Hz, 1H), 4.20 (dd, *J* = 9.3, 1.5 Hz, 1H), 4.06 (dd, *J* = 9.2, 7.7 Hz, 1H), 3.98 (q, *J* = 7.5 Hz, 1H), 3.48 (s, 2H), 3.41 (dd, *J* = 7.5, 1.6 Hz, 2H), 1.44 (s, 9H), 0.88 (s, 9H). ^{13}C NMR (126 MHz, $CDCl_3$) δ 171.4, 170.9, 154.7, 140.9, 140.4, 133.0, 129.4, 128.0, 114.7, 80.8, 65.3, 61.1, 45.1, 42.2, 40.2, 35.7, 28.0, 25.6. MS (ESI) m/z = 438.3 $[M + Na]^+$.

5. *tert*-Butyl 2-(4-((*R*)-5-((1-(((*S*)-4-amino-1-(((*S*)-2-(naphthalen-1-ylmethyl)pent-4-en-1-yl)amino)-1,4-dioxobutan-2-yl) carbamoyl)cyclohexyl)amino)-5-oxopent-1-en-3-yl)phenyl) acetate (**22**). Add H_2O_2 (30%, 518 μL, 4.55 mmol) via syringe to a solution of **21** (380 mg, 0.91 mmol) in 12 mL THF-H_2O (3:1) at 0 °C over 1 min. Then, add LiOH (65 mg, 2.73 mmol), and stir the reaction mixture 0 °C for 1 h. Bring the mixture to ambient temperature, and stir overnight. Destroy excess hydrogen peroxide by addition of 560 mg of Na_2SO_3 in 5.0 mL of H_2O (*see* **Note 6**), and remove THF in vacuo at 30 °C. Extract the residue with CH_2Cl_2 to remove Evans' reagent. Pour the resulting aqueous solution into ice cold 0.5 M HCl (10 mL), extract with EtOAc, wash with ice-cold H_2O and brine, and dry over Na_2SO_4. Concentrate in vacuo and purify by silica gel column chromatography (dichloromethane:MeOH from 30:1 to 20:1) to afford (*R*)-3-(4-(2-(*tert*-butoxy)-2-oxoethyl)phenyl)pent-4-enoic acid as white solid (200 mg, yield 75%).

   Add the preactivated ester solution [formed by the reaction of (*R*)-3-(4-(2-(*tert*-butoxy)-2-oxoethyl) phenyl) pent-4-enoic acid (200 mg, 0.70 mmol) from the previous step, HOBt (158 mg, 1.03 mmol), and *N*,*N*′-diisopropylcarbodiimide (DIPCDI, 130 mg, 1.03 mmol) in 5 mL anhydrous DMF, 10 min] to a solution of amine **15** (320 mg, 0.70 mmol) in 5 mL of dry DMF. Stir the resulting mixture at room temperature for 12 h, and evaporate DMF under reduced pressure. Dissolve the residue in 50 mL EtOAc, wash with saturated $NaHCO_3$ aqueous solution, H_2O, and brine, and dry over Na_2SO_4. Concentrate in vacuo and purify by silica gel column chromatography (chloroform:methanol, from 20:1 to 9:1) to provide **22** as a foam (309 mg, yield 60%). ^{1}H NMR (500 MHz, $CDCl_3$) δ 8.03 (d, *J* = 8.5 Hz, 1H), 7.96 (d, *J* = 8.2 Hz, 1H), 7.81 (d, *J* = 7.5 Hz, 1H), 7.73 (t, *J* = 5.5 Hz, 1H), 7.69 (dd, *J* = 5.5 Hz, 1H), 7.48–7.40 (m, 2H), 7.36–7.33 (m, 2H), 7.12 (d, *J* = 8.0 Hz, 2H), 7.02 (d, *J* = 8.0 Hz, 2H), 6.48 (s, 1H), 6.11 (s, 1H), 5.92–5.79 (m, 2H), 5.33 (s, 1H), 5.09–4.95 (m, 4H), 4.67–4.63 (m, 1H), 3.73 (q, *J* = 7.3 Hz, 1H), 3.45 (s, 2H), 3.35–3.29 (m, 1H), 3.18–3.09 (m, 2H), 3.00–2.93 (m, 2H), 2.65–2.57 (m, 2H), 2.39 (dd, *J* = 15.1, 5.0 Hz, 1H), 2.22 (p, *J* = 6.6 Hz, 1H), 2.10 (h, *J* = 7.4 Hz, 2H), 2.01 (d, *J* = 14.0 Hz, 1H), 1.87 (d, *J* = 14.0 Hz, 1H), 1.77–1.71 (m, 1H), 1.65 (td, *J* = 13.1, 3.6 Hz, 1H), 1.54–1.49 (m, 3H), 1.44 (s, 9H), 1.15 (q, *J* = 8.8 Hz, 2H), 1.01–0.94 (m, 1H). ^{13}C NMR (126 MHz, $CDCl_3$) δ 174.0, 173.8, 171.6, 170.8, 170.6, 140.7, 140.4, 136.7, 136.4, 133.9, 133.1, 132.1, 129.5, 128.6, 127.7, 127.4, 126.7, 125.8, 125.4, 125.3, 124.2, 116.9, 115.0, 80.9, 59.5, 53.4, 50.4, 45.2, 43.0, 42.2, 42.0, 38.7, 36.0, 35.5, 35.3, 32.2, 31.1, 28.0, 24.9, 21.2. MS (ESI) m/z = 737.6 $[M + H]^+$.

6. 2-(4-((10*R*,14*S*,18*S*,*E*)-18-(2-Amino-2-oxoethyl)-14-(naphthalen-1-ylmethyl)-8,17,20-trioxo-7,16,19-triazaspiro[5.14]icos-11-en-10-yl)phenyl)acetic acid (**3**). Add via syringe the Grubbs II catalyst (139 mg, 0.16 mmol) in 37 mL dichloromethane to a solution of **22** (300 mg, 0.41 mmol) in 110 mL anhydrous dichloromethane (deoxygenized with argon) (*see* **Note 7**). Then, reflux the solution was under argon for 60 h with TLC monitoring. Evaporate the solvent in vacuo, and purify the residue by silica gel column chromatography (chloroform:EtOAc:MeOH from 2:1:0 to 14:7:1) to provide *tert*-butyl 2-(4-((10*R*,14*S*,18*S*,*E*)-18-(2-amino-2-oxoethyl)-14-(naphthalen-1-ylmethyl)-8,17,20-trioxo-7,16,19-triazaspiro [5.14]icos-11-en-10-yl)phenyl)acetate as brown crude solid.

   Treatment of the crude product from the previous step with a solution of TFA–triethylsilane–H_2O (3.7 mL–0.1 mL–0.2 mL) at room temperature. Stir the reaction mixture at room temperature for 1 h, and evaporate the solvent under reduced pressure, providing a precipitate which was then dried under high vacuum to provide the crude product. Purify the crude solid by the semipreparative C18-reversed phase silica gel column chromatography (acetone–water 1:1 to 2:1) to provide compound **3** as light yellow solid, and then recrystallize it with methanol to afford the product as white solid (110 mg, 60% yield over two steps). ^{1}H NMR (500 MHz, DMSO-d_6) δ 12.26

(s, 1H), 8.40 (s, 1H), 8.24 (d, J = 8.2 Hz, 1H), 8.13 (d, J = 8.5 Hz, 1H), 7.90 (d, J = 8.1 Hz, 1H), 7.76 (d, J = 7.4 Hz, 1H), 7.56–7.39 (m, 6H), 7.18–7.14 (m, 4H), 7.06 (s, 1H), 5.56 (dd, J = 15.0, 9.4 Hz, 1H), 5.41 (t, J = 14.0 Hz, 1H), 4.31–4.27 (m, 1H), 3.88 (t, J = 11.0 Hz, 1H), 3.58 (dd, J = 13.1, 6.4 Hz, 1H), 3.50 (s, 2H), 3.15 (dd, J = 14.0, 5.7 Hz, 1H), 2.88–2.72 (m, 3H), 2.62–2.57 (m, 1H), 2.41–2.34 (m, 2H), 2.16 (brs, 1H), 2.01 (t, J = 16.8 Hz, 2H), 1.88–1.77 (m, 2H), 1.73–1.68 (m, 2H), 1.54–1.45 (m, 4H), 1.23–1.19 (m, 2H). ^{13}C NMR (126 MHz, DMSO-d_6) δ 173.7, 173.1, 172.8, 170.1, 170.0, 143.5, 136.3, 134.0, 133.4, 132.6, 131.8, 129.3, 129.3, 128.5, 127.1, 126.7, 126.5, 125.9, 125.5, 125.3, 124.0, 58.4, 50.0, 45.7, 45.4, 41.9, 40.3, 38.7, 37.4, 35.1, 34.6, 28.4, 25.0, 21.0, 20.7. HRMS (ESI) Calcd for $C_{38}H_{43}N_4O_6$ (M – H)$^-$ 651.3188, found 651.3196. HPLC purity 95.2%, t_R = 15.75 min (*see* **Note 8**).

7. Methyl 2-(4-((10*R*,14*S*,18*S*,*E*)-18-(2-amino-2-oxoethyl)-14-(naphthalen-1-ylmethyl)-8,17,20-trioxo-7,16,19-triazaspiro [5.14]icos-11-en-10-yl)phenyl)acetate (**4**). Add K_2CO_3 (2 mg, 0.016 mmol) to a solution of **3** (5 mg, 0.008 mmol) in 1 mL DMF at room temperature. After 5 min, add MeI (2.4 μL, 0.039 mmol) to the reaction mixture. Then, stir the reaction mixture at room temperature for 6 h, and remove DMF under reduced pressure. Purify the residue by a small silica gel column chromatography to yield the final compound **4**. ^{1}H NMR (500 MHz, DMSO-d_6) δ 8.40 (s, 1H), 8.23 (d, J = 8.1 Hz, 1H), 8.15–8.11 (m, 1H), 7.90 (dd, J = 8.1, 1.5 Hz, 1H), 7.76 (dd, J = 7.4, 2.0 Hz, 1H), 7.56–7.48 (m, 3H), 7.45–7.37 (m, 3H), 7.18–7.14 (m, 4H), 7.05 (d, J = 2.4 Hz, 1H), 5.56 (ddd, J = 14.9, 9.5, 2.0 Hz, 1H), 5.39 (ddd, J = 14.6, 10.8, 3.2 Hz, 1H), 4.29 (dt, J = 8.2, 5.0 Hz, 1H), 3.88 (ddd, J = 12.2, 9.4, 2.3 Hz, 1H), 3.59 (d, J = 10.2 Hz, 6H), 3.15 (dd, J = 13.9, 5.6 Hz, 1H), 2.87–2.71 (m, 3H), 2.59 (ddd, J = 13.2, 8.5, 4.7 Hz, 1H), 2.42–2.34 (m, 2H), 2.16 (s, 1H), 2.00 (dd, J = 25.1, 13.5 Hz, 2H), 1.88–1.66 (m, 4H), 1.56–1.45 (m, 4H), 1.23 (d, J = 17.9 Hz, 2H). ^{13}C NMR (126 MHz, DMSO-d_6) δ 174.2, 173.5, 173.2, 172.1, 170.5, 144.3, 136.8, 134.4, 133.9, 132.4, 132.3, 129.7, 127.5, 127.3, 127.0, 126.4, 126.0, 125.8, 124.5, 58.8, 52.1, 50.5, 45.9, 21.4, 21.2. HRMS (ESI) Calcd for $C_{39}H_{47}N_4O_6$ (M + H)$^+$ 667.3490, found 667.3489.

8. HPLC purity analysis of the final products. The HPLC analysis used the gradient mobile phase starting with 0.1% TFA in water and ending with 0.1% TFA in water and acetonitrile mixture (water with 0.1% TFA: acetonitrile = 1: 1) for 10 min, and then changed to a 10 min gradient starting with 0.1% TFA in water and acetonitrile 1: 1 mixture and ending with 100% acetonitrile. The purity of compounds **3** and **4** should be ≥95%.

3.2 GRB2 Protein Expression and Purification

1. Culture the *E. coli* BL21 DE3 cells with the GRB2-containing pEHISTEV vector in LB medium with 50 μg/mL kanamycin until the OD_{600} was approximately 1.0.
2. Then, induce the protein expression with 500 μM IPTG at 16 °C overnight.
3. Lyse the cells by sonication.
4. Purify the proteins by two steps of chromatography, including Ni-NTA affinity chromatography and size-exclusion chromatography with a HiLoad 26/600 Superdex 200 pg column using an ÄKTA Pure FPLC system. Elute with the Tris buffer for the GRB2 purification.
5. Determine the purity of GRB2 by SDS-PAGE gel analysis to be greater than 95%. Proteins can be aliquoted and stored at −80 °C.

3.3 Fluorescence Anisotropy-Binding Assays

1. Check the note before the assays (*see* **Note 9**). Perform the experiments in the fluorescence anisotropy assay buffer (*see* **Note 10**).
2. Fix the concentration of human $EGFR_{1068}$ phosphopeptide fluorescence tracer at 5 nM.
3. Add different concentrations of full-length human GRB2 (residues 1–217) to the assay buffer giving a final volume of 100 μL.
4. After the addition, cover the assay plate and gently mix on an orbital shaker for 2 h before the data were recorded at room temperature with an excitation wavelength at 485 nM and an emission wavelength at 535 nM.
5. Repeat each experiment (**steps 1–4**) for three times.
6. Express the results as mean ± standard deviation. The parallel fluorescence intensity (I_s), the perpendicular fluorescence intensity (I_p), and the anisotropy (r) are recorded directly by the plate reader. The total intensity (I) and the fraction ligand bound (L_b) have to be calculated by Eqs. 1 and 2 shown below [21, 22].

$$I = 2 \times I_p \times G + I_s \tag{1}$$

G: the G factor. It was 0.993 for the instrument used.

$$L_b = \frac{r - r_{min}}{\lambda \times (r_{max} - r) + (r - r_{min})} \tag{2}$$

r_{min}: the average anisotropy value for the fluorescently labeled phosphotyrosine peptide.

r_{max}: the average anisotropy value for the fluorescently labeled phosphotyrosine peptide with saturated GRB2.

$$\lambda = \frac{I_{bound}}{I_{unbound}}$$

I_{bound}: the average intensity value for the fluorescently labeled phosphotyrosine peptide with saturated GRB2.

I_{unbound}: the average intensity value for the fluorescently labeled phosphotyrosine peptide.

7. Import the above data to GraphPad Prism 8.0, and analyze the K_D value of the interaction between GRB2 and human $EGFR_{1068}$ phosphopeptide by the nonlinear regression equation (Eq. 3) shown below.

$$Y = \frac{(K_D + X + [\text{fluorescent} - \text{peptide}]) - \sqrt{(K_D + X + [\text{fluorescent} - \text{peptide}])^2 - 4 \times X \times [\text{fluorescent} - \text{peptide}]}}{2} \quad (3)$$

$Y = L_b \times$ [fluorescent-peptide]. [fluorescent-peptide], the concentration was 5 nM.

$X =$ [GRB2]

3.4 Fluorescence Anisotropy Competitive Inhibition Assays

1. Measure the anisotropy at room temperature with an excitation wavelength at 485 nM and an emission wavelength at 535 nM.
2. Set the final reaction volume to 100 μL. Incubate 200 nM human GRB2 (residues 1–217) with 2.5 nM *N*-terminally fluorescein-labeled human $EGFR_{1068}$ phosphopeptide for 30 min at room temperature, and then add different concentrations of the peptide or the compounds in the fluorescence anisotropy assay buffer. The negative control (equivalent to 0% inhibition) refers to 2.5 nM $EGFR_{1068}$ phosphopeptide fluorescence tracer and 200 nM GRB2 in assay buffer without the inhibitor. The positive control (equivalent to 100% inhibition) refers to only 2.5 nM $EGFR_{1068}$ phosphopeptide fluorescence tracer in the assay buffer.
3. Cover each assay plate black and gently mix on an orbital shaker at room temperature for 2.5 h to reach equilibrium before the anisotropy values are read.
4. After measuring the anisotropy values, correct the background of the tested inhibitors by subtracting the raw intensity values of the sample background wells (all components except probe) from the raw intensity values of the corresponding test wells (all components).
5. Determine the IC_{50} values by GraphPad Prism 8.0. Derive the K_i values from the IC_{50} values. The equation used is $K_i = [I]_{50}/([L]_{50}/K_D + [P]_0/K_D + 1)$ (Where $[I]_{50}$ denotes the concentration of the free inhibitor at 50% inhibition, $[L]_{50}$ is the concentration of the free labeled ligand at 50% inhibition, $[P]_0$ is the concentration of the free protein at 0% inhibition, and K_D is the dissociation constant of the protein–ligand

complex). This equation is used to derive K_is for competitive inhibition assays (https://bioinfo-abcc.ncifcrf.gov/IC50_Ki_Converter/index.php) [23, 24].

6. Perform all of the experiments in triplicate and in the presence of 1% DMSO for inhibitors. Assay each compound at least by three independent experiments, and express the results as mean ± standard deviation.

3.5 PAMPA Assays

1. Use the 96-well filtration plates as the artificial membrane support and the receiver plate (*see* **Note 11**). Wet the filter material in each well of the filtration plate with 7.5 μL of the artificial membrane solution.
2. Securely place the filtration plate on top of a donor plate which was prefilled with the donor solution (500 μM compound solution, 280 μL) in the PAMPA phosphate buffer (pH 7.4) in each well.
3. Quickly add equal volumes of the blank receiving solution (phosphate buffer, pH = 7.4) to the wells of the filtration plate.
4. Incubate the stacked donor–receiver plates at room temperature for 5 h with gentle circular shaking. After incubation, assay the receiving solution against the concentrations of the initial donor solution using HPLC (*see* **Note 12**).

   HPLC conditions. The HPLC conditions for the PAMPA assays are identical with that for the compound purity analysis described below. Set the DAD detector to 280 nM. Inject the samples (100 μL) into the HPLC column. Apply the gradient elution: 100% H_2O (0.1% TFA) to H_2O (0.1% TFA): MeCN = 50:50 from 0 to 10 min; and H_2O (0.1% TFA): MeCN = 50:50 to 100% MeCN from 10 to 15 min. The flow rate is 1.5 mL/min. The column has to be equilibrated to each starting mobile phase for approximately 10 min between runs.
5. The results of the artificial membrane permeability are expressed as the percent transport (%T) using Eq. 4.

$$\%T = 100 \times \frac{A_R \times V_R}{A_{D0} \times V_D} \tag{4}$$

where A_{D0} and A_R are the HPLC peak areas of the initial donor solution and the receiving solution after incubation, V_R and V_D are the volumes of the receiving and donor solution, respectively.

The $\%T$ is related to the apparent permeability coefficient P_{app} using Eq. 5.

$$P_{app} = \frac{V_D \times V_R}{(V_D + V_R) \times S \times t} \ln \left[\frac{100 \times V_D}{100 \times V_D - \%T \times (V_D + V_R)} \right] \tag{5}$$

where S is the surface area of the artificial membrane, and t is the incubation time.

4 Notes

1. Compound **8** is a key intermediate, and its synthetic route was reported previously [25]. In our early attempts, 4-pentenoic acid was reacted with pivaloyl chloride to form the mixed anhydride which was then coupled with lithium oxazolidinone to generate (*R*)-3-(pent-4-enoyl)-4-phenyl-1,3-oxazolidin-2-one (**23**) in 60% yield (*see* Fig. 4). However, our synthesis between **23** and 1-(bromomethyl) naphthalene was unsuccessful. In addition to the literature LiHMDS/THF condition, we tried NaH/THF, LDA/THF, LiHMDS/HMPT, NaHMDS/THF, and *n*-BuLi/THF conditions. We then developed an alternative synthetic route for **8** (*see* Fig. 2). The Horner–Wadsworth–Emmons (HWE) reaction between 1-naphthaldehyde and ethyl 2-(diethoxyphosphoryl) acetate gives **5** in 94% yield. The olefin bond of **5** can be saturated under the Pd/C-H_2 condition, and the ethyl ester group can be hydrolyzed under the basic condition to offer **6** in 83% yield over two steps. The carboxylic acid group of **6** can be converted to acyl chloride, and the reaction with Evans' (*S*)-(+)-4-phenyl-2-oxazolidinone chiral auxiliary affords **7** in 90% yield over two steps. Subsequent alkylation with allyl iodide under the NaHMDS condition offers **8** as a single isolated diastereomer after silica gel column chromatography.
2. It is critical to run this reaction under the anhydrous condition.
3. $LiAlH_4$ needs be added slowly and portion-wise.
4. In the literature, phthalimide was used to react with compound **9** through the Mitsunobu reaction and then the hydrazine-mediated cleavage liberated the primary amin **11** [25]. The Gabriel synthesis is chosen in our study. The alcohol group in **9** can be activated by tosylation and then reacted with Boc_2NH. The deprotection of the Boc-protecting groups results in amine **11**. With **11** in hand, key intermediate **15** can successfully be synthesized by following the literature method [25].

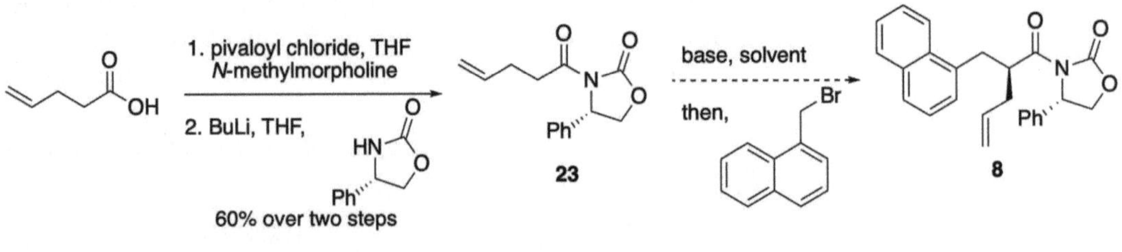

Fig. 4 The early synthetic attempts toward compound **8**

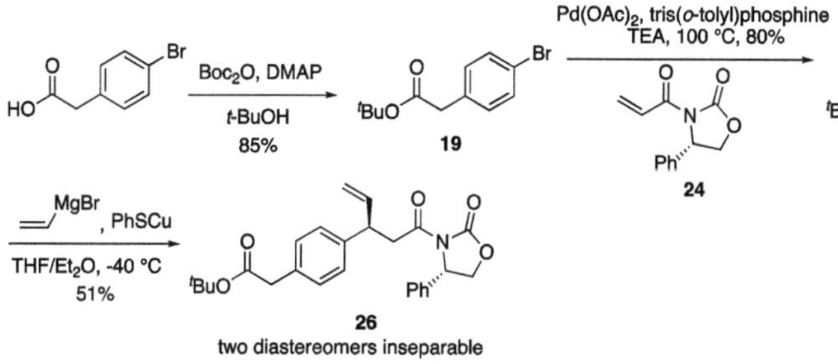

Fig. 5 The attempted synthetic route for compound **25**. (Adapted with permission from ref. 19. Copyright 2021 Elsevier)

5. In our early synthesis, the commercially available Evans' chiral auxiliary, (*S*)-(+)-4-phenyl-2-oxazolidinone, was first attempted to synthesize **26** (*see* Fig. 5) [25, 26]. However, two diastereomers of **26** with a diastereometric ratio (dr) of 3:1 were inseparable by column chromatography. We then screened other Evans' chiral auxiliaries, including (*S*)-4-benzyl-2-oxazolidinone, (*S*)-(−)-4-isopropyl-2-oxazolidinone, (*S*)-(−)-4-benzyl-5,5-dimethyl-2-oxazolidinone, and (*S*)-4-*tert*-butyl-2-oxazolidinone. The chiral auxiliary (*S*)-4-*tert*-butyl-2-oxazolidinone (**17**) offered the highest diastereomer selectivity for the desired product **21** (dr = 100:6) and was used in this synthesis. Compound **20** can undergo 1,4-addition with vinylmagnesium bromide under the PhSCu condition [25] to afford **21** with reproducible yield and high diastereoselectivity. This reaction needs to be operated under the anhydrous condition.
6. It is critical to decompose H_2O_2 before the rotary evaporation in this step. More than one equivalent of Na_2SO_3 are used to destroy excess H_2O_2.
7. This intramolecular ring-closing metathesis (RCM) reaction should be conducted in a quite diluted reaction solution (0.002 M/L in this case) to avoid intermolecular olefin metathesis from occurring during the reaction.
8. The final product **3** is obtained as a single E isomer. The trans configuration of the alkene can be determined by ^{1}H NMR which shows the coupling constant (*J*) of 15 Hz for two vicinal alkene protons.
9. The inhibitory activity of new GRB2 inhibitors for disruption of the protein–protein interaction (PPI) between GRB2 and phosphotyrosine-containing sequence is measured by fluorescence anisotropy competitive inhibition experiments similar to the previously reported GRB2 fluorescence polarization (FP) assay [27]. *N*-terminally His$_6$-tagged human full-length GRB2 (residues 1–217) is overexpressed in *E. coli* and purified. It is

Table 1
Fluorescence anisotropy competitive inhibition assays to determine IC_{50}s of 1, 3, 4, and $EGFR_{1068}$ phosphopeptide for disruption of the interaction between full-length GRB2 and FITC-Ahx-PVPEpYINQSVPKRK-NH_2

Compounds	$IC_{50} \pm$ SD (nM)	$K_i \pm$ SD (nM)
$EGFR_{1068}$ phosphopeptide	1560 ± 219	400 ± 27
1	350 ± 40	16 ± 0.84
3	740 ± 140	140 ± 15
4	>20,000	

The K_i values are derived by the procedure described in Subheading 3.4, **step 5**. The data is expressed as mean ± standard deviation ($n = 3$)
$EGFR_{1068}$ phosphopeptide: Ac-PVPEpYINQSVPKRK-NH_2

known that GRB2 is preferred to bind with pY1068 of EGFR [28]. *N*-terminally fluorescein (FITC)-labeled human $EGFR_{1068}$ phosphopeptide, FITC-Ahx-PVPEpYINQSVPKRK-NH_2 (Ahx: 6-aminohexanoic acid), can be used in this assay. The dissociation constant (K_D) of this EGFR/GRB2 PPI in fluorescence anisotropy-binding experiments is around 191 nM and consistent with the reported K_D [29, 30]. The fluorescence anisotropy competitive inhibition assays was used to evaluate the inhibitory activities of **1**, **3**, and **4**, which was compared with that of the $EGFR_{1068}$ phosphopeptide. Our results are shown in Table 1. Compound **3** displays a K_i of 140 nM and is two-fold more potent than the EGFR1068 phosphopeptide 14-mer.

10. It is important to add 0.01% Triton X-100 and 100 μg/mL γ-globulin to the fluorescence anisotropy assay to minimize non-specific interactions. These two assay components prevent aggregation, ensure solubility and stability of assay components, reduce protein adsorption to assay plate wells, and reduce the assay background. The critical micelle concentration (CMC) of Triton X-100 is 0.02%. The Triton X-100 concentration used in the assay needs to be below the CMC value to avoid formation of micelles.
11. The PAMPA assay was used to assess the permeability of **3** through the artificial membrane. This is a useful system to examine the passive cell permeability of the compounds. Compounds **3** (500 μM) was placed on the donor side of the membrane.
12. After 5-h incubation at room temperature, the amounts of **3** in the receiving solution are quantified by HPLC analyses. The percent transport (%T) and the apparent permeability coefficient (P_{app}) were calculated using the previously reported equations [31, 32]. The PAMPA results show that **3** has good

Table 2
The PAMPA results of compounds 1 and 3

Compound	%$T \pm$ SD	$P_{app} \pm$ SD ($cm \cdot s^{-1} \times 10^{-6}$)
1	0	0
3	24.3 ± 2.5	19.2 ± 2.0

permeability through the artificial membrane, while **1** displays poor permeability in this assay. The results are shown in Table 2.

Acknowledgments

This work was supported by 2017 and 2018 Moffitt Cancer Center Molecular Medicine (MM) Program Innovation Funds and 2019 Moffitt Cancer Center Lung Cancer Center for Excellence DeBartolo Thoracic Research Funds. We also thank the Moffitt Chemical Biology Core for use of the NMR and mass spectrometry facilities supported by National Cancer Institute grant P30-CA76292.

References

1. Cheng AM, Saxton TM, Sakai R, Kulkarni S, Mbamalu G, Vogel W, Tortorice CG, Cardiff RD, Cross JC, Muller WJ, Pawson T (1998) Mammalian Grb2 regulates multiple steps in embryonic development and malignant transformation. Cell 95:793–803
2. Lemmon MA, Schlessinger J (2010) Cell signaling by receptor tyrosine kinases. Cell 141: 1117–1134
3. Bivona TG (2019) Dampening oncogenic RAS signaling. Science 363:1280–1281
4. Ren R (2005) Mechanisms of BCR-ABL in the pathogenesis of chronic myelogenous leukaemia. Nat Rev Cancer 5:172–183
5. Fretz H, Furet P, Garcia-Echeverria C, Schoepfer J, Rahuel J (2000) Structure-based design of compounds inhibiting Grb2-SH2 mediated protein-protein interactions in signal transduction pathways. Curr Pharm Des 6: 1777–1796
6. Sawyer TK, Bohacek RS, Dalgarno DC, Eyermann CJ, Kawahata N, Metcalf CA III, Shakespeare WC, Sundaramoorthi R, Wang Y, Yang MG (2002) SRC homology-2 inhibitors: peptidomimetic and nonpeptide. Mini-Rev Med Chem 2:475–488
7. Burke TR Jr (2006) Development of Grb2 SH2 domain signaling antagonists: a potential new class of antiproliferative agents. Int J Pept Res Ther 12:33–48
8. Kraskouskaya D, Duodu E, Arpin CC, Gunning PT (2013) Progress towards the development of SH2 domain inhibitors. Chem Soc Rev 42:3337–3370
9. Morlacchi P, Robertson FM, Klostergaard J, McMurray JS (2014) Targeting SH2 domains in breast cancer. Future Med Chem 6:1909–1926
10. Gao Y, Voigt J, Wu JX, Yang D, Burke TR Jr (2001) Macrocyclization in the design of a conformationally constrained Grb2 SH2 domain inhibitor. Bioorg Med Chem Lett 11: 1889–1892
11. Wei C-Q, Gao Y, Lee K, Guo R, Li B, Zhang M, Yang D, Burke TR Jr (2003) Macrocyclization in the design of Grb2 SH2 domain-binding ligands exhibiting high potency in whole-cell systems. J Med Chem 46:244–254
12. Yao Z-J, King CR, Cao T, Kelley J, Milne GW, Voigt JH, Burke TR Jr (1999) Potent inhibition of Grb2 SH2 domain binding by non-phosphate-containing ligands. J Med Chem 42:25–35

13. Shi ZD, Lee K, Liu H, Zhang M, Roberts LR, Worthy KM, Fivash MJ, Fisher RJ, Yang D, Burke TR Jr (2003) A novel macrocyclic tetrapeptide mimetic that exhibits low-picomolar Grb2 SH2 domain-binding affinity. Biochem Biophys Res Commun 310:378–383
14. Wei CQ, Li B, Guo R, Yang D, Burke TR Jr (2002) Development of a phosphatase-stable phosphotyrosyl mimetic suitably protected for the synthesis of high-affinity Grb2 SH2 domain-binding ligands. Bioorg Med Chem Lett 12:2781–2784
15. Zhang M, Luo Z, Liu H, Croce CM, Burke TR Jr, Bottaro DP (2014) Synergistic anti-leukemic activity of imatinib in combination with a small molecule Grb2 SH2 domain binding antagonist. Leukemia 28:948–951
16. Vidal M, Gigoux V, Garbay C (2001) SH2 and SH3 domains as targets for anti-proliferative agents. Crit Rev Oncol Hematol 40:175–186
17. Burke TR Jr, Lee K (2003) Phosphotyrosyl mimetics in the development of signal transduction inhibitors. Acc Chem Res 36:426–433
18. Bradshaw JM, Waksman G (2002) Molecular recognition by SH2 domains. Adv Protein Chem 61:161–210
19. Xiao T, Sun L, Zhang M, Li Z, Haura EB, Schonbrunn E, Ji H (2021) Synthesis and structural characterization of a monocarboxylic inhibitor for GRB2 SH2 domain. Bioorg Med Chem Lett 51:128354
20. Lang K, Park J, Hong S (2010) Development of bifunctional aza-bis(oxazoline) copper catalysts for enantioselective Henry reaction. J Org Chem 75:6424–6435
21. Plante JP, Burnley T, Malkova B, Webb ME, Warriner SL, Edwards TA, Wilson AJ (2009) Oligobenzamide proteomimetic inhibitors of the p53-hDM2 protein-protein interaction. Chem Commun 34:5091–5093
22. Yeo DJ, Warriner SL, Wilson AJ (2013) Monosubstituted alkenyl amino acids for peptide “stapling”. Chem Commun 49:9131–9133
23. Nikolovska-Coleska Z, Wang R, Fang X, Pan H, Tomita Y, Li P, Roller PP, Krajewski K, Saito NG, Stuckey JA, Wang S (2004) Development and optimization of a binding assay for the XIAP BIR3 domain using fluorescence polarization. Anal Biochem 332:261–273
24. Cer RZ, Mudunuri U, Stephens R, Lebeda FJ (2009) IC50-to-Ki: a web-based tool for converting IC50 to Ki values for inhibitors of enzyme activity and ligand binding. Nucleic Acids Res 37:W441–W445
25. Gao Y, Wei CQ, Burke TR Jr (2001) Olefin metathesis in the design and synthesis of a globally constrained Grb2 SH2 domain inhibitor. Org Lett 3:1617–1620
26. Burke TR Jr, Liu D-G, Gao Y (2000) Use of a Heck reaction for the synthesis of a new α-azido phosphotyrosyl mimetic suitably protected for peptide synthesis. J Org Chem 65: 6288–6291
27. Luzy J-P, Chen H, Gril B, Liu W-Q, Vidal M, Perdereau D, Burnol A-F, Garbay C (2008) Development of binding assays for the SH2 domain of Grb7 and Grb2 using fluorescence polarization. J Biomol Screen 13:112–119
28. Batzer AG, Rotin D, Urena JM, Skolnik EY, Schlessinger J (1994) Hierarchy of binding sites for Grb2 and Shc on the epidermal growth factor receptor. Mol Cell Biol 14:5192–5201
29. Lemmon MA, Ladbury JE, Mandiyan V, Zhou M, Schlessinger J (1994) Independent binding of peptide ligands to the SH2 and SH3 domains of Grb2. J Biol Chem 269: 31653–31658
30. Chook YM, Gish GD, Kay CM, Pai EF, Pawson T (1996) The Grb2-mSos1 complex binds phosphopeptides with higher affinity than Grb2. J Biol Chem 271:30472–30478
31. Wohnsland F, Faller B (2001) High-throughput permeability pH profile and high-throughput alkane/water log P with artificial membranes. J Med Chem 44:923–930
32. Zhu C, Jiang L, Chen T-M, Hwang K-K (2002) A comparative study of artificial membrane permeability assay for high throughput profiling of drug absorption potential. Eur J Med Chem 37:399–407

Part IV

Engineering SH2 Domains

Chapter 16

Engineering of SH2 Domains for the Recognition of Protein Tyrosine *O*-Sulfation Sites

Sean Paul Waldrop, Wei Niu, and Jiantao Guo

Abstract

Protein engineering has brought advances to industrial processes, biomaterials, nanotechnology, biosensors, and biomedical applications. This chapter will focus on the engineering of Src Homology 2 domains (SH2) to act as an antibody mimetic for the recognition of sulfotyrosine-containing peptides or proteins. In comparison to anti-sulfotyrosine antibodies, SH2 mutants have much smaller size and can be heterologously expressed and purified in large quantity at low cost. This chapter will describe the use of phage display to identify a sulfotyrosine-binding SH2 mutant and the subsequent enrichment of sulfotyrosine-containing peptides in complex biological samples.

Key words Src Homology 2 domain, Sulfotyrosine, Protein tyrosine *O*-sulfation, Phage display, Protein engineering

1 Introduction

SH2 domains are a class of conserved protein modules of approximately 100 residues. The human genome encodes a total of 121 SH2 domains in 111 proteins. They mainly function as regulatory modules of intracellular signaling cascades by interacting specifically with phosphotyrosine (pTyr)-containing targets in a protein sequence-specific manner [1–5]. The core structure of the SH2 domain comprises a central hydrophobic antiparallel β-sheet, flanked by 2 short α-helices [6]. The pTyr-binding site is a deep and positively charged pocket [7], and the "specificity determine region" is usually a hydrophobic pocket that recognizes the side chain of surrounding residues (mainly the pTyr + 3 residue plus lesser contributions from nearby residues [8]; pTyr is defined as position 0) with nearly the same level of binding free-energy change (ΔG°) [7]. Previous work has shown that SH2 domains were highly evolvable toward their pTyr-containing substrates [9]. In this chap-

Teresa Carlomagno and Maja Köhn (eds.), *SH2 Domains: Functional Modules and Evolving Tools in Biology*, Methods in Molecular Biology, vol. 2705, https://doi.org/10.1007/978-1-0716-3393-9_16,

ter, we describe the engineering of a SH2 domain to improve its affinity toward sulfotyrosine (sTyr)-containing peptides (sulfopeptides) or proteins (sulfoproteins) [10, 11].

1.1 Protein Tyrosine O-Sulfation and Research Tools

First observed in fibrinogen in 1954 [12], protein tyrosine O-sulfation is a posttranslational modification in which a sulfate group is added to tyrosine residues of proteins under the catalytic activities of tyrosylprotein sulfotransferases (TPSTs) [13, 14]. It has been linked to both essential physiological processes and the development of diseases [15–20]. Despite studies suggesting that roughly 7% of mammalian proteins may contain sulfotyrosine there is a relatively low amount of reported sulfoproteins. Given the potential prevalence and low discovery rate [21] of sulfoproteins along with the importance of their function, efforts have been focused on development methods for the detection and enrichment of sulfoproteins.

Radioactive S^{35} labeling has been used to label and then detect sulfoproteins. However, labeling of cysteine and methionine can be problematic, and distinguishing between sulfated carbohydrates and tyrosine residues can be difficult [22]. Mass spectrometry has had some success in identifying sulfopeptides and sulfoproteins [22–30]. At the same time, mass spectrometry analysis has also faced some difficulties due to poor ionization efficiency and fragmentation leading to loss of the sulfate moiety [25]. Therefore, several approaches to enrich sulfoproteins before mass spectrometry analysis have been developed, including the use of metal affinity columns [26], imprinted polymers [31], weak anion exchange columns [32], and antibodies [24, 33]. Among them, anti-sulfotyrosine antibodies showed the best selectivity. However, these antibodies show relatively low affinity, making their use in proteomics type research untenable [34]. Our group chose to create an anti-sulfotyrosine antibody mimetic by engineering SH2 domain through phage display. These antibody mimetics showed nanomolar affinity toward sulfotyrosine and showed the ability to label sulfoproteins on the cell membrane [34]. More recently our group has reported new SH2 domain mutants that not only have high affinity for sulfotyrosine but show selectivity toward sulfotyrosine over phosphotyrosine [11].

1.2 Protein Engineering Through Directed Evolution

In order to improve or create desired activities of proteins, directed evolution is often used. It is a process by which a gene of interest is modified, expressed, and evaluated for desired function, and selected for the most promising candidate [35]. There are several methods via which this process can be conducted, including focused mutagenesis, randomized mutagenesis, and recombination [35]. Focused mutagenesis is usually based on prior knowledge of specific residues that are responsible for the desired function of the protein. The typical approach uses degenerate codons at positions of interest to introduce variation. Randomized mutagenesis allows for the sampling of different genetic variants without the need of defining regions of the gene that must be varied. It can be carried out in several ways, such as using mutagens or error prone polymerase chain reaction (PCR). Recombination method rearranges the sequence of the gene of interest either via homologous recombination or sequence homology-independent recombination.

Following alteration of gene of interest, a method for identifying mutants with desired activity must be applied. This process is either carried out by screening or selection. Selection links the target function either to the direct separation of the mutant or the survival of organisms carrying that specific mutant [35]. Screening attempts to find the desirable phenotype from the mutant pool. A common method used for affinity-based selection is phage display [36]. It uses coat proteins of bacteriophages and fuses them to the protein of interest. The protein mutants with the highest affinity for the immobilized target can then be selected and their amino acid sequence be determined. This method allows for the rapid maturation of high affinity mutants and will be discussed in detail in this chapter for the selection of SH2 mutants for sulfotyrosine detection and enrichment.

2 Materials

2.1 Phage Display and Biopanning

1. 37 °C and/or 30 °C incubator.
2. 37 °C and/or 30 °C shaker.
3. Refrigerated centrifuge.
4. Autoclave for sterilization.
5. Fluorometer or fluorescence plate reader.
6. UV-Vis spectrometer.
7. LB liquid medium.
8. LB agar plates.
9. Ampicillin (Amp), 100 mg/mL (1000×) stock solution in sterile H_2O, stored at −20 °C.
10. Kanamycin (Kan), 50 mg/mL (1000×) stock solution in sterile H_2O, stored at −20 °C.
11. Tetracycline (Tet), 12 mg/mL (1000×) stock solution in 37.5% ethanol, stored at −20 °C).
12. *E. coli* TOP10F' cells (ThermoFisher, for all cloning and phage propagation experiments).
13. Neutravidin-coated 96-well plate.
14. Biotin-conjugated sulfopeptide (commercial vendors).
15. Biotin-conjugated phosphopeptide (commercial vendors).

16. Biotin-conjugated unmodified peptide (commercial vendors).
17. 20% polyethylene glycol (PEG)/2.5 M NaCl solution.
18. Phage wash buffer: PBS, 0.05% Tween-20 pH 7.4.
19. Phage blocking buffer: PBS, 0.01% BSA, 0.05% Tween-20.
20. Trypsin.
21. Hyperphage.
22. M13KO7 phage.
23. pSEX vector (Progen).
24. 2YT media.

2.2 Phage ELISA

1. Sulfuric acid.
2. TMB substrate for ELISA.
3. Anti M13/HRP antibody.
4. Neutravidin-coated 96-well plate.

2.3 Protein Purification and Characterization

1. BL21(DE3) cells.
2. LB broth.
3. IPTG.
4. Sonication buffer: 20 mM potassium phosphate buffer (pH 7.5), 500 mM NaCl, 5% glycerol, and 0.5% Triton-100.
5. Bradford assay kit.
6. Ni Sepharose protein purification resin.
7. SDS-PAGE gel.
8. Bovine serum albumin standards (0.125, 0.25, 0.5, 0.75, 1 mg/mL).
9. SDS loading buffer, 2×.
   - 2 mL of 1 M Tris–HCl, pH 6.8.
   - 4 mL of 10% SDS.
   - 2 mL glycerol.
   - 4 mg bromophenol blue.
   - Add water to a volume of 9.5 mL.
   - Store at room temperature for up to 12 months.
   - Immediately before use, mix 5 μL β-mercaptoethanol with 95 μL above buffer.
10. 2-Mercaptoethanol.
11. SDS-PAGE gel.
12. 10× Tris/Glycine/SDS.
13. Prestained protein ladder.

14. Fluorescence polarization buffer: 20 mM potassium phosphate, pH 7.35, 100 mM NaCl, 2 mM DTT, 0.1% bovine gamma globulin (BGG).
15. FITC-conjugated sulfopeptide (commercial vendors).
16. FITC-conjugated phosphopeptide (commercial vendors).
17. FITC-conjugated unmodified peptide (commercial vendors).

2.4 Enrichment of a Sulfoprotein

1. Bovine serum albumin.
2. Sulfoprotein.
3. Ni-sepharose 6 Fast Flow affinity resin.
4. Wash buffer 1 (PBS pH 7.4, 0.1% NP-40, 20 mM imidazole).
5. Wash buffer 2 (PBS pH 7.4, 0.1% NP-40, 50 mM imidazole).
6. Elution buffer (PBS pH 7.4, 0.1% NP-40, 200 mM phenyl sulfate).
7. HEK293T cells.

3 Methods

3.1 Phage Library Preparation

Here, four residues (R35, S37, E38, and T39) that interact with the phosphate group of pTyr are chosen for randomization (Fig. 1). These four residues are randomized to all 20 amino acids using the NNK degenerate codon. A hyperphage display system [36] is utilized for selection of the mutant library. Hyperphage provides several advantages to regular phage display systems. This system displays library proteins on all 5 copies of the pIII coat protein and displays the proteins on a higher percentage of the phage being expressed.

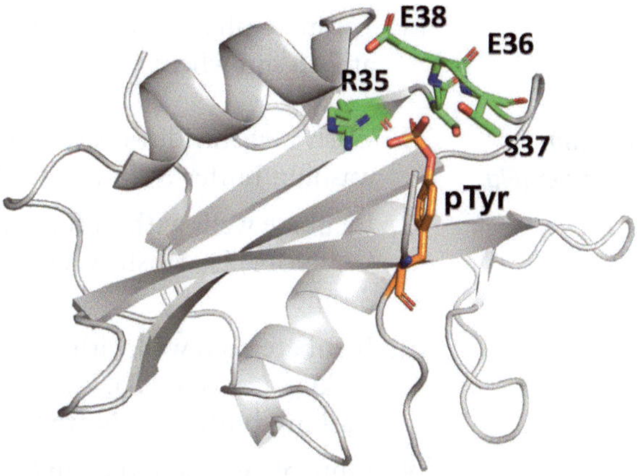

Fig. 1 Crystal structure of Src SH2 domain in complex with a phosphopeptide (PDB, 1SPS). The phosphotyrosine is shown in orange

1. Incubate TOP10F' cells that contain SH2 mutant library in 5 mL 2YT media (with 5 μg/mL Tetracycline, 100 μg/mL Ampicillin, and 100 mM of glucose) overnight at 30 °C, with shaking at 250 rpm.
2. Subculture cells at 1:100 dilution in 5 mL 2YT media (with 5 μg/mL Tetracycline, 100 μg/mL Ampicillin, and 100 mM of glucose) and incubated at 37 °C and 250 rpm until OD_{600} reaches 0.5.
3. Infect cells with 20 times multiplicity of phages (*see* **Note 1**). The cells are incubated for 30 min at 37 °C without shaking to facilitate infection, followed by shaking at 250 rpm for 30 min at 37 °C.
4. Pellet cells at 3220 *g* for 10 min at 4 °C, remove all media, and resuspend pellet with fresh 2YT media with 50 μg/mL Kanamycin and 100 μg/mL Ampicillin (*see* **Note 2**).
5. Allow resuspended cells to grow overnight at 30 °C with shaking at 250 rpm.
6. Transfer cells into centrifuge tubes and centrifuge at 16,000 *g* at 4 °C for 10 min (*see* **Note 3**).
7. Transfer supernatant into a fresh centrifuge tube (*see* **Note 4**).
8. Precipitate phage from supernatant using 20% PEG/2.5 M NaCl solution at 1/5 volume of overall solution and leaving on ice for 1 h (*see* **Note 5**).
9. Pellet precipitated phage by centrifugation at 16,000 *g* for 10 min at 4 °C. Remove supernatant and centrifuge again at 5000 *g* for 3 min at 4 °C. Remove all remaining liquid from and resuspend phage pellet in phage blocking buffer at 1/100 volume of initial culture volume.
10. Centrifuge at 21,000 *g* for 5 min at 4 °C and transfer supernatant to a new vial.
11. Estimate phage forming units (pfu) via measuring absorbance at 268 nm where OD = 1 is equivalent to 5×10^{12} pfu/mL.

3.2 Biopanning-Positive Selection (Fig. 2)

1. Wash Neutravidin-coated 96-well plate with 200 μL of phage washing buffer 3 times (*see* **Note 6**).
2. Incubate wells with sulfopeptide of interest for 1 h with rocking (*see* **Note** 7). Wash wells with 200 μL phage washing buffer 3 times.
3. Incubate each well with 100 μL of harvested phage for 1 h with rocking (*see* **Note 8**). Wash wells 10 times with 200 μL of phage washing buffer (*see* **Note 9**).
4. Elute phage using trypsin (a) or phenyl sulfate (b; *see* **Note 10**).

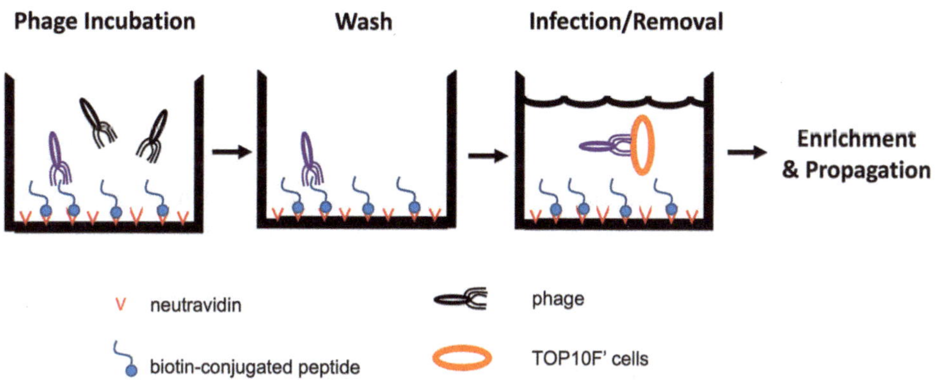

Fig. 2 Positive biopanning. Phages are incubated with immobilized peptide. Unbound phages are removed with washing. Bound phages are eluted and used to infect TOP10F' cells for propagation and additional rounds of biopanning

   (a) Add 100 μL of trypsin (10 ng/μL) for 30 min at 37 °C.
   (b) Add 100 μL of 200 mM phenyl sulfate and allow to incubate for 10 min at 37 °C.

5. Infect TOP10F' cells at OD_{600} = 0.4 with eluted phage. Incubate for 30 min at 37 °C without shaking, followed by shaking for 30 min at 37 °C and 250 rpm.
6. Pellet cells at 3220 *g* at 4 °C for 10 min and remove all supernatant. Resuspend cells in 1 mL of 2YT media and take 5 μL to perform an enrichment test (*see* Subheading 3.2, **step 9**).
7. Do a 1/20 dilution of the resuspended cells in 2YT media containing 5 μg/mL Tetracycline, 100 μg/mL Ampicillin, and 100 mM of glucose. Allow cells to grow overnight at 30 °C with shaking at 250 rpm.
8. Repeat procedure under Subheading 3.1 to carry out another round of positive selection.
9. For each round of positive selection, an enrichment test can be conducted:
   (a) Make serial dilutions with 5 μL of sample recovered in **step 6**.
   (b) Plate dilutions on agar plates containing 5 μg/mL Tetracycline, 100 μg/mL Ampicillin, and 100 mM of glucose. Incubate plates overnight at 37 °C.
   (c) Compare number of colonies grown between test and control wells (*see* **Note 11**).

3.3 Biopanning-Negative Selection

1. Wash Neutravidin-coated 96-well plate with 200 μL of phage washing buffer 3 times.
2. Incubate wells with control peptide of interest (e.g., phospho-peptide or unmodified peptide) for 1 h with rocking (*see* **Note 12**). Wash wells with 200 μL of phage washing buffer 3 times.
3. Incubate wells with 100 μL of harvested phage for 1 h. Unbound phages were collected and used to infect TOP10 F' cells (OD_{600} = 0.4). Incubate for 30 min at 37 °C without shaking, followed by shaking for 30 min at 37 °C and 250 rpm.
4. Pellet cells at 3220 *g* at 4 °C for 10 min and remove all supernatant. Resuspend cells in 1 mL of 2YT media and take 5 μL to perform an enrichment test (*see* Subheading 3.2, **step 9**).
5. Do a 1/20 dilution of the resuspended cells in 2YT media containing 5 μg/mL Tetracycline, 100 μg/mL Ampicillin, and 100 mM of glucose. Allow cells to grow overnight at 30 °C with shaking at 250 rpm.
6. Repeat procedure under Subheading 3.1 to carry out another round of positive selection.

3.4 Phage ELISA

1. Wash Neutravidin-coated 96-well plate with 200 μL of phage washing buffer 3 times.
2. Incubate one set of wells with control peptide (e.g., phospho-peptide) for 1 h with rocking and incubate another set of wells with sulfopeptide for 1 h with rocking. Wash all wells with 200 μL of phage washing buffer 3 times.
3. Incubate each well with 100 μL of harvested phage for 1 h with rocking. Wash wells with 200 μL of phage washing buffer 5 times.
4. Add 100 μL of anti-M13/HRP antibody at a 1/3000 dilution in phage blocking buffer and incubate for 30 min with rocking. Wash wells with 200 μL of phage washing buffer 5 times.
5. Add 100 μL of TMB and incubate for 15 min at room temperature with rocking followed immediately by the addition of 50 μL of 1 M sulfuric acid.
6. Detect HRP activity via measuring absorbance at 450 nM (*see* **Note 13**).

3.5 Protein Expression and Characterization

1. Isolate and sequence individual mutants after three rounds of biopanning.
2. Clone mutants into protein expression vector (e.g., pET30b) plasmid (*see* **Note 14**).

3. Transform BL21(DE3) cells with protein expression plasmids and culture overnight in LB broth containing 50 μg/mL Kanamycin at 37 °C with shaking at 250 rpm.
4. Subculture (1/100 dilution) into new LB broth containing 50 μg/mL Kanamycin. Allow cells to grow to reach OD_{600} of 0.5 before adding IPTG (0.25 mM) to induce protein expression. Incubate cells for 12–18 h at room temperature with shaking at 250 rpm.
5. Centrifuge cells for 10 min at 5000 *g*, remove supernatant, and resuspend cells in sonication buffer (with volume equivalent to 4 times of the weight of the cell pellet; *see* **Note 15**).
6. Sonicate resuspended cells at 60% intensity for 10 s on and 30 s off for 6 min. Centrifuge at 21,000 *g* for 30 min at 4 °C.
7. Load supernatant onto Ni-sepharose column and follow standard purification protocols under nondenaturing conditions [37].
8. Conduct SDS-PAGE analysis to ensure proper protein expression and purification.
9. Buffer exchange the purified proteins into fluorescence polarization buffer (without BGG) and measure protein concentration using Bradford Assay.
10. Add BGG to protein solution to a final concentration of 0.1%. Create a series of serial dilutions of protein solution.
11. Dilute FITC-conjugated peptide to a final concentration of 10 nM with fluorescence polarization buffer (with 01% BGG; *see* **Note 16**).
12. Add FITC-conjugated peptide solution to black 96-well plate and serial dilution protein solutions. Allow solutions to incubate for 20 min in the dark at room temperature. Read fluorescence with plate reader equipped with standard filter cube (excitation = 485 nM, BP = 20 nM; emission = 528 nM) [38].
13. Analyze data and using graphing program (e.g., PRISM) to fit to Hill equation for the determination of K_d values [39]. The mathematic equation is described, as follows:

$$\text{Protein Bound Fraction (\%)} = B_{\max}[\text{protein}]^h / \left(K_{D^h} + [\text{protein}]^h\right)$$

In the above equation, $B_{\max}$ is the maximum specific binding; h is the Hill slope; [protein] is the concentration of protein.

3.6 Enrichment of a Sulfopeptide

1. Load his-tagged SH2 mutant of interest onto Ni-sepharose column. Incubate for 10 min and then wash with 5-column volumes of wash buffer 1 (PBS pH 7.4, 0.1% NP-40, 20 mM imidazole).
2. Incubate column with sulfopeptide-containing sample (e.g., 100:1 BSA:sulfopeptide mixture) for 2 h (*see* **Note 17**).
3. Wash with two 25 column volumes of wash buffer 1 (PBS pH 7.4, 0.1% NP-40, 20 mM imidazole) followed by nine 25 column volumes of wash buffer 2 (PBS pH 7.4, 0.1% NP-40, 50 mM imidazole).
4. Elute bound peptides with four column volumes of elution buffer.
5. Subject eluted fraction to SDS-PAGE and/or Western blot analysis.

4 Notes

1. Either hyperphage or M13KO7 phage can be used at this point depending on biopanning protocol being used.
2. Eight times the volume of original culture should be used to resuspend pellet.
3. Cells should be producing phage will pIII fused with protein of interest at this point.
4. Phages are soluble in the supernatant.
5. Concentration of components of precipitation solution can be adjusted to increase the precipitation of phage.
6. Neutravidin is used to reduce nonspecific binding when performing selection process.
7. Sulfopeptide is synthesized based on the sequence of native substrate of SH2 domains. A biotin tag enables the attachment of peptide to neutravidin-coated wells. Concentration of sulfopeptide used may be varied to alter stringency of biopanning.
8. Concentration of phage should be 1–5 × 10^3 greater than library diversity to ensure all mutants can be selected against. The library diversity for this example is 1.05 × 10^6.
9. Decrease number of washes to decrease stringency of selection. Increasing concentration of Tween-20 in wash buffer can increase stringency of selection.
10. Trypsin recognizes site between pIII coat protein and SH2 mutant. Phenyl sulfate provides a competitive substrate to bind into binding pocket and displace bound sulfopeptide. If phenyl sulfate is used to elute hyperphage, a following trypsin digestion is needed for the hyperphage to infect cells for next panning round.

11. Control wells that were not incubated with sulfopeptide, and/or incubated with control peptide should be included to determine enrichment factor when carrying out enrichment test.
12. Selection against control is conducted to remove mutants with affinity for nonsulfated peptide. This aims to increase the selectivity of mutants.
13. Comparison of signal intensity between sulfopeptide-containing wells and control peptide wells can give a qualitative measurement of selectivity. Control wells without any peptide can provide qualitative measurements for affinity. ELISA can be carried out with individual mutants isolated from library or with entire library.
14. pET30b plasmid includes a C-terminal his tag for purification.
15. Samples containing pure and peptide mixtures should always be kept on ice.
16. FITC-conjugated peptide can be a sulfopeptide, phosphopeptide, or unmodified peptide, depending on the goal of the experiment.
17. Different peptide mixtures can be tested to determine capability of enrichment in different samples.

Acknowledgments

This work was supported by National Institute of Health (grant 1R01GM138623 and 1R01GM147785 to J.G. and W.N.).

References

1. Pawson T, Gish GD (1992) SH2 and SH3 domains: from structure to function. Cell 71(3):359–362. https://doi.org/10.1016/0092-8674(92)90504-6
2. Anderson D, Koch CA, Grey L et al (1990) Binding of SH2 domains of phospholipase Cγ1, GAP, and Src to activated growth factor receptors. Science 250(4983):979–982. https://doi.org/10.1126/science.2173144
3. Moran MF, Koch CA, Anderson D et al (1990) Src homology region 2 domains direct protein-protein interactions in signal transduction. Proc Natl Acad Sci U S A 87(21):8622–8626. https://doi.org/10.1073/pnas.87.21.8622
4. Filippakopoulos P, Mueller S, Knapp S (2009) SH2 domains: modulators of nonreceptor tyrosine kinase activity. Curr Opin Struct Biol 19(6):643–649. https://doi.org/10.1016/j.sbi.2009.10.001
5. Wagner MJ, Stacey MM, Liu BA et al (2013) Molecular mechanisms of SH2- and PTB-domain-containing proteins in receptor tyrosine kinase signaling. Cold Spring Harb Perspect Biol 5(12):a008987/008981–a008987/008919. https://doi.org/10.1101/cshperspect.a008987
6. Machida K, Mayer BJ (2005) The SH2 domain: versatile signaling module and pharmaceutical target. Biochim Biophys Acta, Proteins Proteomics 1747(1):1–25. https://doi.org/10.1016/j.bbapap.2004.10.005
7. Waksman G, Kumaran S, Lubman O (2004) SH2 domains: role, structure and implications for molecular medicine. Expert Rev Mol Med 6(3):1–18
8. Songyang Z, Shoelson SE, Chaudhuri M et al (1993) SH2 domains recognize specific phosphopeptide sequences. Cell 72(5):767–778

9. Kaneko T, Huang H, Cao X et al (2012) Superbinder SH2 domains act as antagonists of cell signaling. Sci Signal 5(243):ra68. https://doi.org/10.1126/scisignal.2003021
10. Ju T, Niu W, Cerny R et al (2013) Molecular recognition of sulfotyrosine and phosphotyrosine by the Src homology 2 domain. Mol Biosyst 9(7):1829–1832. https://doi.org/10.1039/c3mb70061e
11. Lawrie J, Waldrop S, Morozov A et al (2021) Engineering of a small protein scaffold to recognize sulfotyrosine with high specificity. ACS Chem Biol 16(8):1508–1517. https://doi.org/10.1021/acschembio.1c00382
12. Lee RWH, Huttner WB (1983) Tyrosine-O-sulfated proteins of PC12 pheochromocytoma cells and their sulfation by a tyrosylprotein sulfotransferase. J Biol Chem 258(18):11326–11334
13. Tanaka S, Nishiyori T, Kojo H et al (2017) Structural basis for the broad substrate specificity of the human tyrosylprotein sulfotransferase-1. Sci Rep 7(1):8776–8776. https://doi.org/10.1038/s41598-017-07141-8
14. Teramoto T, Fujikawa Y, Kawaguchi Y et al (2013) Crystal structure of human tyrosylprotein sulfotransferase-2 reveals the mechanism of protein tyrosine sulfation reaction. Nat Commun 4:1572–1579. https://doi.org/10.1038/ncomms2593
15. Sherry DM, Murray AR, Kanan Y et al (2010) Lack of protein-tyrosine sulfation disrupts photoreceptor outer segment morphogenesis, retinal function and retinal anatomy. Eur J Neurosci 32:1461–1472. https://doi.org/10.1111/j.1460-9568.2010.07431.x
16. Farzan M, Mirzabekov T, Kolchinsky P et al (1999) Tyrosine sulfation of the amino terminus of CCR5 facilitates HIV-1 entry. Cell 96(5):667–676. https://doi.org/10.1016/S0092-8674(00)80577-2
17. Sasha Tait A, Dong JF, López JA et al (2002) Site-directed mutagenesis of platelet glycoprotein Ibα demonstrating residues involved in the sulfation of tyrosines 276, 278, and 279. Blood 99:4422–4427. https://doi.org/10.1182/blood.V99.12.4422
18. Kanan Y, Siefert JC, Kinter M et al (2014) Complement factor H, vitronectin, and opticin are tyrosine-sulfated proteins of the retinal pigment epithelium. PLoS One 9:e105409. https://doi.org/10.1371/journal.pone.0105409
19. Gao J, Choe H, Bota D et al (2003) Sulfation of tyrosine 174 in the human C3a receptor is essential for binding of C3a anaphylatoxin. J Biol Chem 278:37902–37908. https://doi.org/10.1074/jbc.M306061200
20. Jiang X, Liu H, Chen X et al (2012) Structure of follicle-stimulating hormone in complex with the entire ectodomain of its receptor. Proc Natl Acad Sci U S A 109:12491–12496. https://doi.org/10.1073/pnas.1206643109
21. Moore KL (2003) The biology and enzymology of protein tyrosine O-sulfation. J Biol Chem 278:24243–24246. https://doi.org/10.1074/jbc.R300008200
22. Seibert C, Sakmar TP (2008) Toward a framework for sulfoproteomics: synthesis and characterization of sulfotyrosine-containing peptides. Biopolymers 90(3):459–477. https://doi.org/10.1002/bip.20821
23. Baeuerle PA, Huttner WB (1987) Tyrosine sulfation is a trans-Golgi-specific protein modification. J Cell Biol 105:2655–2664. https://doi.org/10.1083/jcb.105.6.2655
24. Hoffhines AJ, Damoc E, Bridges KG et al (2006) Detection and purification of tyrosine-sulfated proteins using a novel anti-sulfotyrosine monoclonal antibody. J Biol Chem 281:37877–37887. https://doi.org/10.1074/jbc.M609398200
25. Robinson MR, Moore KL, Brodbelt JS (2014) Direct identification of tyrosine sulfation by using ultraviolet photodissociation mass spectrometry. J Am Soc Mass Spectrom 25:1461–1471. https://doi.org/10.1007/s13361-014-0910-3
26. Balderrama GD, Meneses EP, Orihuela LH et al (2011) Analysis of sulfated peptides from the skin secretion of the Pachymedusa dacnicolor frog using IMAC-Ga enrichment and high-resolution mass spectrometry. Rapid Commun Mass Spectrom 25:1017–1027. https://doi.org/10.1002/rcm.4950
27. Monigatti F, Hekking B, Steen H (2006) Protein sulfation analysis-a primer. Biochim Biophys Acta 1764. https://doi.org/10.1016/j.bbapap.2006.07.002
28. Önnerfjord P, Heathfield TF, Heinegård D (2004) Identification of tyrosine sulfation in extracellular leucine-rich repeat proteins using mass spectrometry. J Biol Chem 279:26–33. https://doi.org/10.1074/jbc.M308689200
29. Ward CM, Andrews RK, Smith AI et al (1996) Mocarhagin, a novel cobra venom metalloproteinase, cleaves the platelet von Willebrand factor receptor glycoprotein Ibalpha. Identification of the sulfated tyrosine/anionic sequence Tyr-276-Glu-282 of glycoprotein Ibalpha as a binding site for von Willebr. Biochemistry 35:4929–4938. https://doi.org/10.1021/bi952456c

30. Yu Y, Hoffhines AJ, Moore KL et al (2007) Determination of the sites of tyrosine O-sulfation in peptides and proteins. Nat Methods 4:583–588. https://doi.org/10.1038/nmeth1056
31. Shinde S, Bunschoten A, Kruijtzer JAW et al (2012) Imprinted polymers displaying high affinity for sulfated protein fragments. Angew Chem Int Ed 51(33):8326–8329. https://doi.org/10.1002/anie.201201314
32. Robinson MR, Brodbelt JS (2016) Integrating weak anion exchange and ultraviolet photodissociation mass spectrometry with strategic modulation of peptide basicity for the enrichment of sulfopeptides. Anal Chem 88(22):11037–11045
33. Kehoe JW, Velappan N, Walbolt M et al (2006) Using phage display to select antibodies recognizing post-translational modifications independently of sequence context. Mol Cell Proteomics 5(12):2350–2363. https://doi.org/10.1074/mcp.M600314-MCP200
34. Ju T, Niu W, Guo J (2016) Evolution of Src homology 2 (SH2) domain to recognize sulfotyrosine. ACS Chem Biol 11(9):2551–2557. https://doi.org/10.1021/acschembio.6b00555
35. Packer MS, Liu DR (2015) Methods for the directed evolution of proteins. Nat Rev Genet 16(7):379–394. https://doi.org/10.1038/nrg3927
36. Smith GP (1985) Filamentous fusion phage: novel expression vectors that display cloned antigens on the virion surface. Science 228(4705):1315–1317
37. Spriestersbach A, Kubicek J, Schaefer F et al (2015) Purification of His-tagged proteins. Methods Enzymol 559:1–15. https://doi.org/10.1016/bs.mie.2014.11.003. Laboratory methods in enzymology: protein, part D.
38. Lynch BA, Loiacono KA, Tiong CL et al (1997) A fluorescence polarization based src-SH2 binding assay. Anal Biochem 247(1):77–82. https://doi.org/10.1006/abio.1997.2042
39. Rossi AM, Taylor CW (2011) Analysis of protein-ligand interactions by fluorescence polarization. Nat Protoc 6(3):365–387. https://doi.org/10.1038/nprot.2011.305

Chapter 17

Engineering SH2 Domains with Tailored Specificities and Affinities

Gregory D. Martyn, Gianluca Veggiani, and Sachdev S. Sidhu

Abstract

The Src Homology 2 (SH2) domain is an emerging biotechnology with applications in basic science, drug discovery, and even diagnostics. The SH2 domains rapid uptake into different areas of research is a direct result of the wealth of information generated on its biochemical, biological, and biophysical role in mammalian cell biology. Functionally, the SH2 domain binds and recognizes specific phosphotyrosine (pTyr) residues in the cell to mediate protein–protein interactions (PPIs) that govern signal transduction networks. These signal transduction networks are responsible for relaying growth and stress state signals to the cell's nucleus, ultimately effecting a change in cell biology. Protein engineers have been able to increase the affinity of SH2 domains for pTyr while also tailoring the domains' specificity to unique amino acid sequences flanking the pTyr residue. In this way, it has been possible to develop unique SH2 variants for use in affinity-purification coupled to mass spectrometry (AP-MS) experiments, microscopy, or even synthetic biology. This chapter outlines methods to tailor the affinity and specificity of virtually any human SH2 domain using a combination of rational engineering and phage-display approaches.

Key words Protein engineering, SH2 superbinders, Phage-display, Structural biology, Synthetic biology

Abbreviations

PTM Posttranslational modification
pTyr Phosphotyrosine
SH2 Src-Homology 2

1 Introduction

1.1 Structure and Function of the Human SH2 Domain

The SH2 domain is a small modular domain (approximately 12 kDa; 100 amino acids) that is comprised of a central β-sheet flanked by a pair of α-helices [1] (Fig. 1). The domain interacts with its ligands in a pTyr and sequence specific manner using a binding

Teresa Carlomagno and Maja Köhn (eds.), *SH2 Domains: Functional Modules and Evolving Tools in Biology*, Methods in Molecular Biology, vol. 2705, https://doi.org/10.1007/978-1-0716-3393-9_17,

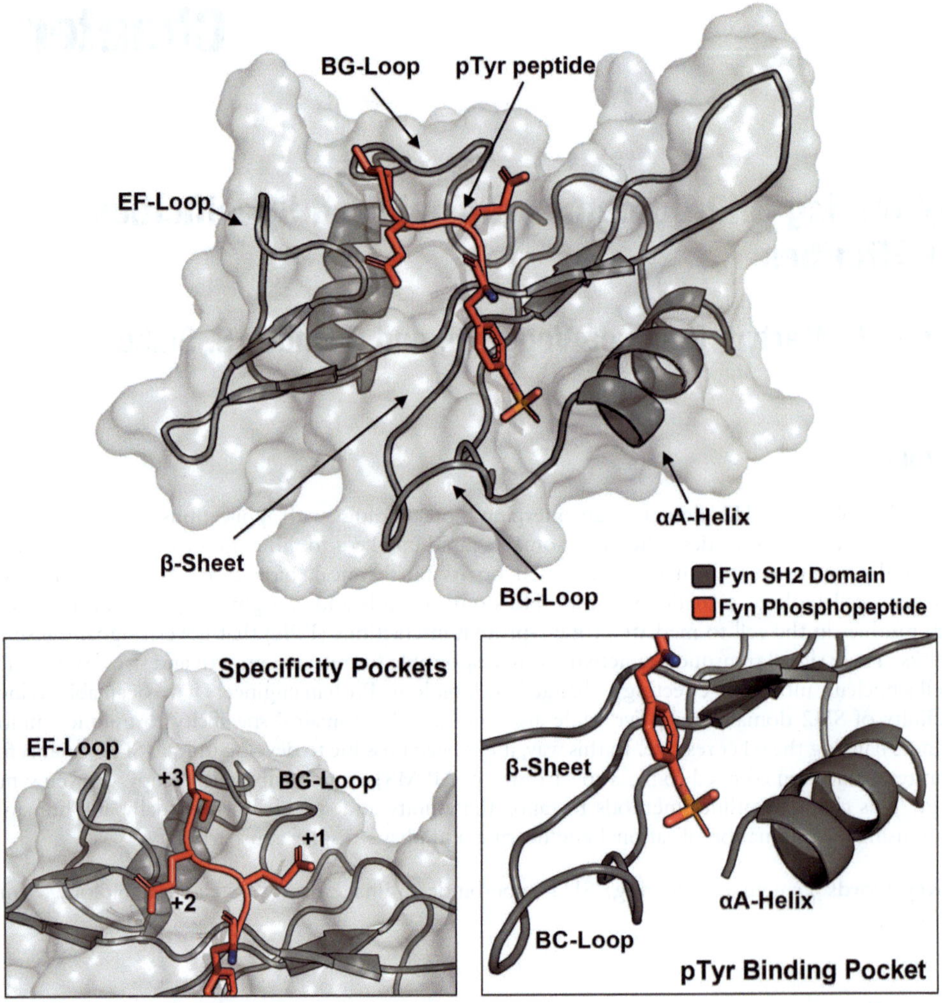

Fig. 1 Structure of the Fyn SH2 domain (PDB ID: 1AOT; gray) bound to a pTyr peptide (red). Secondary structure is annotated in the top panel with the location and composition of the specificity determining pocket and the pTyr-binding pocket highlighted below. Note that the SH2 domain binds its ligands in a two-pronged binding mode using the specificity determining pockets and pTyr-binding pocket to recognize specific pTyr-containing sequences

pocket that enables direct interaction with the phosphoryl moiety of the pTyr residue (pTyr-binding pocket), and a series of specificity determining pockets that interact with residues C-terminal to the pTyr. The pTyr-binding pocket is formed by the αA-Helix, BC-loop, and one side of the central β-sheet, whereas the specificity determining regions are formed by the EF- and BG-loops [2]. This two-pronged binding mode means that the 122 different human SH2 domains bind different intracellular pTyr-containing ligands with modest affinities (0.1–10 μM) [3].

The SH2 domain is often associated with other modular binding domains (Src Homology 3; SH3, pleckstrin homology domains; PH) or enzymatic (kinase/phosphatase) domains [4]. Therefore, the SH2 domain is responsible for localizing a protein to a precise subcellular location, in response to a specific signal, at the correct time to execute a specific function. Such an arrangement enables multiple modular binding domains to engage with different regions of a ligand, and their weak interaction strength ensures the ability to quickly react to stimuli, resulting in discrete signaling outputs that control and regulate cell signaling [5]. There are tens of thousands of different pTyr signals at any one time in a cell and are frequently targets for alteration in diseases such as cancer [6]. The PPIs involved in pTyr-mediated signal transduction are transient. Therefore, understanding how these signal transduction networks are operating at a systems level all the way down to the minute effects on individual proteins involved in these networks is essential for the study of cell biology [7], drug discovery efforts [8] as well as the creation of potential diagnostics [9].

1.2 Engineering the Affinity of Human SH2 Domains

Using phage-display, the human SH2 domain of the Tyrosine-protein Kinase Fyn protein was engineered to bind target pTyr-ligands with high-affinity [10]. Specifically, 3 residues in the αA-helix, 5 residues in the BC loop and 7 residues on the central β-sheet in the backside of the pTyr-binding pocket were randomized (Fig. 2a, b), obtaining a large Fyn SH2 variant phage-display library that was used in affinity selections against target pTyr peptides. Following affinity selections, a Fyn variant characterized by two hydrophilic to hydrophobic mutations in the backside of the pTyr-binding pocket, and one mutation in the BC loop was obtained. This Fyn variant displayed an increased affinity for pTyr-targets of ~380-fold relative to the wild-type domain and was named Fyn superbinder (superFyn; sFyn). The mutations identified in sFyn were grafted into the SH2 domains of the Src tyrosine-protein kinase (Fig. 2c) and the growth factor receptor-bound protein 2 (Grb2) to develop two additional SH2 domain superbinders (superSrc; sSrc and superGrb2; sGrb2). More recently, we used phage-display to engineer the affinity of the SH2 domain of the human Tyrosine-protein kinase Fes [11]. Like sSrc, we randomized 2 residues in the αA-helix, all 6 residues comprising the BC loop and 5 residues in the backside of the pTyr-binding pocket (Fig. 2d, e). Following affinity selections against a pTyr-peptide, we obtained superFes (sFes) which contained 5 mutations all located in the BC loop, while the hydrophobic backside residues were maintained (Fig. 2f).

Although using phage-display libraries to engineer the binding affinity of SH2 domains has proved fruitful, such a method is laborious, time consuming and impractical for engineering the

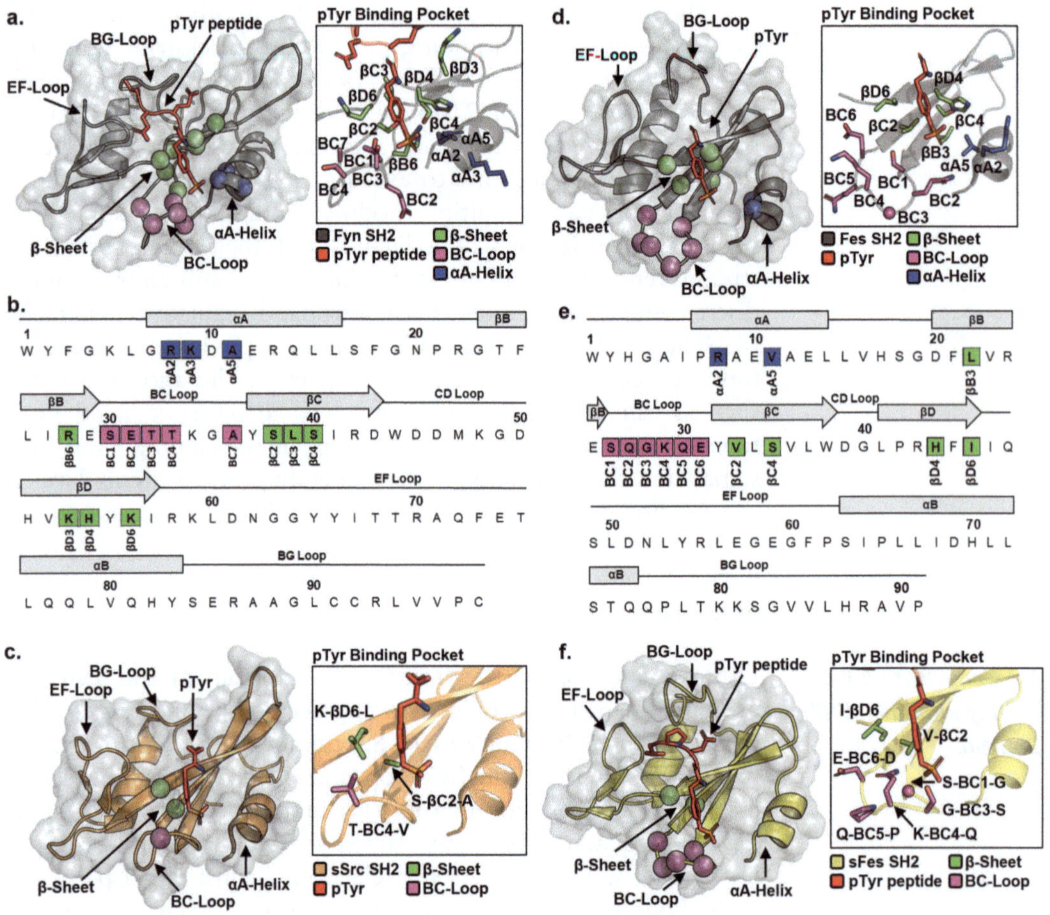

Fig. 2 Design of phage-display libraries to increase affinity using the Fyn and Fes SH2 domains and selection of sSrc and sFes superbinder SH2 variants. (**a**) Structure of the Fyn SH2 domain (PDB ID: 1AOT; gray) bound to a pTyr peptide (red) with secondary structures annotated. The positions of residues randomized in the phage-display library are colored blue (αA-helix), magenta (BC-loop), and green (β-sheet; backside of pTyr-binding pocket) and are represented as spheres highlighting the Cα coordinates of each residue. The inset (right) is a close-up view of the pTyr-binding pocket with side chains of residues subjected to mutagenesis depicted as previously described. (**b**) Sequence of the Fyn SH2 domain (UniProt ID: P06241; residues 149–246) with secondary structure shown and positions randomized highlighted in colors described in (**a**). (**c**) Structure of sSrc (PDB ID: 4F5B; orange) bound to a pTyr residue (red) with secondary structure annotated. The Fyn SH2 phage-display library depicted in (**a**) was used to select Fyn variants with high affinity. These substitutions were grafted into the Src SH2 domain (UniProt ID: P12931; residues 151–248) to produce sSrc with 1 substitution in the BC-loop and 2 substitutions in the backside of the pTyr-binding pocket (right inset). (**d**) Structure of the Fes SH2 domain (PDB ID: 1WQU; gray) bound to a pTyr peptide (red) with secondary structures annotated. Since no structure of wt Fes in complex with a pTyr peptide is available, the Fes SH2 domain was aligned to the structure of the Fyn SH2 domain (PDB ID: 1AOT) using Cα coordinates and the ALIGN function in PyMol. Depicted here is the alignment with just the pTyr residue. The inset (right) is a close-up view of the pTyr-binding pocket with side chains of residues depicted in colors described. (**e**) Sequence of the Fes SH2 domain (UniProt ID: P07332; residues 460–549) with secondary structure shown and positions randomized highlighted as described in (**a**). (**f**) Structure of sFes (PDB ID: 7T1K; yellow) bound to a pTyr residue (red) with secondary structure annotated. The inset (right) depicts residues in the BC-loop with substitutions that confer increased binding affinity in sFes. Note that hydrophobic positions in the backside of the pTyr binding pocket (green) were not substituted when affinity selections were performed with the Fes SH2 phage-display library

affinity of all 122 human SH2 domains. In combination with phage-display, we recently developed a novel approach that takes advantage of (i) computational analysis of human SH2 domains, (ii) structural information, and (iii) rational protein engineering to develop a general strategy for enhancing the binding affinity of human SH2 domains without the need for additional phage-displayed SH2 domain libraries [11]. We successfully used this strategy to increase, by two orders of magnitude, the affinity of 17 human SH2 domains. However, given the modularity of such an approach, we anticipate that this method could be applicable to any SH2 domain.

Our strategy was developed by using the following algorithm:

Bioinformatic Analysis

1. Downloading all 122 human SH2 domain sequences from the Universal Protein resource (UniProt) [12] and their available associated structures from the Protein Data Bank (PDB) (Table 1).
2. Aligning the sequences using the Constraint-Based Alignment Tool (COBALT) [13] from the National Center for Biotechnology Information (NCBI) and visualizing the alignment in Jalview [14] or Excel (Fig. 3).
3. Identifying the position of the αA2 residue, BC loop and βC2/βD6 backside residues using known sequence information from the Fyn and Src SH2 domains.
4. Structural alignments and mapping residue positions. The Protein Data Bank (PDB) [15] and UniProt databases is queried for SH2 domain structures (individual structures or part of a larger protein). The position of the αA2 residue, BC loop and βC2/βD6 backside residues are verified by aligning the query structure(s) with the reference structure from the Src (PDB ID: 1HCT) or Fyn (PDB ID: 1AOT) SH2 domains. These alignments are performed in PyMol [16] for residues spanning the SH2 domain boundaries, as defined by UniProt sequence, using the Cα coordinates (Fig. 4, *see* **Note 1**).

Rational Protein Engineering

Using the sequence alignment generated in the previous section (Fig. 3), the present approach relies on grafting mutations identified via phage-display in sSrc and sFes into other SH2 domains:

1. The entire BC loop of sFes (GQSQPD) and corresponding βC2/βD6 (V/I) are grafted into the SH2 domain of choice to produce a superbinder variant ($sSH2^F$, where F indicates that the domain contains sFes superbinding motif).

Table 1
SH2 domain-containing proteins and their associated gene names

Name	Gene names	Uniprot ID	Sequence	Length	Known structure
3BP2	SH3BP2, 3BP2, RES4-23	P78314	VFVNTTESCEVERLFKATSPRGEPQDGLYCIRNSSTKSGKVL VVWDETSNKVRNYRIFEKDSKFYLEGEVLFVSVGSMVEHYH THVLPSHQSLLLRHPY	99	Yes
ABL1	ABL1, ABL, JTK7	P00519	WYHGPVSRNAAEYLLSSGINGSFLVRESESSPGQRSISLRYEG RVYHYRINTASDGKLYVSSESRFNTLAELVHHHSTVADGLI TTLHYPA	91	Yes
ABL2	ABL2, ABLL, ARG	P42684	WYHGPVSRSAAEYLLSSLINGSFLVRESESSPGQLSISLRYEG RVYHYRINTTADGKVYVTAESRFSTLAELVHHHSTVADGL VTTLHYPA	91	Yes
BCAR3	BCAR3, NSP2, SH2D3B, UNQ271/PRO308	O75815	WYHGRIPRQVSENLVQRDGDFLVRDSLSSPGNFVLTC QWKNLAQHFKINRTVLRLSEAYSRVQYQFEMESFDSIPGL VRCYVGNRRPISQQSGAIIFQPI	100	No
BLK	BLK	P51451	WFFRSQGRKEAERQLLAPINKAGSFLIRESETNKGAFSLSVKD VTTQGELIKHYKIRCLDEGGYYISPRITFPSLQALVQH YSKKGDGLCQRLTLPC	97	No
BLNK	BLNK, BASH, SLP65	Q8WV28	WYAGACDRKSAEEALHRSNKDGSFLIRKSSGHDSKQPYTL VVFFNKRVYNIPVRFIEATKQYALGRKKNGEEYFGSVAEII RNHQHSPLVLIDSQNNTKDSTRLKYAV	108	No
BMX	BMX	P51813	WFAGNISRSQSEQLLRQKGKEGAFMVRNSSQVGMYTVSLF SKAVNDKKGTVKHYHVHTNAENKLYLAENYCFDSIPKLIH YHQHNSAGMITRLRHPV	97	Yes
BTK	BTK, AGMX1, ATK, BPK	Q06187	WYSKHMTRSQAEQLLKQEGKEGGFIVRDSSKAGKYTVSVFAK STGDPQGVIRHYVVCSTPQSQYYLAEKHLFSTIPELINYH QHNSAGLISRLKYPV	97	Yes

CBL	CBL, CBL2, RNF55	P22681	QPWSSLLRNWNSLAVTHPGYMAFLTYDEVKARLQKFIHKPG SYIFRLSCTRLGQWAIGYVTADGNILQTIPHNKPLF QALIDGFREGFYLFPDGRNQNPDLTG	103	Yes
CBLB	CBLB, RNF56, Nbla00127	Q13191	QPWGSILRNWNFLAVTHPGYMAFLTYDEVKARLQKYSTKPG SYIFRLSCTRLGQWAIGYVTGDGNILQTIPHNKPLFQALIDG SREGFYLYPDGRSYNPDLTG	103	Yes
CBLC	CBLC, CBL3, RNF57	Q9ULV8	QPWPTLLKNWQLLAVNHPGYMAFLTYDEVQERLQAC RDKPGSYIFRPSCTRLGQWAIGYVSSDGSILQTIPANKPL SQVLLEGQKDGFYLYPDGKTHNPDLTE	103	Yes
CHIN	CHN1, ARHGAP2, CHN	P15882	EFHGMISREAADQLLIVAEGSYLIRESQRQPGTYTLALRFG SQTRNFRLYYDGKHFVGEKRFESIHDLVTDGLITLYIE TKAAEYIA	87	No
CHIO	CHN2, ARHGAP3, BCH	P52757	EFHGIISREQADELLGGVEGAYILRESQRQPGCYTLALRFGN QTLNYRLFHDGKHFVGEKRFESIHDLV	69	No
CISH	CISH, G18	Q9NSE2	WYWGSITASEARQHLQKMPEGTFLVRDSTHPSYLFTLSVK TTRGPTNVRIEYADSSFRLDSNCLSRPRILAFPDVVSLVQHY	82	No
CLNK	CLNK, MIST	Q7Z7G1	WYIGEYSRQAVEEAFMKENKDGSFLVRDCSTKSKEEPYVLAVF YENKVYNVKIRFLERNQQFALGTGLRGDEKFDSVEDIIEH YKNFPIILIDGKDKTGVHRKQCHLTQPL	111	No
CRK	CRK	P46108	WYWGRLSRQEAVALLQGQRHGVFLVRDSSTSPGDYVLSVSEN SRVSHYIINSSGPRPPVPPSPAQPPPGVSPSRLRIGDQEFD SLPALLEFYKIHYLDTTTLIEPV	106	Yes
CRKL	CRKL	P46109	WYMGPVSRQEAQTRLQGQRHGMFLVRDSSTCPGDYVL SVSENSRVSHYIINSLPNRRFKIGDQEFDHLPALLEFYKIH YLDTTTLIEPA	89	Yes
CSK	CSK	P41240	WFHGKITREQAERLLYPPETGLFLVRESTNYPGDYTLC VSCDGKVEHYRIMYHASKLSIDEEVYFENLMQLVEH YTSDADGLCTRLIKPK	90	Yes
DAPP1	DAPP1, BAM32, HSPC066	Q9UN19	WYHGNLTRHAAEALLLSNGCDGSYLLRDSNETTGLYSL SVRAKDSVKHFHVEYTGYSFKFGFNEFSSLKDFVKHFAN QPLIGSETGTLMVLKHPY	95	No

(continued)

Table 1
(continued)

Name	Gene names	Uniprot ID	Sequence	Length	Known structure
FER	FER, TYK3	P16591	WYHGAIPRIEAQELLKKQGDFLVRESHGKPGEYVLSVYSDG QRRHFIIQYVDNMYRFEGTGFSNIPQLIDHHYTTKQVITKK SGVVLLNPI	91	Yes
FES	FES, FPS	P07332	WYHGAIPRAEVAELLVHSGDFLVRESQGKQEYVLSVLWDGLP RHFIIQSLDNLYRLEGEGFPSIPLLIDHLLSTQQPLTKKSG VVLHRAV	90	Yes
FGR	FGR, SRC2	P09769	WYFGKIGRKDAERQLLSPGNPQGAFLIRESETTKGAYSLSIRD WDQTRGDHVKHYKIRKLDMGGYYITTRVQFNSVQELVQH YMEVNDGLCNLLIAPC	98	No
FRK	FRK, PTK5, RAK	P42685	WFFGAIGRSDAEKQLLYSENKTGSFLIRESESQKGEFSL SVLDGAVVKHYRIKRLDEGGFFLTRRRIFSTLNEFVSHYTK TSDGLCVKLGKPC	93	No
FYN	FYN	P06241	WYFGKLGRKDAERQLLSFGNPRGTFLIRESETTKGAYSLSIRD WDDMKGDHVKHYKIRKLDNGGYYITTRAQFETLQQLVQH YSERAAGLCCRLVVPC	98	Yes
GRAP	GRAP	Q13588	WYSGRISRQLAEEILMKRNHLGAFLIRESESSPGEFSVSVNYGD QVQHFKVLREASGKYFLWEEKFNSLNELVDFYRTTTIAKK RQIFLRDEEPL	95	No
GRAP2	GRAP2, GADS, GRB2L, GRID	O75791	WFHEGLSRHQAENLLMGKEVGFFIIRASQSSPGDFSI SVRHEDDVQHFKVMRDNKGNYFLWTEKFPSLNKLVD YYRTNSISRQKQIFLRDRT	92	Yes
GRAPL	GRAPL	Q8TC17	WYSGRISRQLAEEILMKRNHLGAFLIRESESSPGEFSVSVNNRA QRGPCLGPKSHSRLG	59	No
GRB10	GRB10, GRBIR, KIAA0207	Q13322	WFHGRISREESHRIIKQQGLVDGLFLLRDSQSNPKAFVL TLCHHQKIKNFQILPCEDDGQTFFSLDDGNTKFSDLIQL VDFY	82	Yes

GRB14	GRB14	Q14449	WFHHKISRDEAQRLIIQQGLVDGVFLVRDSQSNPKTFVLSMSHGQKIKHFQIIPVEDDGEMFHTLDDGHTRFTDLIQLVEFYQLNKGVLPCKLKHYC	97	Yes
GRB2	GRB2, ASH	P62993	WFFGKIPRAKAEEMLSKQRHDGAFLIRESESAPGDFSLSVKFGNDVQHFKVLRDGAGKYFLWVVKFNSLNELVDYHRSTSVSRNQQIFLRDIE	93	Yes
GRB7	GRB7	Q14451	WFHGRISREESQRLIGQQGLVDGLFLVRESQRNPQGFVLSLCHLQKVKHYLILPSEEEGRLYFSMDDGQTRFTDLLQLVEFHQLNRGILPCLLRHCC	97	Yes
HCK	HCK	P08631	WFFKGISRKDAERQLLAPGNMLGSFMIRDSETTKGSYSLSVRDYDPRQGDTVKHYKIRTLDNGGFYISPRSTFSTLQELVDHYKKGNDGLCQKLSVPC	98	Yes
HSH2D	HSH2D, ALX	Q96JZ2	WFHGAISREDAENLLESQPLGSFLIRVSHSHVGYTLSYKAQSSCCHFMVKLLDDGTFMIPGEKVAHTSLDALVTFHQQKPIEPRRELLTQPC	92	Yes
ITK	ITK, EMT, LYK	Q08881	WYNKSISRDKAEKLLLDTGKEGAFMVRDSRTAGTYTVSVFTKAVVSENNPCIKHYHIKETNDNPKRYYVAEKYVFDSIPLLINYHQHNGGGLVTRLRYPV	100	No
JAK1	JAK1, JAK1A, JAK1B	P23458	GCHGPICTEYAINKLRQEGSEEGMYVLRWSCTDFDNILMTVTCFEKSEQVQGAQKQFKNFQIEVQKGRYSLHGSDRSFPSLGDLMSHLKKQILRTDNISFMLKRCC	106	Yes
JAK2	JAK2	O60674	HGPISMDFAISKLKKAGNQTGLYVLRCSPKDFNKYFLTFAVERENVIEYKHCLITKNENEEYNLSGTKKNFSSLKDLLNCYQ	82	Yes
JAK3	JAK3	P52333	QCHGPITLDFAINKLKTGGSRPGSYVLERSPQDFDSFLLTVCVQNPLGPDYKGCLIRRSPTGTFLLVGLSRPHSSLRELLATCWDGGLHVDGVAVTLTSCC	101	No
KSYK_N	SYK	P43405	FFFGNITREEAEDYLVQGGMSDGLYLLRQSRNYLGGFALSVAHGRKAHHYTIERELNGTYAIAGGRTHASPADLCHYHSQESDGLVCLLKKPF	93	No
KSYK_C	SYK	P43405	WFHGKISREESEQIVLIGSKTNGKFLIRARDNNGSYALCLLHEGKVLHYRIDKDKTGKLSIPEGKKFDTLWQLVEHYSYKADGLLRVLTVPC	92	Yes

(continued)

Table 1
(continued)

Name	Gene names	Uniprot ID	Sequence	Length	Known structure
LCK	LCK	P06239	WFFKNLSRKDAERQLLAPGNTHGSFLIRESESTAGSFSL SVRDFDQNQGEVVKHYKIRNLDNGGFYISPRITFPGLHEL VRHYTNASDGLCTRLSRPC	98	Yes
LCP2	LCP2	Q13094	WYVSYITRPEAEAALRKINQDGTFLVRDSSKKTTTNPYVLMVL YKDKVYNIQIRYQKESQVYLLGTGLRGKEDFLSVSDIIDYF RKMPLLLIDGKNRGSRYQCTLTHAA	109	No
LYN	LYN, JTK8	P07948	WFFKDITRKDAERQLLAPGNSAGAFLIRESETLKGSFSL SVRDFDPVHGDVIKHYKIRSLDNGGYYISPRITFPCISDMIKH YQKQADGLCRRLEKAC	98	No
MATK	MATK, CTK, HYL	P42679	WFHGKISGQEAVQQLQPPEDGLFLVRESARHPGDYVLCVSFG RDVIHYRVLHRDGHLTIDEAVFFCNLMDMVEH YSKDKGAICTKLVRPK	90	Yes
NCK1	NCK1, NCK	P16333	WYYGKVTRHQAEMALNERGHEGDFLIRDSESSPNDFSVSLKA QGKNKHFKVQLKETVYCIGQRKFSTMEELVEHYKKAPIF TSEQGEKLYLVKHL	95	Yes
NCK2	NCK2, GRB4	O43639	WYYGNVTRHQAECALNERGVEGDFLIRDSESSPSDFSVSLKA SGKNKHFKVQLVDNVYCIGQRRFHTMDELVEHYKKAPIF TSEHGEKLYLVRALQ	96	Yes
P55G_C	PIK3R3	Q92569	WFVEDINRVQAEDLLYGKPDGAFLIRESSKKGCYACSVVADGE VKHCVIYSTARGYGFAEPYNLYSSLKELVLHYQQTSL VQHNDSLNVRLAYPV	95	No
P55G_N	PIK3R3	Q92569	WYWGDISREEVNDKLRDMPDGTFLVRDASTKMQGDYTLTL RKGGNNKLIKIYHRDGKYGFSDPLTFNSVVELINHYHHE SLAQYNPKLDVKLMYPV	96	No
P85A_N	PIK3R1, GRB1	P27986	WYWGDISREEVNEKLRDTADGTFLVRDASTKMHGDYTLTL RKGGNNKLIKIFHRDGKYGFSDPLTFSSVVELINHYRNESLA QYNPKLDVKLLYPV	96	Yes

P85B_C	PIK3R2	O00459	WYVGKINRTQAEEMLSGKRDGTFLIRESSQRGCYAC SVVVDGDTKHCVIYRTATGFGFAEPYNLYGSLKELVLH YQHASLVQHNDALTVTLAHPV	95	No
P85B_N	PIK3R2	O00459	WYWGDISREEVNEKLRDTPDGTFLVRDASSKIQGEYTLTL RKGGNNKLIKVFHRDGHYGFSEPLTFCSVVDLINHYRHE SLAQYNAKLDTRLLYPV	96	No
PLCG1_C	PLCG1, PLC1	P19174	WYHASLTRAQAEHMLMRVPRDGAFLVRKRNEPNSYAISF RAEGKIKHCRVQQEGQTVMLGNSEFDSLVDLISYYEKHPL YRKMKLRYPI	89	No
PLCG1_N	PLCG1, PLC1	P19174	WFHGKLGAGRDGRHIAERLLTEYCIETGAPDGSFLVRESETF VGDYTLSFWRNGKVQHCRIHSRQDAGTPKFFLTDNLVFD SLYDLITHYQQVPLRCNEFEMRLSEPV	108	No
PLCG2_C	PLCG2	P16885	WYYDSLSRGEAEDMLMRIPRDGAFLIRKREGSDSYAITFRA RGKVKHCRINRDGRHFVLGTSAYFESLVELVSYYEKHSL YRKMRLRYPV	90	No
PLCG2_N	PLCG2	P16885	WFHKKVEKRTSAEKLLQEYCMETGGKDGTFLVRESETFPND YTLSFWRSGRVQHCRIRSTMEGGTLKYYLTDNLTFSSIYALI QHYRETHLRCAEFELRLTDPV	104	No
PTK6	PTK6, BRK	Q13882	WFFGCISRSEAVRRLQAEGNATGAFLIRVSEKPSADYVLSVRD TQAVRHYKIWRRAGGRLHLNEAVSFLSLPELVNYHRAQSL SHGLRLAAPC	93	Yes
PTN11_C	PTPN11, PTP2C, SHPTP2	Q06124	WFHGHLSGKEAEKLLTEKGKHGSFLVRESQSHPGDFVL SVRTGDDKGESNDGKSKVTHVMIRCQELKYDVGGGERFD SLTDLVEHYKKNPMVETLGTVLQLKQPL	105	No
PTN11_N	PTPN11, PTP2C, SHPTP2	Q06124	WFHPNITGVEAENLLLTRGVDGSFLARPSKSNPGDFTL SVRRNGAVTHIKIQNTGDYYDLYGGEKFATLAEL VQYYMEHHGQLKEKNGDVIELKYPL	97	Yes
PTN6_C	PTPN6, HCP, PTP1C	P29350	WYHGHMSGGQAETLLQAKGEPWTFLVRESLSQPGDFVLSVL SDQPKAGPGSPLRVTHIKVMCEGGRYTVGGLETFDSLTDL VEHFKKTGIEEASGAFVYLRQPY	104	Yes

(continued)

Table 1 (continued)

Name	Gene names	Uniprot ID	Sequence	Length	Known structure
PTN6_N	PTPN6, HCP, PTP1C	P29350	WFHRDLSGLDAETLLKGRGVHGSFLARPSRKNQGDFSL SVRVGDQVTHIRIQNSGDFYDLYGGEKFATLTELVE YYTQQQGVLQDRDGTIIHLKYPL	97	No
RASA1_C	RASA1, GAP, RASA	P20936	WFHGKISKQEAYNLLMTVGQVCSFLVRPSDNTPGDYSLYF RTNENIQRFKICPTPNNQFMMGGRYYNSIGDIIDHYRKEQI VEGYYLKEPV	91	Yes
RASA1_N	RASA1, GAP, RASA	P20936	WYHGKLDRTIAEERLRQAGKSGSYLIRESDRRPGSFVLSFL SQMNVVNHFRIIAMCGDYYIGGRRFSSLSDLIGYYSH VSCLLKGEKLLYPV	92	No
RIN1	RIN1	Q13671	WLQLQANAAAALHMLRTEPPGTFLVRKSNTRQCQALCM RLPEASGPSFVSSHYILESPGGVSLEGSELMFPDLVQLICA YCHTRDILLLPLQLPR	95	No
RIN2	RIN2, RASSF4	Q8WYP3	WLQLSLSEEEAAEVLQAQPPGIFLVHKSTKMQKKVLSL RLPCEFGAPLKEFAIKESTYTFSLEGSGISFADLFRLIAFYCI SRDVLPFTLKLPY	94	No
RIN3	RIN3	Q8TB24	WLQLSLGQAEVARILHRVVAGMFLVRRDSSSKQLVLCVHFP SLNESSAEVLEYTIKEEKSILYLEGSALVFEDIFRLIAFYC VSRDLLPFTLRLPQ	96	No
SH21A	SH2D1A, DSHP, SAP	O60880	VYHGKISRETGEKLLLATGLDGSYLLRDSESVPGVYCLCVL YHGYIYTYRVSQTETGSWSAETAPGVHKRYFRKIKNLISAF QKPDQGIVIPLQYPVEK	99	Yes
SH21B	SH2D1B, EAT2	O14796	YYHGRLTKQDCETLLLKEGVDGNFLLRDSESIPGVLCLC VSFKNIVYTYRIFREKHGYYRIQTAEGSPKQVFPSLKELI SKFEKPNQGMVVHLLKPI	97	Yes

SH22A	SH2D2A, SCAP, TSAD, VRAP	Q9NP31	WFHGFITRREAERLLEPKPQGCYLVRFSESAVTFVLTYRSRTCC RHFLLAQLRDGRHVVLGEDSAHARLQDLLLHYTAHPLSP YGETLTEPL	92	No
SH23A	SH2D3A, NSP1, UNQ175/ PRO201	Q9BRG2	WYHGLLSRQKAEALLQQNGDFLVRASGSRGGNPVISCRWRG SALHFEVFRVALRPRPGRPTALFQLEDEQFPSIPALVHSYM TGRRPLSQATGAVVSRPV	100	No
SH24A	SH2D4A, PPP1R38, SH2A	Q9H788	WFHGILTLKKANELLLSTGMPGSFLIRVSERIKGYALSYL SEDGCKHFLIDASADAYSFLGVDQLQHATLADLVE YHKEEPITSLGKELLLYPC	94	No
SH24B	SH2D4B	Q5SQS7	WFHGIISREDAEALLENMTEGAFLVRVSEKIWGYTLSYRL QKGFKHFLVDASGDFYSFLGVDPNRHATLTDLVDFHKEEII TVSGGELLQEPC	93	No
SH2B1	SH2B1, KIAA1299, SH2B	Q9NRF2	WFHGMLSRLKAAQLVLTGGTGSHGVFLVRQSETRRGEYVL TFNFQGKAKHLRLSLNEEGQCRVQHLWFQSIFDMLEHF RVHPIPLESGGSSDVVLVSYV	99	Yes
SH2B3	SH2B3, LNK	Q9UQQ2	WFHGPISRVKAAQLVQLQGPDAHGVFLVRQSETRRGEYVL TFNFQGIAKHLRLSLTERGQCRVQHLHFPSVVDMLHHF QRSPIPLECGAACDVRLSSYV	99	No
SH2D3	SH2D3C, NSP3, UNQ272/ PRO309/PRO34088	Q8N5H7	WYHGRIPREVSETLVQRNGDFLIRDSLTSLGDYVLTCRWRN QALHFKINKVVVKAGESYTHIQYLFEQESFDHVPALVRYH VGSRKAVSEQSGAIIYCPV	100	No
SH2D5	SH2D5	Q6ZV89	WAFAGISRPCALALLRRDVLGAFLLWPELGASGQWCL SVRTQCGVVPHQVFRNHLGRYCLEHLPAEFPSLEAL VENHAVTERSLFCPLDMGRLNPTY	97	No
SH2D6	SH2D6	Q7Z4S9	WYSGNCDRYAVESALLHLQKDGAYTVRPSSGPHGSQPFTLA VLLRGRVFNIPIRRLDGGRHYALGREGRNREELFSSVAAM VQHFMWHPLPLVDRHSGSRELTCLLFPT	109	No
SH2D7	SH2D7	A6NKC9	WFHGFITRKQTEQLLRDKALGSFLIRLSDRATGYILSYRGSD RCRHFVINQLRNRRYIISGDTQSHSTLAELVHHYQEA QLEPFKEMLTAAC	92	No

(continued)

Table 1 (continued)

Name	Gene names	Uniprot ID	Sequence	Length	Known structure
SHB	SHB	Q15464	WYHGAISRGDAENLLRLCKECSYLVRNSQTSKHDYSLSLRSN QGFMHMKLAKTKEKYVLGQNSPPFDSVPEVIH YYTTRKLPIKGAEHLSLLYPV	95	No
SHC1	SHC1, SHC, SHCA	P29353	WFHGKLSRREAEALLQLNGDFLVRESTTTPGQYVLTGLQSG QPKHLLLVDPEGVVRTKDHRFESVSHLISYHMDNHLPII SAGSELCLQQPV	92	Yes
SHC2	SHC2, SCK, SHCB	P98077	WYHGRMSRRAAERMLRADGDFLVRDSVTNPGQYVL TGMHAGQPKHLLLVDPEGVVRTKDVLFESISHLIDHHL QNGQPIVAAESELHLRGVV	92	No
SHC3	SHC3, NSHC, SHCC	Q92529	WYQGEMSRKEAEGLLEKDGDFLVRKSTTNPGSFVLTGMHNG QAKHLLLVDPEGTIRTKDRVFDSISHLINHHLESSLPIVSAG SELCLQQPV	92	No
SHC4	SHC4, SHCD, UNQ6438/ PRO21364	Q6S5L8	CYHGKLSRKAAESLLVKDGDFLVRESATSPGQYVLSGLQGG QAKHLLLVDPEGKVRTKDHVFDNVGHLIRYHMDNSLPII SSGSEVSLKQPV	92	No
SHD	SHD	Q96IW2	WFHGPLNRADAESLLSLCKEGSYLVRLSETNPQDCSLSL RSSQGFLHLKFARTRENQVVLGQHSGPFPSVPELVLH YSSRPLPVQGAEHLALLYPV	96	No
SHE	SHE	Q5VZ18	WYHGAISRAEAESRLQPCKEAGYLVRNSESGNSRYSIALK TSQGCVHIIVAQTKDNKYTLNQTSAVFDSIPEVVH YYSNEKLPFKGAEHMTLLYPV	96	No
SHF	SHF	Q7M4L6	WYHGAISRTDAENLLRLCKEASYLVRNSETSKNDFSLSLK SSQGFMHMKLSRTKEHKYVLGQNSPPFSSVPEIVHHYA SRKLPIKGAEHMSLLYPV	96	No
SHIP1	INPP5D, SHIP, SHIP1	Q92835	WNHGNITRSKAEELLSRTGKDGSFLVRASESISRAYALCVL YRNCVYTYRILPNEDDKFTVQASEGVSMRFFTKLDQLIEF YKKENMGLVTHLQYPV	97	Yes

SHIP2	INPPL1, SHIP2	O15357	WYHRDLSRAAAEELLARAGRDGSFLVRDSESVAGAFALCVL YQKHVHTYRILPDGEDFLAVQTSQGVPVRRFQTLGELIGL YAQPNQGLVCALLLPV	97	Yes
SLAP1	SLA, SLAP, SLAP1	Q13239	WLFEGLGRDKAEELLQLPDTKVGSFMIRESETKKGFYSLSVRH RQVKHYRIFRLPNNWYYISPRLTFQCLEDLVNHYSE VADGLCCVLTTPC	92	No
SLAP2	SLA2, C20orf156, SLAP2	Q9H6Q3	WLYEGLSREKAEELLLLPGNPGGAFLIRESQTRRGSYSLSVRL SRPASWDRIRHYRIHCLDNGWLYISPRLTFPSLQALVDH YSELADDICCLLKEPC	98	Yes
SOCS1	SOCS1, SSI1, TIP3	O15524	FYWGPLSVHGAHERLRAEPVGTFLVRDSRQRNCFFALSVKMA SGPTSIRVHFQAGRFHLDGSRESFDCLFELLEHYVAAP RRMLGAPLRQRRVRPL	96	No
SOCS2	SOCS2, CIS2, SSI2, STATI2	O14508	WYWGSMTVNEAKEKLKEAPEGTFLIRDSSHSDYLLTISVK TSAGPTNLRIEYQDGKFRLDSIICVKSKLKQFDSVVHLID YYVQMCKDKRTGPEAPRNGTVHLYLTKPL	109	Yes
SOCS3	SOCS3, CIS3, SSI3	O14543	FYWSAVTGGEANLLLSAEPAGTFLIRDSSDQRHFFTLSVKTQSG TKNLRIQCEGGSFSLQSDPRSTQPVPRFDCVLKLVHH YMPPPGAPSFPSPPTE	97	No
SOCS4	SOCS4, SOCS7	Q8WXH5	CYWGVMDKYAAEALLEGKPEGTFLLRDSAQEDYLFSVSF RRYSRSLHARIEQWNHNFSFDAHDPCVFHSPDITGLLEH YKDPSACMFFEPLLSTPL	96	Yes
SOCS5	SOCS5, CIS6, CISH5, CISH6, KIAA0671	O75159	CYWGVMDRYEAEALLEGKPEGTFLLRDSAQEDYLFSVSF RRYNRSLHARIEQWNHNFSFDAHDPCVFHSSTVTGLLEH YKDPSSCMFFEPLLTISL	96	No
SOCS7	SOCS7, NAP4, SOCS6	O14512	WYWGPMNWEDAEMKLKGKPDGSFLVRDSSDPRYILSLSF RSQGITHHTRMEHYRGTFSLWCHPKFEDRCQSVVEFIK RAIMHSKNGKFLYFLRSRVPGLPPTPVQLLYPV	110	No
SPT6H	SUPT6H, KIAA0162, SPT6H	Q7KZ85	YIKRVIAHPSFHNINFKQAEKMMETMDQGDVIIRP SSKGENHLTVTWKVSDGIYQHVDVREEGKENAFSLGATL WINSEEFEDLDEIVARYVQPMASFARDLLNHKY	107	Yes

(continued)

Table 1 (continued)

Name	Gene names	Uniprot ID	Sequence	Length	Known structure
SRC	SRC, SRC1	P12931	WYFGKITRRESERLLLNAENPRGTFLVRESETTKGAYCL SVSDFDNAKGLNVKHYKIRKLDSGGFYITSRTQFNSLQQL VAYYSKHADGLCHRLTTVC	98	Yes
SRMS	SRMS, C20orf148	Q9H3Y6	WYFSGVSRTQAQQLLLSPPNEPGAFLIRPSESSLGGYSLSVRA QAKVCHYRVSMAADGSLYLQKGRLFPGLEELLTYYKAN WKLIQNPLLQPC	93	No
STAP1	STAP1, BRDG1	Q9ULZ2	ACFYTVSRKEATEMLQKNPSLGNMILRPGSDSRNYSITI RQEIDIPRIKHYKVMSVGQNYTIELEKPVTLPNLFSVIDYF VKETRGNLRPFICSTDENTGQEPS	104	Yes
STAP2	STAP2, BKS	Q9UGK3	YMMSEVLAKEEARRALETPSCFLKVSRLEAQLLLE RYPECGNLLLRPSGDGADGVSVTTRQMHNGTHVVRHYK VKREGPKYVIDVEQPFSCTSLDAVVNYFVSHTKKAL VPFLLDE	116	Yes
STAT1	STAT1	P42224	WNDGCIMGFISKERERALLKDQQPGTFLLRFSESSREGAITF TWVERSQNGGEPDFHAVEPYTKKELSAVTFPDIIRNYK VMAAENIPENPLKYLYPN	98	Yes
STAT2	STAT2	P52630	WNDGRIMGFVSRSQERRLLKKTMSGTFLLRFSESSEGGITC SWVEHQDDDKVLIYSVQPYTKEVLQSLPLTEIIRHYQLL TEENIPENPLRFLYPR	96	No
STAT3	STAT3, APRF	P40763	WNEGYIMGFISKERERAILSTKPPGTFLLRFSESSKEGGVTF TWVEKDISGKTQIQSVEPYTKQQLNNMSFAEIIMGYKIMDA TNILVSPL	91	No
STAT4	STAT4	Q14765	WIDGYVMGFVSKEKERLLLKDKMPGTFLLRFSESHLGGITF TWVDHSESGEVRFHSVEPYNKGRLSALPFADILRDYK VIMAENIPENPLKYLYPD	96	No
STAT5A	STA5A, STAT5	P42229	WNDGAILGFVNKQQAHDLLINKPDGTFLLRFSDSEIGGITIA WKFDSPERNLWNLKPFTTRDFSIRSLADRLGDLSYLIYVFPD RPKDEVFSKYYTPV	98	No

STAT5B	STA5B	P51692	WNDGAILGFVNKQQAHDLLINKPDGTFLLRFSDSEIGGITIA WKFDSQERMFWNLMPFTTRDFSIRSLADRLGDLNYLI YVFPDRPKDEVYSKYYTPV	98	No
STAT6	STAT6	P42226	WFDGVLDLTKRCLRSYWSDRLIIGFISKQYVTSLLLNEPDG TFLLRFSDSEIGGITIAHVIRGQDGSPQIENIQPFSAKDLSI RSLGDRIRDLAQLKNLYPKKPKDEAFRSHYKPE	116	No
TEC	TEC, PSCTK4	P42680	WYCRNMNRSKAEQLLRSEDKEGGFMVRDSSQPGLYTVSL YTKFGGEGSSGFRHYHIKETTTSPKKYYLAEKHAFGSIPEIIE YHKHNAAGLVTRLRYPV	99	No
TENS1	TNS1, TNS	Q9HBL0	WYKPEISREQAIALLKDQEPGAFIIRDSHSFRGAYGLAMK VSSPPPTIMQQNKKGDMTHELVRHFLIETGPRG VKLKGCPNEPNFGSLSALVYQHSIIPLALPCKLVIPN	110	No
TENS3	TNS3, TEM6, TENS1, TPP	Q68CZ2	WYKADISREQAIAMLKDKEPGSFIVRDSHSFRGAYGLAMKVA TPPPSVLQLNKKAGDLANELVRHFLIECTPKGVRLKGC SNEPYFGSLTALVCQHSITPLALPCKLLIPE	111	No
TENS4	TNS4, CTEN, PP14434	Q8IZW8	WFKPNITREQAIELLRKEEPGAFVIRDSSSYRGSFGLALKVQE VPASAQSRPGEDSNDLIRHFLIESSAKGVHLKGADEEPYFG SLSAFVCQHSIMALALPCKLTIPQ	108	No
TNS2	TNS2, KIAA1075, TENC1	Q63HR2	WYKPHLSRDQAIALLKDKDPGAFLIRDSHSFQGAYGLALKVA TPPPSAQPWKGDPVEQLVRHFLIETGPKGVKIKGCPSEPYFG SLSALVSQHSISPISLPCCLRIPS	108	Yes
TXK	TXK, PTK4, RLK	P42681	WYHRNITRNQAEHLLRQESKEGAFIVRDSRHLGSYTI SVFMGARRSTEAAIKHYQIKKNDSGQWYVAERHAFQSIPELI WYHQHNAAGLMTRLRYPV	97	Yes
TYK2	TYK2	P29597	GIHGPLLEPFVQAKLRPEDGLYLIHWSTSHPYRLILTVA QRSQAPDGMQSLRLRKFPIEQQDGAFVLEGWGRSFPSVREL	80	No
VAV	VAV1, VAV	P15498	WYAGPMERAGAESILANRSDGTFLVRQRVKDAAEFAISIKYN VEVKHIKIMTAEGLYRITEKKAFRGLTELVEFYQQN SLKDCFKSLDTTLQFPF	95	Yes

(continued)

Table 1 (continued)

Name	Gene names	Uniprot ID	Sequence	Length	Known structure
VAV2	VAV2	P52735	WFAGNMERQQTDNLLKSHASGTYLIRERPAEAERFAI SIKFNDEVKHIKVVEKDNWIHITEAKKFDSLLELVEYYQCH SLKESFKQLDTTLKYPY	95	Yes
YES	YES1, YES	P07947	WYFGKMGRKDAERLLLNPGNQRGIFLVRESETTKGAYSLSIRD WDEIRGDNVKHYKIRKLDNGGYYITTRAQFDTLQKLVKH YTEHADGLCHKLTTVC	98	No
ZAP70_C	ZAP70, SRK	P43403	WYHSSLTREEAERKLYSGAQTDGKFLLRPRKEQGTYALSLI YGKTVYHYLISQDKAGKYCIPEGTKFDTLWQLVE YLKLKADGLIYCLKEAC	92	Yes
ZAP70_N	ZAP70, SRK	P43403	FFYGSISRAEAEEHLKLAGMADGLFLLRQCLRSLGGYVLSL VHDVRFHHFPIERQLNGTYAIAGGKAHCGPAELCEF YSRDPDGLPCNLRKPC	93	Yes

The UniProt identifier code, amino acid sequence, and length of each SH2 domain are reported. Additionally, indicated is the presence or absence of known structures in the PDB

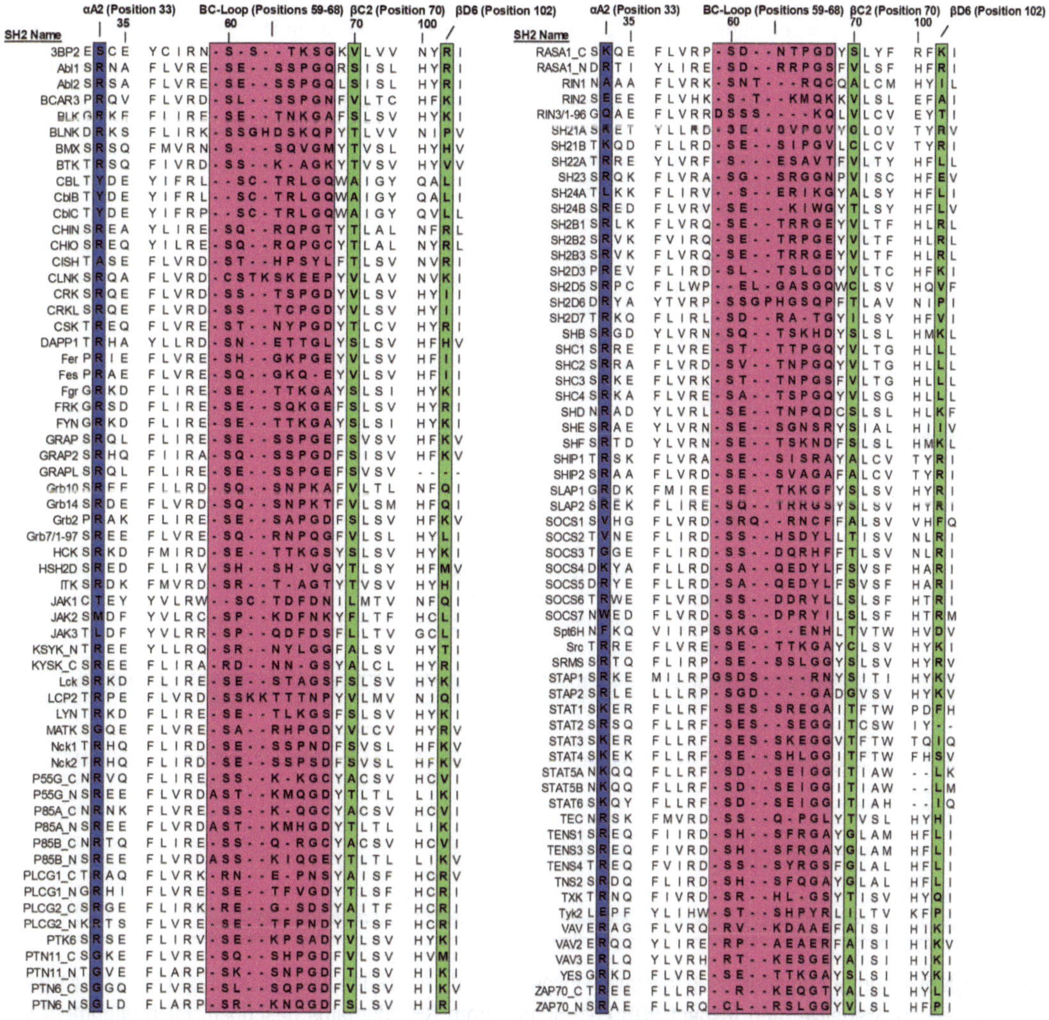

Fig. 3 Sequence alignment of all 122 human SH2 domains. The primary amino acid sequences of all human SH2 domains were downloaded from UniProt, aligned using the COBALT algorithm from NCBI and visualized in Excel. The alignment was performed using the entire amino acid sequence from each domain. Here are shown sections of the alignment covering parts of the αA-helix (αA2-position; blue), BC-loop (magenta), and pTyr backside (green). The colored positions are essential for using the grafting algorithm described in the Methods section

2. The entire BC loop of sSrc (SETVKGA) and corresponding βC2/βD6 (A/L) are grafted into the SH2 domain of choice to produce an additional superbinder variant (sSH2^{S}, where S indicates that the domain contains sSrc superbinding motif) (*see* **Note 2**).
3. If the domain of interest does not have an Arginine (Arg; R) in the αA2 position, this position is substituted with Arg either alone or with the sFes and sSrc motif grafts to produce a superbinder variant (sSH2$^{wt/X\text{-}\alpha A2\text{-}R}$, sSH2$^{sFes/X\text{-}\alpha A2\text{-}R}$, or sSH2$^{sSrc/X\text{-}\alpha A2\text{-}R}$).

PyMol Align Function → see https://pymol.org/pymol-command-ref.html#align for in depth description of align function.

align mobile/segment/chain/residue/atom, target/segment/chain/residue/atom

align = command recognized by PyMol API.
mobile = object that to be aligned.
target = object to be aligned to.
segment/chain = see below.
residue = amino acids of interest in object.
atom = specific atoms of interest. Default is all.
c = carbon (main chain)
n = nitrogen (main chain)
o = oxygen (main chain)
ca = alpha carbon

Example Alignment

Input:

align /1AOT/A/F/7-105/ca, /1HCT/B/B/148-245/ca
mobile = 1AOT (Fyn SH2 domain).
target = 1HCT (Src SH2 domain)..
segment/chain = see below.
residue = amino acids of SH2 defined by UniProt sequences.
atom = ca (alpha carbon).

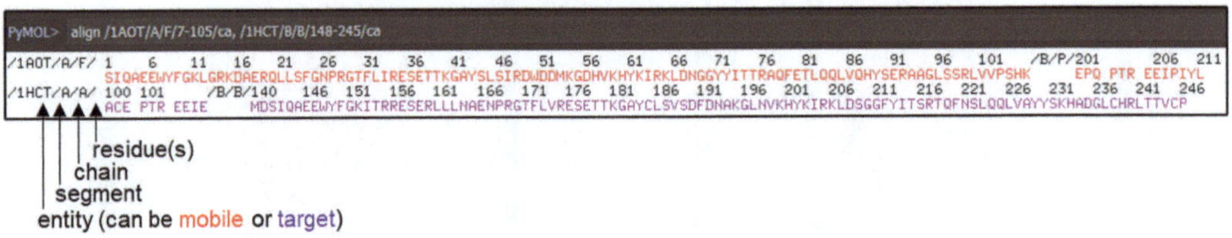

Output:

Object 1AOT is now aligned with 1HCT based on Cα co-ordinates

Fig. 4 PyMol alignment function used to visually inspect SH2 domain secondary structure. PyMol software was used to import, visualize, and inspect PDB coordinates from SH2 domains described. Each alignment was performed one at a time using Cα coordinates for each domain. Note that additional documentation and description of the alignment algorithm can be found in the URL provided at the top of the figure

Biophysical Characterization of SH2 Variants

The SH2 variants are cloned into a vector (i.e., pET21) [10] suitable for recombinant expression in *Escherichia coli (E. coli)*. The SH2 variants are expressed as recombinant protein in bacteria and purified using appropriate methodologies [17]. We recommend cloning and expressing SH2 domains containing about 10 additional amino acids flanking both the N-terminus and the C-terminus of the domain. In this way, the expressed domain will likely assume the correct and typical fold of SH2 domains [17].

1. The literature is queried for established SH2-pTyr-peptide interactions [3, 18–20] and the peptides of interest are either synthesized or purchased as N-terminally biotinylated (Bio)

peptides (**Bio-linker-X_n-pTyr-X_m**; linker is 6-aminohexanoic acid (Ahx), X is any amino acid, n/m are number of any amino acids (typically 4/6, respectively) flanking the pTyr residue).

2. The wild-type and SH2 variants are assayed for affinity using an enzyme-linked immunosorbent assay (ELISA). Alternatively, fluorescence polarization, biolayer interferometry (BLI), surface plasmon resonance (SPR), or other affinity measurement techniques can be used.

As the SH2 domain fold is highly conserved, this algorithm to rationally engineer the affinity of a human SH2 domain can also be applied to nonhuman SH2 domains. While the described algorithm has not been tested in all human SH2 domains, it has been successful in engineering human SH2 domains with different degrees of relatedness to one another and therefore provides a general and modular framework for enhancing the affinity of virtually any SH2 domain.

1.3 Engineering the Specificity of Human SH2 Domains

The human Fyn SH2 domain was also engineered using phage-display to have exquisite specificity for unique pTyr-peptides [21, 22]. In these studies, Fyn SH2 phage-display libraries were created by randomizing residues in the EF- and BG-loops, which define the specificity pockets and selecting for variants with high specificity for unique pTyr-peptides (Fig. 5). One of the cloned phage-display libraries had an insertion of three additional residues in the BG-loop which were also mutated to mimic variable loop lengths observed in other human SH2 domains. The Fyn variants displayed tailored specificities for unique pTyr-peptides, therefore demonstrating that by changing the sequence of both the EF- and BG-loops, it is possible to tailor the specificity preferences of SH2 domains.

However, the variability in size and amino acid composition of the EF/BG loops and flanking regions means that primary amino acid sequence alignments are not optimal to define the boundaries of these loops in other human SH2 domains. Therefore, it is advisable to select an SH2 domain with a defined structure and identify the positions of the EF- and BG-loops by structural alignment. The following is a brief description of how to define the EF/BG loops and isolate SH2 variants with tailored specificities using phage-display:

Bioinformatics

1. The PDB and UniProt databases are queried for a structure of an SH2 domain of interest, preferably in complex with a pTyr-peptide.

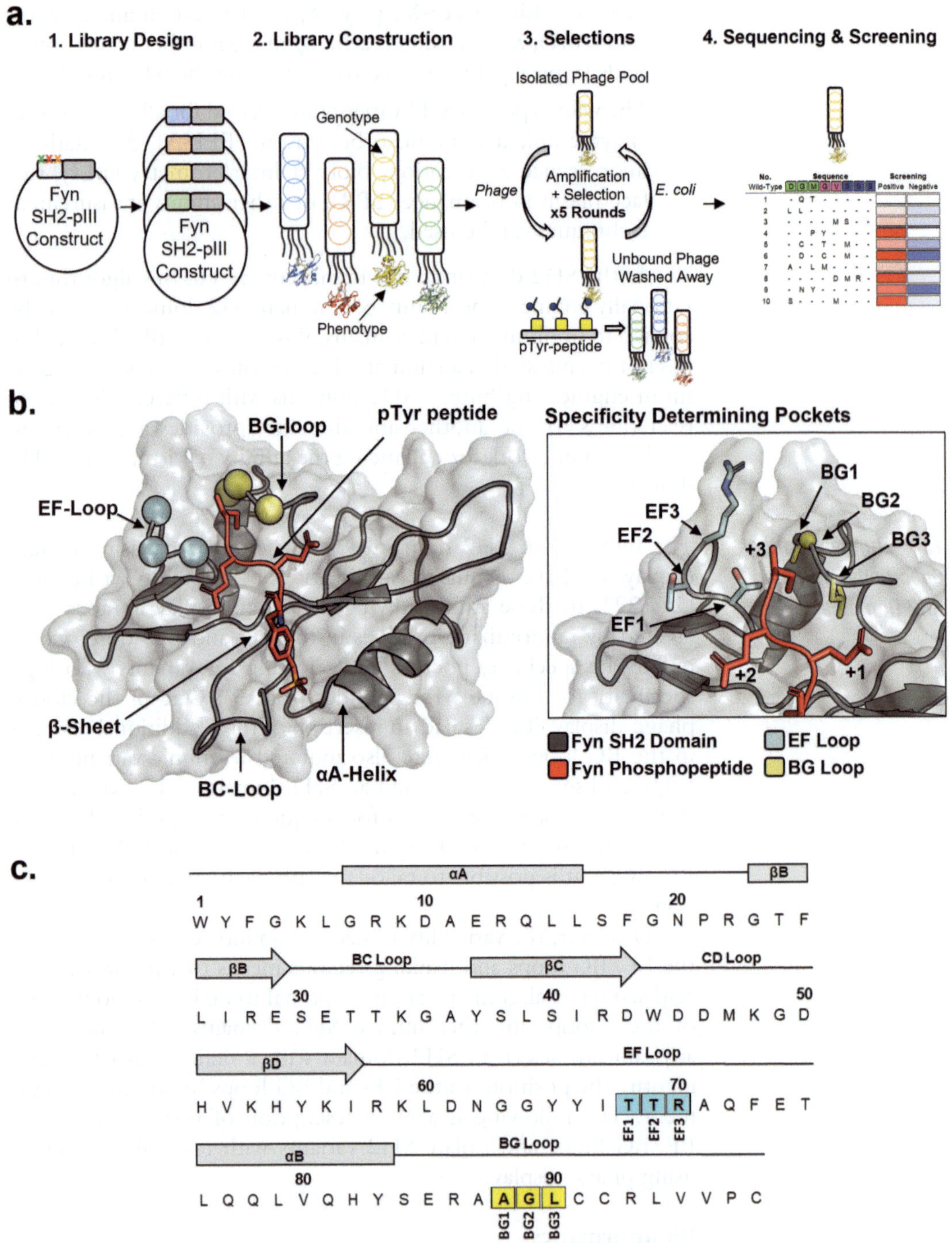

Fig. 5 Schematic of phage-display workflow and design of phage-display libraries to tailor specificity using the Fyn SH2 domain. (**a**) Schematic of phage-display library design, construction selections, and analysis. The phage-display library is synthesized and used in affinity selections against pTyr-peptides of interest. Eluted phage is used to infect *E. coli* and analyze the ssDNA in the viral capsid using a simple PCR reaction. Sanger sequencing and bioinformatic analysis of the isolated SH2 domain variants enables the identification of unique mutations in each unique SH2 domain clone. (**b**) Structure of the Fyn SH2 domain (PDB ID: 1AOT; gray) bound to a pTyr-peptide (red) with secondary structures annotated. The positions of residues randomized in the

2. The query SH2 domain structure is aligned with that of Fyn (PDB ID: 1AOT). Alignments can be performed in PyMol using the Cα coordinates for residues spanning the SH2 domain boundaries (Fig. 4).
3. EF- and BG-loop residues of the query SH2 domain(s) are identified by visually inspecting the structural alignment.

Phage-Display Library Design and Construction

1. The SH2 domain(s) of interest are cloned in frame with a C-terminal phage major coat protein III to enable monovalent display of the domain on the surface of phage particles.
2. Three mutagenic oligonucleotides (Oligo1–3) are designed with annealing sections of 15 base pairs (bp) on either side of the region to be randomized. Oligo1 should have annealing sections around the EF-loop with all 3 residues randomized with NNK (N = A/C/G/T and K = G/C) degenerate codons (total length of primer is 39 bp). Oligo2 should have annealing sections around the BG-loop with all 3 residues randomized with NNK degenerate codons (total length of primer is 39 bp). Oligo3 should have annealing sections around the EF-loop with all 3 residues randomized with NNK degenerate codons plus an additional 2 NNK degenerate codons (BG-loop length is 5 amino acids; total length of primer is 45 bp) (*see* **Note 3**).
3. A Kunkel reaction is performed with the ssdU-DNA template [23] to create library1 (mixture of Oligo1 and Oligo2) and library2 (mixture of Oligo1 and Oligo3).
4. A phage-displayed library of Fyn SH2 variants is produced using established methods [24, 25].

Affinity Selections

1. The N-terminally biotinylated peptide of interest is synthesized or purchased. Since the specificity-determining pocket of SH2 domains primarily recognizes amino acids C-terminal to the pTyr moiety, these residues will likely dictate the interaction between the Fyn SH2 variant and the pTyr-peptide. We suggest using an Ahx linker between the pTyr-peptide and the N-terminal biotin to maximize peptide accessibility. Alternatively, other linkers, such as polyethylene glycol (PEG) can be used.

Fig. 5 (continued) phage-display libraries are colored cyan (EF-loop) and yellow (BG-loop) and are represented as spheres highlighting the Cα coordinates of each residue. The inset (right) is a close-up view of the specificity determining pockets with side chains of residues depicted in colors described. Note the proximity to (within <10 Å) and orientation toward side chains of residues C-terminal (residues +1/2/3) to the pTyr residue. (**c**) Sequence of the Fyn SH2 domain (UniProt ID: P06241; residues 149–246) with secondary structure shown and positions randomized are highlighted as in (**a**)

2. Bio-pTyr-peptides are immobilized onto solid support coated with streptavidin (e.g., ELISA plate, magnetic beads), and following the blocking of unspecific binding sites, the Fyn SH2 variant libraries can be incubated in presence of the pTyr peptide of interest. Nonbinding Fyn SH2 variants are removed by washing the solid support and interacting Fyn SH2 variants are eluted and amplified overnight to enrich the phage pool of pTyr-peptide binding Fyn SH2 variants. This process should be repeated four to five times.
3. After four to five rounds of affinity selections, isolated phage can be used to infect *E. coli* cells and isolate individual colonies to send for DNA sequencing.

Biophysical Characterization of SH2 Variants

1. The DNA sequences of the Fyn SH2 variants are analyzed and the binding affinities are estimated using phage-ELISA to generate EC_{50} values.
2. The SH2 variants are cloned into an *E. coli* protein expression vector (i.e., pET21) [10] expressed as recombinant protein in bacteria and purified using appropriate methodologies and guidelines [17].
3. The wild-type and SH2 variants are assayed for affinity using a technique of choice such as competitive enzyme-linked immunosorbent assay (ELISA). Other biochemical and biophysical methods can be used.

Note that the EF- and BG-loops do not confer specificity in the same context for all human SH2 domains. For example, the Cbl SH2 domain has a tyrosine kinase binding (TKB) domain and EF-hand domain adjacent to its SH2 domain. Residues N-terminal to pTyr make significant contacts with the four-helix bundle of the TKB domain thereby dictating binding specificity [26]. The Cbl SH2 domain can also interact with the pTyr ligand in the reverse orientation, with N-terminal residues contacting the specificity determining pockets [27]. The N-terminal SH2 domain of phospholipase-Cγ forms additional contacts with some pTyr-peptides at other sites on the SH2 domain [28]. Other noncanonical SH2-pTyr-peptide interactions and binding preferences have been reviewed in more detail previously [29]. Therefore, it is important to understand the structure of the SH2 domain of interest. Instead of engineering a new SH2 domain, it might be advantageous to repurpose ones that have already been engineered (such as the Fyn SH2 domain) to bind to the target pTyr-peptide of interest. However, this methodology can be applied to any SH2 domain to create phage-display libraries and perform affinity selection experiments for sequence specificity.

1.4 The SH2 Domain as an Emerging Biotechnology

The ability to tailor the affinity and specificity of proteins is essential for the creation of useful biotechnologies [30]. While antibodies represent ideal scaffolds with which to engineer novel biotechnologies [31]; they have shown some limitations in recognizing with high-affinity and specificity posttranslational modifications (PTMs), including pTyr [32]. The SH2 domain is a specialized pTyr-specific binding scaffold with extensive research backing up its biological and biochemical function [33]. Indeed, the wild-type or engineered versions of different SH2 domains have been previously used in AP-MS workflows [34, 35], as biosensors for mapping signal transduction pathways [36–38], synthetic biology tools for rewiring signal transduction pathways [39, 40], or even as probes for superresolution microscopy [41]. By tailoring the affinity and/or specificity of the SH2 domain for a particular affinity, target, or even both in combination, one can create custom SH2 variants to investigate the biological system of interest.

2 Materials

2.1 Bioinformatics Software and Miscellaneous

1. Sequence information of all human SH2 domain (https://www.uniprot.org/).
2. Multiple protein sequence alignment tool (NCBI COBALT) (https://www.ncbi.nlm.nih.gov/tools/cobalt/re_cobalt.cgi).
3. Jalview, multiple sequence alignment editing, visualization, and analysis software (https://www.jalview.org/).
4. PyMol software (https://pymol.org/2/).

2.2 Rational Engineering of SH2 Domains

- 3,3′,5,5′-Tetramethylbenzidine (TMB) Microwell Peroxidase Substrate System.
- Agar.
- Bovine serum albumin (BSA).
- Carbenicillin Disodium Salt.
- Dithiothreitol (DTT).
- *E. coli* BL21 competent cells.
- MaxiSorp 384-well plates.
- Phenylmethylsulfonyl fluoride (PMSF).
- Phosphoric acid.
- Protease Inhibitor Tablets (EDTA-free).
- Sodium chloride.
- Tris(hydroxymethyl)aminomethane.
- Tryptone.

- Tween-20.
- UltraPure Glycerol.
- Yeast extract.

2.3 Engineering SH2 Domains Using Phage-Display

- 0.2 cm gap electroporation cuvettes.
- 1 Kb DNA Ladder.
- 3,3′,5,5′-Tetramethylbenzidine (TMB) Microwell Peroxidase Substrate System.
- 10 mM Adenosine 5′-Triphosphate (ATP).
- 10 mM deoxynucleoside triphosphate (dNTP) mix.
- Anti-M13 antibody HRP-conjugated.
- DpnI.
- Exonuclease I (ExoI).
- MaxiSorp 96-well plates.
- MaxiSorp 384-well plates.
- Mini tube 96-well system.
- QIAprep Spin Miniprep kit (includes proprietary PB, PE, and QG buffers).
- Shrimp alkaline phosphatase (SAP).
- Taq polymerase.
- T4 DNA ligase.
- T4 PNK.
- T7 polymerase.
- 1 M Glucose in sterile water for irrigation, filter sterilize.
- 1 M Tris pH 11.0 in sterile water for irrigation, filter sterilize.
- 1 M H_3PO_4 in milli-q water.
- 100 mM HCl in sterile water for irrigation, filter sterilize.
- Carbenicillin 100 mg/mL in milli-q water, filter sterilize.
- Chloramphenicol 10 mg/mL in 2-propanol.
- MLB buffer: 1 M sodium perchlorate, 30% (v/v) 2-propanol.
- MP buffer: dissolve 3.3 g citric acid monohydrate at room temperature in 2 mL sterile water for irrigation. Filter sterilize.
- Kanamycin 25 mg/mL in milli-q water, filter sterilize.
- Phosphate buffer saline (PBS): 8 mM Na_2HPO_4, 1.5 mM KH_2PO_4, 3 mM KCl, 137 mM NaCl. Adjust pH to 7.4 using HCl and autoclave.
- PBT buffer: PBS pH 7.4 containing 0.5% (w/v) BSA and 0.05% (v/v) Tween-20. Filter sterilize.

- PEG/NaCl: 20% (w/v) PEG 8000, 2.5 M NaCl in milli-q water. Mix thoroughly and autoclave.
- TAE buffer: 40 mM Tris-acetate pH 8.0, 1 mM ethylenediaminetetraacetic acid (EDTA).
- Tris buffer saline (TBS): 50 mM Tris–HCl pH 8.0, 150 mM NaCl.
- TBT buffer: TBS pH 8.0 containing 0.5% (w/v) BSA and 0.05% (v/v) Tween-20. Filter sterilize.
- Tetracycline solution: 10 mg/mL in ethanol.
- 10× TM buffer: 0.1 M $MgCl_2$, 0.5 M Tris pH 7.5. Filter sterilize.
- Uridine: 0.25 mg/mL in sterile water for irrigation, filter sterilize.
- *E. coli* CJ236 electrocompetent cells.
- *E. Coli* OmniMAX™ 2 $T1^R$ cells.
- *E. Coli* SS320 cells.
- M13KO7 helper phage.
- 2YT medium: 10 g yeast extract, 16 g tryptone, 5 g NaCl. Adjust the pH of the medium to 7.0 using NaOH, add water to bring the final volume to 1 L, and autoclave.
- 2YT/carb/cam medium: 2YT medium containing 100 μg/mL carbenicillin (carb) and 10 μg/mL chloramphenicol (cam).
- 2YT/carb/kan medium: 2YT medium containing 100 μg/mL carbenicillin (carb) and 25 μg/mL kanamycin (kan).
- 2YT/carb/kan/uridine medium: 2YT medium containing 100 μg/mL carbenicillin (carb), 25 μg/mL kanamycin (kan) and 0.25 μg/mL uridine.
- SOC medium: 5 g yeast extract, 20 g tryptone, 0.5 g NaCl, 0.2 g KCl. Adjust the pH of the medium to 7.0 using NaOH, add water to bring the final volume to 1 L and autoclave. Add 5 mL of autoclaved 2 M $MgCl_2$ and 20 mL of filter sterilized glucose.
- Superbroth medium: 24 g yeast extract, 12 g tryptone, 5 mL glycerol. Adjust the pH of the medium to 7.0 using NaOH, add water to bring the final volume to 900 mL and autoclave. Add 100 mL of autoclaved 0.17 M KH_2PO_4 and 0.72 K_2HPO_4.

3 Methods

3.1 Rational Engineering of SH2 Domains in the Human Proteome

Phage-display enabled the isolation of SH2 domain variants with outstanding binding affinity and tailored specificity [10, 11, 21]. However, the human genome encodes 122 SH2 protein domains, and although developing individual phage-displayed

SH2 domains library is technically possible, such an approach would be laborious, expensive, and time-consuming. To overcome this problem, we recently developed a general and modular strategy to engineer high-affinity variants of human SH2 domains without the need for combinatorial libraries [11]. In fact, by taking advantage of the highly conserved structure of SH2 domains, and of sets of mutations identified by the phage-display, that greatly enhanced the affinity of two SH2 domains with limited sequence identity, we successfully expanded the repertoire of high-affinity SH2 domains.

Our rational engineering strategy requires (i) sequence alignment of SH2 domains of interest with those of two high-affinity phage-display-derived SH2 domain variants (sSrc and sFes), (ii) structural analysis and alignment of SH2 domains, (iii) site-directed mutagenesis to graft sSrc or sFes binding-enhancing mutations into SH2 domains of interest. The method here described requires bioinformatic skills, standard molecular biology procedures and the ability to express and purify recombinant proteins.

3.1.1 Bioinformatic Analysis of the Location of Important pTyr-Interacting Residues

SH2 domains are characterized by high sequence and structural homology, thereby amenable to motif grafting strategies.

1. Download the amino acid sequence of SH2 domains of interest from UniProt [42].
2. Perform multiple sequence alignment using COBALT [13], and visualize the alignment using Jalview software [14]. Assemble aligned sequences into an unrooted phylogenetic tree using fast minimum evolution with a maximum sequence dissimilarity cut-off of 0.9.
3. Download the structure of the SH2 domain of interest, as well as those of Src (PDB ID: 1HCT) or Fyn (PDB ID: 1AOT) from the Protein Data Bank and visualize them using the open source PyMol software [16].
4. Perform structural alignment of the SH2 domains in PyMol. Alternatively, custom Python scripts can be used to perform the structural analysis.
5. Map superSrc or superFes binding-enhancing mutations into the sequence and the structure of the SH2 domain of interest. In addition, we recommend assessing the presence of an Arg residue in position αA2 as this amino acid contributes to coordinate the phosphoryl moiety of pTyr.

3.1.2 Grafting of Phage-Display-Identified Binding-Enhancing Mutations into SH2 Domains

The superSrc and superFes SH2 domains share very limited sequence identity, but both the engineered domains contain critical mutations in their BC loops, with superSrc having two hydrophilic-to-hydrophobic substitutions at positions βC2 and βD6 that greatly contribute to enhance binding affinity. Given the high structural and sequence similarity of SH2 domains is thereby possible to introduce, via site-directed mutagenesis, superSrc and superFes binding-enhancing mutations as well as an Arg residue in position αA2 to increase the affinity of other human SH2 domains.

1. Clone the desired SH2 domain into a bacterial expression vector. We recommend using a vector containing His-tag for facile affinity chromatography purification of the domain, and to fuse at the N-terminus of the SH2 domain of interest for ELISA-binding analysis.
2. Perform site-directed mutagenesis of the SH2 domain of interest by designing mutant oligonucleotides oriented in the inverse direction to amplify the entire template, thus incorporating the required mutation(s). We suggest using inverse PCR [43], as this method is highly robust and allows the introduction of a large number of mutations in one reaction. For each SH2 domain of interest, we suggest cloning a variant containing sSrc binding-enhancing mutations as well as another one containing mutations that improved sFes affinity, thereby maximizing the likelihood of developing a novel SH2 domain superbinder.
3. Following amplification of a linear, double-stranded PCR product containing the desired mutations, remove methylated DNA template by DpnI digestion.
4. Purify the linear, double-stranded, mutated PCR product using a Miniprep kit according to the manufacturer's instructions.
5. Phosphorylate the linear, double-stranded, mutated PCR product using T4 polynucleotide kinase, and then circularize the phosphorylated PCR amplicon by using T4 DNA ligase to obtain a circular plasmid.
6. Transform the obtained circular mutated plasmids into *E. Coli* cells and confirm the success of the mutagenesis reaction by Sanger sequencing.

3.2 Phage-Display

This protocol describes the procedures necessary for constructing phage-display SH2 domain libraries to isolate SH2 domain variants with tailored binding specificity. The genetic diversity is encoded in the libraries using precisely designed degenerate mutagenetic oligonucleotides that enable the fine control over the location and the chemical diversity introduced. Phage-display libraries containing billions of unique SH2 domains sequences are cloned using standard molecular biology techniques, and following electroporation in an *E. coli* host, the genetic library is converted into a phage-displayed SH2 domain library that can be used for isolating SH2 variants targeting a multitude of phosphorylated peptides.

A phagemid is a specialized vector that contains a double-stranded DNA origin of replication (dsDNA ori) for replication in *E. coli* hosts, and a single-stranded DNA (ssDNA) filamentous phage origin of replication (fl ori) that enables ssDNA to be packaged into phage particles. The SH2 domain is fused at the N-terminus to a stII leader peptide that ensures secretion of the SH2 domain in the bacterial periplasm via the *E. Coli* Sec pathway [44, 45], while at the C-terminal the SH2 domain is fused to the truncated M13 bacteriophage minor coat protein 3 (pIII), thus enabling the display of the domain on the surface of phage particles. The expression of the SH2 domain-pIII fusion is controlled by the alkaline phosphatase promoter (*PhoA*), but other promoters have been successfully used to control the expression of SH2 domains [46]. The display of SH2 domains on the surface of phage particles is achieved by infecting *E. coli* cells transformed with SH2 domain-encoding phagemids with helper phage, such as M13KO7, which provides all the additional proteins necessary for the assembly of phage particles.

The library described here is derived from a phagemid encoding the SH2 domain of the tyrosine kinase Fyn (residues 149–246; UniProt ID: P06241) [21]. Genetic diversity is introduced by annealing uracil-containing ssDNA (dU-ssDNA) encoding the SH2 domain-pIII fusion template (purified from an *E. coli dut-/ ung- E. Coli* strain) to synthetic mutagenic oligonucleotides designed to randomize 3 residues in both the EF- and BG-loops, with additional nucleotides added for elongation of the BG-loop. T7 polymerase is used to extend mutagenic oligonucleotides for the synthesis of a complementary DNA strand that is ligated by the activity of T4 DNA ligase, thus obtaining a covalently closed circular dsDNA (CCC-dsDNA) heteroduplex. Following the transformation of CCC-dsDNA into an *E. coli dut+/ung+* host, the uracil-containing DNA strand is degraded [47], resulting in the replication and propagation of the randomized DNA strand.

The mutagenic oligonucleotides are designed with complementarity to the sequences immediately preceding and following the region subjected to mutagenesis, with the oligonucleotide sequence between the two annealing regions that can be of any desired sequence and length, therefore enabling facile and precise control over the explored sequence space. Provided the absence of overlap, such an approach allows the simultaneous introduction of mutations in multiple regions in the template dU-ssDNA. Furthermore, this strategy provides a simple way to vary the length of loops that make contact with the pTyr by simply using pools of degenerate oligonucleotides, thus generating SH2 domain variants with extended binding surfaces [21].

3.2.1 Preparation of dU-ssDNA Template

Highly pure dU-dsDNA phagemid template, extracted from phage particles, is essential for successful and efficient library construction. We recommend using 20 μg of dU-dsDNA template for the construction of one library containing 10^9–10^{10} unique SH2 domain variants.

1. Transform the phagemid vector encoding the SH2 domain-pIII fusion into competent *E. coli* CJ236 (or analogous *dut-/ ung-* bacterial host). The phagemid template here described contains a β-lactamase gene (Bla) that confers resistance to ampicillin and carbenicillin, enabling the selection of successfully transformed *E. Coli* cells by simply plating them onto LB agar plates containing 100 μg/mL carbenicillin and overnight incubation at 37 °C.
2. Pick a single colony of *E. coli* CJ236 harboring the phagemid vector and inoculate 1 mL of 2YT/carb/cam medium. Chloramphenicol is used to select the F′ episome in *E. coli* CJ236 cells, while carbenicillin enables the selection of phagemid-containing *E. coli* cells. Following incubation of the culture for 30 min at 37 °C, with 200 rpm shaking, add 1 μL of M13KO7 helper phage (10^{10} pfu/mL), and allow infection to proceed at 37 °C, 200 rpm for 2 h.
3. Add 25 μg/mL kanamycin to the culture media to select for bacteria coinfected with M13KO7 helper phage. Kanamycin is used to select cells superinfected by the helper phage. Incubate culture for 6 h at 37 °C, with 200 rpm shaking.
4. Transfer the culture to a baffled 250 mL flask containing 30 mL 2YT/carb/kan/uridine medium and incubate at 37 °C and 200 rpm for 20 h. Uridine is supplemented to culture media to ensure the incorporation of uracil bases in place of thymine.
5. Transfer the culture to a 50 mL tube and pellet bacteria by centrifuging at 7871 g, 4 °C for 10 min.
6. Collect the phage-containing supernatant in a new 50 mL tube and add to the solution 6 mL PEG/NaCl solution to precipitate phage particles. Incubate the solution on ice for 20 min.
7. Phage can be precipitated by centrifugation at 17,709 g at 4 °C for 30 min. Discard the supernatant and further precipitate phage at 17,709 g 4 °C for another 3 min.
8. Carefully remove residual supernatant and resuspend precipitated phage with 500 μL sterile PBS pH 7.4. Transfer phage to a 1.5 mL microcentrifuge tube, and spin the tube at 17,709 g, 4 °C for 5 min to completely remove bacterial cell debris. Transfer the phage-containing supernatant to a new 1.5 mL microcentrifuge tube.
9. Lyse phage particles by adding 7 μL of MP buffer to the phage-containing solution and incubating the mixture at room temperature for 5 min.
10. Apply sample to a QIAprep spin column, and spin at 7871 g, 25 °C for 30 s. Discard the flow-through.
11. Extract DNA by adding 700 μL MLB buffer to the QIAprep spin column, and spin at 7871 g, 25 °C for 30 s. Discard the flow-through.

12. Add 700 μL MLB buffer and incubate the column at 25 °C for 1 min. Spin the column at 7871 g, 25 °C for 30 s. Discard the flow-through. The extracted dU-ssDNA will be bound to the column's matrix.
13. Precipitate dU-ssDNA onto the column's matrix by adding 700 μL of PE buffer (Qiagen) and spinning the column at 7871 g, 25 °C for 30 s. Discard flow-through.
14. Further remove residual proteins and salt by adding 700 μL PE buffer (Qiagen). Spin the column 7871 g, 25 °C for 30 s. Discard flow-through.
15. Remove excess ethanol (contained in the PE buffer) by spinning the column at 7871 g, 25 °C for 30 s.
16. Transfer the column into a new 1.5 mL microcentrifuge tube and add to the center of the column 50 μL sterile water for irrigation. Incubate the column at 37 °C for 10 min to ensure highly efficient dU-ssDNA elution. Spin the column at 17,709 g, 25 °C for 1 min, and save the eluate (containing purified dU-ssDNA).
17. To ensure complete recovery of column-bound dU-ssDNA, perform an additional elution step by adding 50 μL of sterile water to the middle of the column and incubate it at 37 °C for 10 min. Spin the column at 17,709 g, 25 °C for 1 min. Collect the eluate and combine it with previously eluted dU-ssDNA.
18. Analyze the purity of dU-ssDNA by loading 5 μL onto a 1% agarose gel and determine spectrophotometrically its concentration. An absorbance value of 1.0 at 260 nm indicates a concentration of 33 ng/μL dU-ssDNA. The presented procedure typically yields dU-ssDNA concentrations in the 250–500 ng/μL range.
19. Aliquot 20 μg dU-ssDNA into 1.5 mL microcentrifuge tubes. Store samples at −20 °C. Avoid repeated freeze and thaw cycles.

3.2.2 Synthesis of CCC-dsDNA

CCC-dsDNA is synthesized via a three-step procedure that requires (i) mutagenic oligonucleotide phosphorylation, (ii) annealing of mutagenic oligonucleotides dU-ssDNA template encoding the SH2 domain-pIII fusion and (iii) introduction of genetic diversity via T7 DNA polymerase-mediated extension of mutagenic oligonucleotides and T4 DNA ligase-mediated ligation of the obtained product to form CCC-dsDNA. The protocol described here requires 20 μg of highly pure dU-ssDNA to generate a library containing 10^{9}–10^{10} unique Fyn SH2 domain clones.

1. Phosphorylate each mutagenic oligonucleotide by assembling the following reaction in a PCR tube

0.6 μg	mutagenic oligonucleotide
2 μL	10× TM buffer

2 μL	10 mM ATP
1 μL	100 mM DTT
2 μL	T4 polynucleotide kinase (20 units)

Add sterile irrigation water to bring the final reaction volume to 20 μL. Incubate the reaction mixture at 37 °C for 1 h.

For obtaining SH2 domain libraries with extended loops and binding surfaces, it is possible to follow the same procedure, albeit using pools of mutagenic oligonucleotides.

2. Immediately after mutagenic oligonucleotides phosphorylation, anneal mutagenic primers to 20 μg purified dU-ssDNA encoding the SH2 domain template. Since two main regions in SH2 domain are diversified, two mutagenic, nonoverlapping, oligonucleotides are used simultaneously.

20 μg	dU-ssDNA
25 μL	10× TM buffer
20 μL	each phosphorylated oligonucleotide

Add sterile water to bring the final reaction volume to 250 μL. Incubate the reaction mixture at 90 °C for 3 min, 50 °C for 3 min, and 25 °C for 5 min.

3. Following mutagenic oligonucleotides annealing to the dU-ssDNA template, extend in vitro synthesis of CCC-dsDNA by assembling the following reaction mixture:

250 μL	annealed oligonucleotide/dU-ssDNA mixture
10 μL	10 mM ATP
10 μL	10 mM dNTP mix
15 μL	100 mM DTT
5 μL	T4 DNA ligase (2000 units)
3 μL	T7 DNA polymerase (30 units)

Incubate the reaction mixture overnight at 25 °C.

4. Purify synthesized CCC-dsDNA using the QIAquick gel extraction kit (Qiagen) by adding to the reaction mixture 1 mL of buffer QG (Qiagen) and applying the sample to two QIAquick spin columns. Spin the columns at 7871 g, 25 °C for 1 min. Discard the flow-through.

5. Precipitate and desalt DNA by adding 750 μL PE buffer (Qiagen) to each column and spinning them at 17709 g, 25 °C for 30 s. Discard flow-through.

6. To ensure greater salt removal, add 750 μL PE buffer (Qiagen) to each column and incubate at 25 °C for 1 min. Spin columns at 17,709 g, 25 °C for 30 s. Discard flow-through.

7. Transfer the columns into two new 1.5 mL microcentrifuge tubes and remove excess ethanol (contained in the PE buffer) by centrifugation at 17,709 g, 25 °C for 1 min.

8. Transfer the columns into two new 1.5 mL microcentrifuge tubes and elute CCC-dsDNA by adding to the center of each column 30 μL of sterile water. Incubate the column at 37 °C for 10 min and elute CCC-dsDNA by centrifugation at 17,709 g, for 1 min at 25 °C.
9. Combine the CCC-dsDNA eluted from each column and determine the concentration of the CCC-dsDNA heteroduplex by measuring absorbance at 260 nm ($A_{260} = 1.0$ for 50 ng/μL CCC-dsDNA).
10. Analyze the efficiency of the mutagenesis reaction by loading 5 μL CCC-dsDNA alongside the dU-ssDNA template onto a 1% agarose gel, as the CCC-dsDNA is characterized by different electrophoretic mobility relative to the dU-ssDNA template [24].

3.2.3 Conversion of CCC-dsDNA into a Phage-Displayed SH2 Domain Library

The final step required to construct highly diverse phage-displayed SH2 domain libraries is the transformation of the CCC-dsDNA heteroduplex into an *E. coli* dut+/ung+ host containing the F′ episome to enable M13 bacteriophage infection and propagation, as well as degradation of the dU-ssDNA template. In the presented procedure, we use the *E. coli* SS320 strain that enables high-efficiency electroporation and phage production [48].

1. Chill purified CCC-dsDNA heteroduplex and a 0.2 cm gap electroporation cuvette on ice.
2. Prewarm 25 mL SOC medium at 37 °C (20 mL in a 250 mL baffled flask and 5 mL in a 15 mL tube).
3. Thaw on ice a 350 μL aliquot of electrocompetent *E. coli* SS320 cells (~3×10^{11} cfu/mL) and add to them purified CCC-dsDNA. Gently mix by pipetting several times, without introducing bubbles.
4. Transfer the CCC-dsDNA:*E. Coli* cells mixture to the electroporation cuvette, gently tap it to remove bubbles and wipe its sides with a paper tissue. Electroporate according to the manufacturer's instruction. For the BTX ECM-630 electroporation system (BTX), we use the following settings: 2.5 kV field strength, 125 Ω resistance, and 50 μF capacitance.
5. Immediately after electroporation, rescue bacterial cells by adding SOC medium prewarmed into a 15 mL tube to the cuvette and transfer the solution to the 250 mL baffled flask containing the remaining SOC medium. Rinse the cuvette with SOC medium multiple times to ensure the collection of all transformed cells. Incubate cells at 37 °C, for 30 min with 200 rpm shaking.
6. To determine library diversity, transfer 10 μL from the 250 mL baffled flask and perform tenfold serial dilutions into 90 μL 2YT medium in a 96-well plate. Plate 5 μL of each dilution,

using a multichannel pipette, onto an LB agar plate containing 100 μg/mL carbenicillin (selecting phagemid-containing cells), an LB agar plate containing 25 μg/mL kanamycin (selecting cells infected with the M13KO7 helper phage) and an LB agar plate containing 10 μg/mL tetracycline (selecting the F′ episome) to determine the titer of *E. coli* SS320 cells. Grow bacteria by incubating plates at 37 °C, overnight.

7. Transfer the 25 mL culture containing electroporated *E. coli* SS320 cells into a 2 L baffled flask containing 500 mL 2YT/carb/kan medium. Incubate culture overnight at 37 °C and 200 rpm shaking.
8. Following overnight amplification of phage, pellet bacteria by transferring the culture to a 1 L bottle and centrifuging 10 min, at 7871 g, 4 °C.
9. Pick a single colony of *E. coli* OmniMAX™ 2 T1^{R} cells from a fresh LB agar plate containing 10 μg/mL tetracycline and inoculate 25 mL 2YT/tet medium. Incubate the culture at 37 °C with 200 rpm shaking.
10. Following bacterial cell precipitation, transfer the supernatant (containing the phage-displayed SH2 domain library) to a new 1 L bottle containing 120 mL PEG/NaCl solution prechilled on ice. Mix the solution by inversion and incubate it on ice for 20 min.
11. Precipitate phage by centrifugation at 12,298 g, 4 °C for 30 min. Decant the supernatant and spin down phage particles at 12,298 g, 4 °C for additional 3 min.
12. Remove the remaining supernatant with a pipette and gently resuspend phage with 20 mL prechilled filter-sterilized TBT buffer. Since SH2 domain phage-displayed libraries are used in affinity selections with phosphorylated tyrosine-containing targets, we recommend avoiding the use of phosphate-containing buffers. Transfer the solution into a 50 mL tube and remove bacterial cell debris by centrifugation at 17,709 g, for 10 min at 4 °C.
13. Transfer the supernatant to a new 50 mL tube and precipitate the SH2 domain phage-displayed library a second time by adding 5 mL of ice-cold PEG/NaCl solution. Mix the tube by inversion and incubate it on ice for 20 min.
14. Precipitate phage at 17,709 g, 4 °C for 30 min. Decant the supernatant and further spin down at 17,709 g, 4 °C for additional 3 min. Remove the remaining supernatant with a pipette and resuspend the phage pellet with 5 mL of filter sterilized TBT buffer.

15. Determine the concentration of the phage library by infecting mid-log phase (OD_{600} = 0.6–0.8) *E. coli* OmniMAX™ 2 $T1^R$ cells. Dilute 10 μL phage library into 90 μL 2YT medium and perform tenfold serial dilutions. Use 10 μL of each phage serial dilution to infect 90 μL of mid-log phase *E. coli* OmniMAX™ 2 $T1^R$ cells. Incubate at 37 °C, with 200 rpm shaking for 30 min.
16. Using a multichannel pipette, plate 5 μL of each dilution onto an LB agar plate containing 100 μg/mL carbenicillin, an LB agar plate containing 25 μg/mL kanamycin and an LB agar plate containing 10 μg/mL tetracycline. Incubate plates overnight at 37 °C.
17. Add to the resuspended phage-display library glycerol to a final concentration of 40%, aliquot and freeze it using liquid nitrogen. Store the library at −80 °C.

3.2.4 Selection of High-Affinity SH2 Domain Variants

Phage-displayed SH2 domain libraries have been used to isolate high-affinity SH2 domains targeting a multitude of tyrosine-phosphorylated targets [10, 11, 21], as well as tyrosine-sulfated peptides [46, 49]. Typically, affinity selections are performed by immobilizing biotinylated peptides containing tyrosine-modified residues onto streptavidin-coated immunoplates. To maximize the likelihood of isolating SH2 domain variants recognizing the post-translationally modified tyrosine residues, rather than amino acids flanking the PTM site, we recommend using a negative selection step during which the phage-displayed SH2 domain library is incubated with a peptide having identical sequence to the target but lacking the modified tyrosine residue.

1. Coat 8 wells of two MaxiSorp immunoplates with 100 μL of streptavidin at 2 μg/mL in TBS pH 8.0 and incubate the plate at 4 °C, overnight.
2. Remove the coating solution and block wells by adding 300 μL/well of TBS pH 8.0 containing 0.5% (w/v) BSA. Incubate immunoplates at 25 °C for 1 h with gentle shaking.
3. Remove the blocking solution and wash wells three times with 300 μL/well of TBS pH 8.0 containing 0.05% (v/v) Tween-20 (TBT buffer).
4. Add to each streptavidin-coated well 100 μL of biotinylated peptides at 2 μg/mL in TBS pH 8.0 and incubate plates at 25 °C, with gentle shaking for 30 min. While one immunoplate will be incubated with the biotinylated pTyr-containing peptide, the other plate will be incubated with the same peptide lacking tyrosine phosphorylation.
5. Remove the excess nonphosphorylated peptide solution and wash wells three times with 300 μL/well of TBT buffer.

6. Add 100 μL of library phage solution in TBS buffer containing 0.05% (v/v) Tween-20 and 0.05% (w/v) BSA to each well and incubate the plate at 25 °C for 1 h with gentle shaking. These wells will be used to deplete from the phage pool SH2 domain variants that bind to streptavidin or the peptide independently of tyrosine phosphorylation. We recommend using a phage concentration in each well exceeding the library diversity by 1000-fold.
7. Remove the excess pTyr-peptide solution and wash wells three times with 300 μL/well of TBT buffer.
8. Transfer unbound phage from the plate immobilized with nonphosphorylated peptides to the one containing the target pTyr-peptide. Incubate the plate at 25 °C for 1 h with gentle shaking.
9. Remove the phage solution and wash each well with 300 μL/well TBT buffer. The stringency of the washing step can be modified by adjusting the number of washing steps performed, the duration of each step, and/or the concentration of Tween-20 used in the buffer.
10. Elute bound phage-displayed SH2 domain variants by adding 100 μL/well of 100 mM HCl. Incubate at 25 °C for 5 min with gentle shaking. Transfer the HCl solution to a 1.5 mL microcentrifuge tube containing 1 M Tris pH 11 to neutralize the pH of the solution.
11. Use eluted phage particles to infect 10 volumes of mid-log phase OmniMAX™ 2 T1^R cells ($OD_{600} = 0.6–0.8$) growing in 2YT/tet medium. Incubate culture at 37 °C with 200 rpm shaking for 30 min (*see* **Note 4**).
12. Perform tenfold serial dilutions of phage infected Omni-MAX™ 2 T1^R cells and plate 5 μL of each dilution onto an LB agar plate containing 100 μg/mL carbenicillin to determine the titer of eluted phage. Incubate plates overnight at 37 ° C.
13. Following infection of OmniMAX™ 2 T1^R cells with eluted phage, superinfect cells by adding M13KO7 helper phage to a final concentration of 10^{10} pfu/mL. Incubate at 37 °C the culture for 1 h with shaking at 200 rpm.
14. Transfer superinfected *E. coli* OmniMAX™ 2 T1^R cells to a 250 mL baffled flask containing 25 volumes of 2YT/carb/kan. Incubate the culture overnight at 37 °C with 200 rpm shaking.
15. Transfer the culture to a 50 mL tube and remove bacterial cells by centrifugation for 10 min at 7871 g and 4 °C.
16. Transfer the supernatant-containing phage to a new 50 mL tube containing ice-cold PEG/NaCl solution. Mix the tube by inversion and incubate on ice for 20 min.

17. Precipitate phage for 30 mi at 17,709 g and 4 °C. Remove the supernatant and precipitate phage particles for additional 3 mi at 17,709 g and 4 °C. Remove the remaining supernatant with a pipette and resuspend the phage pellet with 1 mL of filter-sterilized TBT buffer.
18. Transfer phage to a 1.5 mL microcentrifuge tube and remove bacterial cell debris by centrifugation at 17,709 g, for 5 min and 4 °C. Transfer the phage-containing supernatant to a new 1.5 mL microcentrifuge tube and determine the phage concentration spectrophotometrically.
19. Repeat the selection cycle. Significant enrichment is observed typically after 3 rounds of selection with most selection campaigns lasting for four to five rounds in total.
20. Pick individual clones for analysis of SH2 domain variants identity via DNA sequencing and analysis of binding specificity via phage enzyme-linked immunosorbent assay (ELISA).

3.2.5 Binding Analysis of Selected SH2 Domain Variants by Phage ELISA and Sequencing of Specific Phage Clones

To identify SH2 domains specifically recognizing pTyr-peptides, binding assays (such as phage ELISA) that use the culture medium containing unique SH2 domain variants displayed on the surface of phage particles can be used. The procedure described below is optimized for the use of 384-well MaxiSorp immunoplates. Following binding analysis, the identity of SH2 domain variants can be deduced by sequencing of the phagemid DNA encapsulated in the phage particles.

1. Pick single colonies harboring selected phage-displayed SH2 domain variants and inoculate 96-well microtubes containing 400 μL 2TY/carb medium and M13KO7 helper phage to the final concentration of 10^{10} pfu/mL. Produce phage particles displaying unique SH2 domain variants by incubating microtubes overnight at 37 °C with 200 rpm shaking.
2. Coat MaxiSorp immunoplate wells with 25 μL of streptavidin at 2 μg/mL in TBS pH 8.0 and incubate the plate at 4 °C, overnight.
3. Remove the coating solution and block wells by adding 60 μL/well of TBS pH 8.0 containing 0.5% (w/v) BSA. Incubate immunoplates at 25 °C for 1 h with gentle shaking.
4. Remove the blocking solution and wash wells three times with 90 μL/well TBT buffer.
5. Add to each streptavidin-coated well 25 μL of biotinylated peptides at 2 μg/mL in TBS pH 8.0 and incubate plates at 25 °C, with gentle shaking for 30 min. In order to assess binding specificity, we recommend adding to adjacent wells biotinylated nonphosphorylated peptides and biotinylated pTyr-containing peptides.

6. Remove the excess peptide solution and wash wells three times with 90 μL/well of TBT buffer.
7. Centrifuge 96-well microtubes at 1968 g, 4 °C for 10 min to remove bacteria cells. Transfer phage-containing supernatant to a 96-well nonbinding microplate and dilute it threefold with TBT buffer.
8. Transfer 25 μL of each diluted phage supernatant to immunoplate wells and incubate at 25 °C for 30 min with gentle shaking. In addition to binding to the target pTyr-peptide and the same peptide lacking tyrosine phosphorylation, binding should also be assayed using wells coated with streptavidin but not incubated with biotinylated peptides (negative control).
9. Remove phage solution and wash each well four times with 90 μL TBT buffer.
10. Add to each well 25 μL horseradish peroxidase conjugate anti-M13 antibody diluted 1:3000 in TBT buffer. Incubate the immunoplate at 25 °C for 30 min with gentle shaking.
11. Remove the excess of anti-M13 antibody and wash each well six times with 90 μL TBT buffer.
12. Add to each well 25 μL of freshly prepared TMB substrate. Incubate the plate at 25 °C with gentle shaking and allow chromogenic development for 5 to 10 min.
13. Quench the colorimetric reaction by adding to each well 25 μL of 1 M H_3PO_4. Read the immunoplate spectrophotometrically at 450 nm.
14. To determine binding specificity, subtract absorbance values obtained from wells blocked with BSA only and compare the measured absorbance intensity of wells coated with pTyr-peptide to that of wells coated the nonphosphorylated target or streptavidin.
15. In order to perform Sanger DNA sequencing in a 96-well format by using the same phage-containing supernatant used for screening of specific SH2 domain variants, assemble a PCR reaction mixture as follows:

2.5 μL	10× ThermoPol buffer (New England Biolabs)
0.25 μL	10 mM dNTP mix
0.25 μL	each PCR primer (FWD and REV)
2 μL	phage supernatant
0.5 μL	Taq polymerase (2.5 units)

Add sterile water to bring the final reaction volume to 25 μL.

16. Amplify the DNA fragment using the following PCR program: 5 min at 95 °C, 25 cycles of amplification (30 s at 95 °C, 15 s at 55 °C, 60 s at 68 °C), 5 min at 68 °C, and storage at 4 °C.

17. Assess the presence of PCR amplicons by loading 5 μL of each reaction onto a 1% agarose gel.
18. To degrade the residual PCR primers and dephosphorylate dNTPs that could interfere with the Sanger sequencing reaction, we recommend using the ExoI and SAP enzymes. Assemble a clean-up reaction mixture by mixing:

    0.5 μL SAP (2.5 units)
    0.3 μL ExoI (3 units)

    Add sterile irrigation water to bring the final reaction volume to 2 μL, and dispense the reaction mixture, without changing the buffer, to each well of the 96-well PCR plate containing PCR amplicons.
19. Analyze the ExoI and SAP-treated PCR products by Sanger DNA sequencing.

4 Notes

1. There are multiple alignment algorithms available in PyMol. For alignments with human SH2 domains, the ALIGN function does a reasonably good job for all human SH2 domains.
2. Since Fes and Src SH2 domains share only 35% sequence similarity, grafting the BC-loops and βC2/βD6 residues of both superbinder domains into the desired SH2 domain will maximize the likelihood of identifying variants with enhanced affinity.
3. In this example, we used NNK degenerate oligonucleotides that allow one to encode all 20 natural amino acids, while reducing the frequency of stop codons incorporation. However, other randomization strategies can also be used. Moreover, we present a synthetic library design that requires only three mutagenic primers, but larger pools of diverse oligonucleotides can be used to increase the diversity of the chemical space introduced in the SH2 domain.
4. Volumes refer to volume of phage eluted after selections. Typically, one-half of eluted phage is used to infect cells for the next round, while the other half is mixed with glycerol (40% v/v; final concentration), frozen using liquid nitrogen, and stored at −80 °C.

References

1. Waksman G, Kominos D, Robertson SC et al (1992) Crystal structure of the phosphotyrosine recognition domain of SH2 of v-src complexed with tyrosine-phosphorylated peptides. Nature 358:646–653
2. Kaneko T, Huang H, Zhao B et al (2010) Loops govern SH2 domain specificity by controlling access to binding pockets. Sci Signal 3:ra34. https://doi.org/10.1126/scisignal.2000796
3. Hause RJ, Leung KK, Barkinge JL et al (2012) Comprehensive binary interaction mapping of SH2 domains via fluorescence polarization reveals novel functional diversification of ErbB

receptors. PLoS One 7:e44471. https://doi.org/10.1371/journal.pone.0044471
4. Haslam NJ, Shields DC (2012) Peptide-binding domains: are limp handshakes safest? Sci Signal 5:pe40. https://doi.org/10.1126/scisignal.2003372
5. Thibault P, Michnick SW, Kanshin E et al (2015) A cell-signaling network temporally resolves specific versus promiscuous phosphorylation. Cell Rep 10:1202–1214. https://doi.org/10.1016/j.celrep.2015.01.052
6. Liu BA, Engelmann BW, Nash PD (2012) The language of SH2 domain interactions defines phosphotyrosine-mediated signal transduction. FEBS Lett 586:2597–2605. https://doi.org/10.1016/j.febslet.2012.04.054
7. Lim WA, Pawson T (2010) Phosphotyrosine signaling: evolving a new cellular communication system. Cell 142:661–667. https://doi.org/10.1016/j.cell.2010.08.023. Phosphotyrosine
8. Li L, Tibiche C, Fu C et al (2012) The human phosphotyrosine signaling network: evolution and hotspots of hijacking in cancer. Genome Res 22:1222–1230. https://doi.org/10.1101/gr.128819.111
9. Doll S, Gnad F, Mann M (2019) The case for proteomics and phospho-proteomics in personalized cancer medicine. Proteomics Clin Appl 13:1800113. https://doi.org/10.1002/prca.201800113
10. Kaneko T, Huang H, Cao X et al (2012) Superbinder SH2 domains act as antagonists of cell signaling. Sci Signal 5:ra68. https://doi.org/10.1126/scisignal.2003021
11. Martyn GD, Veggiani G, Kusebauch U et al (2022) Engineered SH2 domains for targeted phosphoproteomics. ACS Chem Biol 17: 1472–1484. https://doi.org/10.1021/acschembio.2c00051
12. Schneider M, Lane L, Boutet E et al (2009) The UniprotKB/Swiss-Prot knowledgebase and its plant proteome annotation program. J Proteome 72:567–573. https://doi.org/10.1016/j.jprot.2008.11.010.The
13. Papadopoulos JS, Agarwala R (2007) COBALT: constraint-based alignment tool for multiple protein sequences. Bioinformatics 23:1073–1079. https://doi.org/10.1093/bioinformatics/btm076
14. Waterhouse AM, Procter JB, Martin DMA et al (2009) Jalview Version 2-a multiple sequence alignment editor and analysis workbench. Bioinformatics 25:1189–1191. https://doi.org/10.1093/bioinformatics/btp033
15. Burley SK, Berman HM, Bhikadiya C et al (2019) RCSB Protein Data Bank: biological macromolecular structures enabling research and education in fundamental biology, biomedicine, biotechnology and energy. Nucleic Acids Res 47:D464–D474. https://doi.org/10.1093/nar/gky1004
16. Schrodinger L, DeLano W (2020) PyMol. Retrieved from http://www.pymol.org/pymol
17. Liu BA, Ogiue-Ikeda M, Machida K (2017) Expression and production of SH2 domain proteins. In: SH2 domains: methods and protocols. Springer Nature, pp 117–162
18. Tinti M, Kiemer L, Costa S et al (2013) The SH2 domain interaction landscape. Cell Rep 3: 1293–1305. https://doi.org/10.1016/j.celrep.2013.03.001.The
19. Jones RB, Gordus A, Krall JA, MacBeath G (2006) A quantitative protein interaction network for the ErbB receptors using protein microarrays. Nature 439:168–174. https://doi.org/10.1038/nature04177
20. Huang H, Li L, Wu C et al (2007) Defining the specificity space of the human Src homology 2 domain. Mol Cell Proteomics 7:768–784. https://doi.org/10.1074/mcp.M700312-MCP200
21. Veggiani G, Huang H, Yates BP et al (2019) Engineered SH2 domains with tailored specificities and enhanced affinities for phosphoproteome analysis. Protein Sci 28:403–413. https://doi.org/10.1002/pro.3551
22. Liu H, Huang H, Voss C et al (2019) Surface loops in a single SH2 domain are capable of encoding the spectrum of specificity of the SH2 family. Mol Cell Proteomics 18:372–382. https://doi.org/10.1074/mcp.RA118.001123
23. Kunkel TA (1985) Rapid and efficient site-specific mutagenesis without phenotypic selection. Proc Natl Acad Sci U S A 82:488–492. https://doi.org/10.1073/pnas.82.2.488
24. Tonikian R, Zhang Y, Boone C, Sidhu SS (2007) Identifying specificity profiles for peptide recognition modules from phage-displayed peptide libraries. Nat Protoc 2: 1368–1386. https://doi.org/10.1038/nprot.2007.151
25. Huang H, Kaneko T, Sidhu SS, Li SSC (2017) Creation of phosphotyrosine superbinders by directed evolution of an SH2 domain. In: Machida K, Liu BA (eds) SH2 domains methods and protocols, 1555th edn. Springer Nature, New York, pp 225–254
26. Hu J, Hubbard SR (2005) Structural characterization of a novel Cbl phosphotyrosine recognition motif in the APS family of adapter proteins. J Biol Chem 280:18943–18949. https://doi.org/10.1074/jbc.M414157200

27. Ng C, Jackson RA, Buschdorf JP et al (2008) Structural basis for a novel intrapeptidyl H-bond and reverse binding of c-Cbl-TKB domain substrates. EMBO J 27:804–816. https://doi.org/10.1038/emboj.2008.18
28. Stein A, Aloy P (2008) Contextual specificity in peptide-mediated protein interactions. PLoS One 3:e2524. https://doi.org/10.1371/journal.pone.0002524
29. Wagner MJ, Stacey MM, Liu BA, Pawson T (2013) Molecular mechanisms of SH2- and PTB-domain-containing proteins in receptor tyrosine kinase signaling. Cold Spring Harb Perspect Biol 5. https://doi.org/10.1101/cshperspect.a008987
30. Nuttall SD, Walsh RB (2008) Display scaffolds: protein engineering for novel therapeutics. Curr Opin Pharmacol 8:609–615. https://doi.org/10.1016/j.coph.2008.06.007
31. Bostrom J, Lee CV, Haber L, Fuh G (2009) Improving antibody binding affinity and specificity for therapeutic development. In: Therapeutic antibodies: methods and protocols. Springer, pp 353–376
32. Hattori T, Koide S (2019) Next-generation antibodies for post-translational modifications. Curr Opin Struct Biol 51:141–148. https://doi.org/10.1016/j.sbi.2018.04.006.Next-generation
33. Machida K, Liu BA (2017) SH2 domains: methods and protocols. In: Methods in molecular biology, 1555th edn. Springer Nature, New York, pp 1–546
34. Bian Y, Li L, Dong M et al (2016) Ultra-deep tyrosine phosphoproteomics enabled by a phosphotyrosine superbinder. Nat Chem Biol 12:959–966. https://doi.org/10.1038/nchembio.2178
35. Tong J, Cao B, Martyn GD et al (2017) Protein-phosphotyrosine proteome profiling by superbinder-SH2 domain affinity purification mass spectrometry, sSH2-AP-MS. Proteomics 17:1600360. https://doi.org/10.1002/pmic.201600360
36. Freeman J, Kriston-vizi J, Seed B, Ketteler R (2012) A high-content imaging workflow to study Grb2 signaling complexes by expression cloning. J Vis Exp 68:e4382. https://doi.org/10.3791/4382
37. Noguchi T, Ishiba H, Honda K et al (2017) Synthesis of Grb2 SH2 domain proteins for mirror-image screening systems. Bioconjug Chem 28:609–619. https://doi.org/10.1021/acs.bioconjchem.6b00692
38. Sun J, Lei L, Tsai CM et al (2017) Engineered proteins with sensing and activating modules for automated reprogramming of cellular functions. Nat Commun 8:477. https://doi.org/10.1038/s41467-017-00569-6
39. Findlay GM, Smith MJ, Lanner F et al (2013) Interaction domains of Sos1/Grb2 are finely tuned for cooperative control of embryonic stem cell fate. Cell 152:1008–1020. https://doi.org/10.1016/j.cell.2013.01.056
40. Barnea G, Strapps W, Herrada G et al (2008) The genetic design of signaling cascades to record receptor activation. PNAS 105:64–69
41. Farrell MV, Nunez AC, Yang Z et al (2022) Protein-PAINT: Superresolution microscopy with signaling proteins. Sci Signal 15: eabg9782
42. The Uniprot Consortium, Bateman A (2019) UniProt: a worldwide hub of protein knowledge. Nucleic Acids Res 47:D506–D515. https://doi.org/10.1093/nar/gky1049
43. Silva D, Santos G, Barroca M, Collins T (2017) Inverse PCR for point mutation introduction. In: Methods in molecular biology. Springer, pp 87–100
44. Fujimoto K, Fukuda T, Marumoto R (1988) Expression and secretion of human epidermal growth factor by Escherichia coli using enterotoxin signal sequences. J Biotechnol 8:77–86
45. Thie H, Schirrmann T, Paschke M et al (2008) SRP and Sec pathway leader peptides for antibody phage display and antibody fragment production in E. coli. New Biotechnol 25:49–54. https://doi.org/10.1016/j.nbt.2008.01.001
46. Lawrie J, Niu W, Guo J (2019) Engineering of a sulfotyrosine-recognizing small protein scaffold for the study of protein tyrosine O-sulfation. In: Methods in enzymology. Elsevier, pp 67–89
47. Warner HR, Duncan BK, Garrett C, Neuhard J (1981) Synthesis and metabolism of uracil-containing deoxyribonucleic acid in Escherichia coli. J Bacteriol 145:687–695. https://doi.org/10.1128/jb.145.2.687-695.1981
48. Sidhu SS, Lowman HB, Cunningham BC, Wells JA (2000) Phage display for selection of novel binding peptides. Methods Enzymol 328:333–363
49. Ju T, Niu W, Guo J (2016) Evolution of Src homology 2 (SH2) domain to recognize sulfotyrosine. ACS Chem Biol 11:2551–2557. https://doi.org/10.1021/acschembio.6b00555

Part V

SH2-Domain Containing Proteins and Regulation of Activity

Chapter 18

Measuring Protein Tyrosine Phosphatase Activity Dependent on SH2 Domain-Mediated Regulation

Pablo Rios, Azin Kiani, and Maja Köhn

Abstract

Src-homology-2 (SH2) domains bind selectively to phosphotyrosine (pTyr) residues located in target binding proteins; therefore, they are key elements in pTyr-mediated signaling pathways. The binding of an SH2 domain to a pTyr acts as a docking mechanism that attracts proteins into signaling hubs, and in some cases, it can also regulate the catalytic activity of signaling enzymes such as protein kinases or protein phosphatases. Therefore, compounds that selectively bind SH2 domains can be potentially used to modulate the activity of such SH2 domain-containing enzymes. This chapter describes how to measure the regulation of protein tyrosine phosphatase activity through allosteric binding of peptides to SH2 domains, and uses human recombinant protein tyrosine phosphatase SHP2 (Src homology-2 domain-containing protein tyrosine phosphatase 2) purified from bacteria as a case example. The phosphatase activity against the artificial substrate DiFMUP (6, 8-Difluoro-4-Methylumbelliferyl Phosphate) is measured over time in the presence of a peptide that selectively binds and activates SHP2 at different concentrations to determine the half maximal effective concentration (EC_{50}).

Key words Protein Phosphatases, SH2 domain, DiFMUP, Phosphopeptides, Enzyme activity

1 Introduction

Cellular signaling mechanisms are largely governed by a delicate balance of phosphorylation and dephosphorylation of proteins (or other signaling molecules) that is maintained by the interplay of protein kinases and protein phosphatases. As opposing enzymes to protein kinases, protein phosphatases dephosphorylate protein and some nonprotein substrates, and play equal important roles in cell homeostasis.

Among the different families of protein phosphatases in humans, only a few enzymes contain SH2 domains: the two closely related lipid phosphatases SHIP1 and SHIP2 (SH2-containing Inositol 5′-phosphatase 1 and 2, encoded by the genes INPP5D and INPPL1, respectively), the protein tyrosine phosphatase (PTP) Tensin-2 and its inactive homolog Tensin-1, and the two closely

Teresa Carlomagno and Maja Köhn (eds.), *SH2 Domains: Functional Modules and Evolving Tools in Biology*, Methods in Molecular Biology, vol. 2705, https://doi.org/10.1007/978-1-0716-3393-9_18,

related PTPs tyrosine phosphatase nonreceptor type 6 (also known as Src homology-2 domain-containing protein tyrosine phosphatase 1, SHP1, encoded by the PTPN6 gene) and Tyrosine phosphatase nonreceptor type 11 (also known as Src homology-2 domain-containing protein tyrosine phosphatase 2, SHP2, encoded by the PTPN11 gene). This chapter will focus only on the activity modulation of SH2 domain-containing tyrosine phosphatases SHP1 and SHP2.

The SHP1 and SHP2 phosphatases are multidomain proteins containing at their amino terminus two SH2 domains (N-SH2 and C-SH2), followed by a PTP domain and a *C*-terminal tail that includes two tyrosine residues with regulatory functions mediated by phosphorylation [1]. SHP1 and SHP2 stay under basal conditions in an inactive, closed conformation, in which the N-SH2 domain blocks the catalytic site of the PTP domain. In normal physiological conditions, extracellular stimuli activate cell signaling, and recruit them to the plasma membrane through the binding of both SH2 domains to pTyr residues located in receptor tyrosine kinases, cytokine receptors, immune inhibitory receptors and other signaling adaptor proteins, leading to an open conformation on which the N-SH2 domain moves apart and unblocks the active site [2–5]. Thus, binding of the SH2 domains to pTyr motifs provoke conformational changes that render the PTP domain active (Fig. 1a).

SHP1 is expressed predominantly in hematopoietic cells, acting as a negative regulator of signaling pathways that prevents inflammation and autoimmunity [6]. It also exerts tumor suppressor functions, terminating signaling events that are involved in cell survival, proliferation, migration and invasion [7]. On the contrary, some cancers show higher expression of SHP1 [7], and it is being considered a candidate for checkpoint inhibition for immunotherapy [8, 9].

SHP2 is a widely expressed phosphatase that is predominantly involved in cancer promotion and developmental disorders, primarily promoting the full activation of the Ras-MAPK signaling pathway [1, 10]. Although a tumor suppressor function has been also described [11, 12], it is a well-established oncogenic phosphatase, with somatic mutations being found in several tumors (different leukemias in adults, juvenile myelomonocytic leukemia, breast cancer, gastric cancer) [13]. In addition, germline mutations in the PTPN11 gene are found in other diseases as the Noonan syndrome (with SHP2 mutations accounting for 50% of the cases) and in the LEOPARD syndrome [14].

Another important role of SHP2 is the control of T cell activation. Binding of both SH2 domains to the tyrosine-phosphorylated motifs Immunoreceptor Tyr-based Motif (ITIM) and the Immunoreceptor Tyr-based Switch Motif (ITSM) of the immune checkpoint receptor programmed cell death protein-1 (PD-1) activates

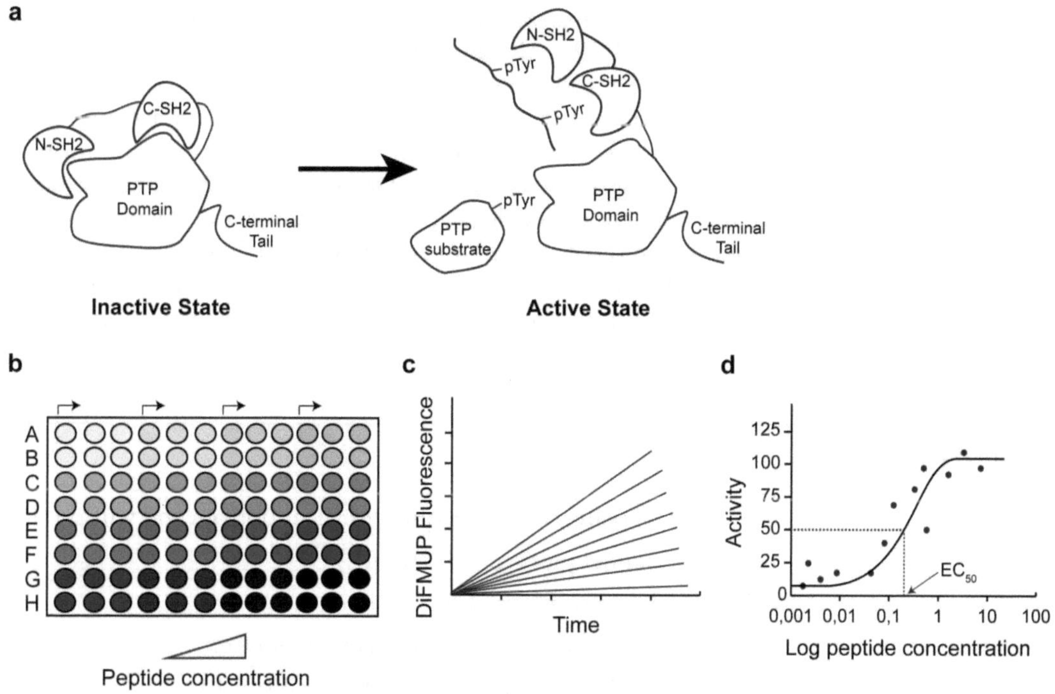

Fig. 1 Measuring PTP activity dependent on SH2 domain-mediated regulation. (**a**) Activation mechanism of SHP2 mediated by a bidentate pTyr-containing peptide or protein. In the active state, substrates of SHP2 can occupy the active site upon release of the N-SH2 domain from the catalytic pocket of SHP2. (**b**) Schematic pipetting layout of the 96-well plate. Sixteen different concentrations of peptide are distributed in the plate as follows: row A and control row B (blank no enzyme) contain the first four peptide concentrations (shown as grey scale) in triplicate (depicted with arrows); row C and control row D (blank no enzyme) contain the next four peptide concentrations, and subsequently for rows E–F and G–H. (**c**) DiFMUP dephosphorylation kinetics recorded for every peptide concentration. After subtracting the corresponding blank control measurement, the data is represented as fluorescence signal over time, and the slope values are determined for every peptide concentration. (**d**) Dose–response curve obtained after nonlinear fitting of the data from (**c**) to calculate the final EC_{50}

SHP2 and leads to suppressed T-cell receptor (TCR) signaling and T cell proliferation. Cancer cells expressing PD-L1, the cell surface ligand of PD-1, use the PD-1/SHP2 pathway as an advantage to escape the immune system [15, 16].

All these facts make SHP1 and SHP2 potential targets for therapy of human diseases, and large efforts are currently underway to develop SHP1 and SHP2 drugs to modulate their function [7, 17, 18]. Hence, a full understanding of their molecular activation mechanism and their biological regulation is essential.

A detailed protocol for measuring SHP2 activation in vitro is given here, and it serves as a guideline for further studies (either activation or inhibition-based) for both SHP1 and SHP2. It is based on the mechanism shown in Fig. 1a, with the natural substrate being replaced in vitro by the unnatural fluorescent phosphatase substrate 6,8-difluoro-4-methylumbelliferyl phosphate (DiFMUP). Although the activator molecule used here is peptide based, molecules of other chemical nature could be used. The catalytic activity of the SHP2 is monitored in a plate reader instrument in 96-well plates. A predetermined concentration of protein phosphatase is incubated with a wide concentration range of the activator molecule (Fig. 1b). The kinetics of the phosphatase reaction are recorded upon addition of a predetermined optimum DiFMUP concentration (Fig. 1c), and the data is further analyzed for calculation of the half maximal effective concentration (EC_{50}) (Fig. 1d).

2 Materials

1. 96-well plates, black opaque with clear bottom and without lid.
2. 12-Channel pipette (10–100 μL).
3. Aluminum or plastic foil.
4. Reagent reservoirs for multichannel pipette.
5. DiFMUP phosphatase substrate (*see* **Note 1**).
6. Plate reader with fluorescence readout mode (excitation at 358 nm, emission at 452 nm) (*see* **Note 2**).
7. Purified human recombinant SHP2 phosphatase (*see* **Note 3**).
8. Reaction buffer: 50 mM Hepes (pH 7.6), 150 mM NaCl, 1 mM EDTA, 0.05% Tween 20 (freshly added), 5 mM DTT (freshly added, as DTT is sensitive to oxidation).
9. MilliQ water.
10. Phosphopeptides dissolved in reaction buffer.
11. Data analysis software (we use Microsoft Excel and GraphPad Prism).

3 Methods

3.1 Preparation of Reagents

The following procedure is described for a biochemical assay in a 96-well plate. Volumes are given in sufficient amount to run an experiment filling the full plate. If necessary, the protocol should be adapted by users based on their experimental needs.

1. Prepare 10 mL of 10% Tween 20 (1 mL of Tween 20 in 9 mL of water).
2. Prepare 5 mL of 1 M DTT stock: dissolve 0.77 g of DTT in 5 mL of water and vortex until it is completely dissolved. Make 0.5 mL aliquots and freeze four of them at −20 °C for later use. Leave one aliquot on ice for direct use.

3. Add 150 μL of 10% Tween 20 and 150 μL of 1 M DTT to 30 mL of reaction buffer and mix with a vortex.
4. Prepare 400 μL of intermediate stock of SHP2 phosphatase at a concentration of 100 nM in reaction buffer.
5. Mix 200 μL of SHP2 100 nM with 3.8 mL of reaction buffer. This will be a 10× stock (5 nM) with respect the final concentration of SHP2 in the assay at 0.5 nM.
6. Dissolve the peptide powder at a concentration of 2 mM in reaction buffer (*see* **Note 4**).
7. Prepare intermediate stocks of the peptide (in this example: 2000 μM, 200 μM, 20 μM, 2 μM, 0.2 μM, 0.02 μM, 0.002 μM) in reaction buffer.
8. Prepare 16 dilutions of the peptide in reaction buffer in separate Eppendorf tubes with a final volume of 70 μL using the different intermediate peptide stocks prepared in **step** 7. These dilutions will be 10× concentrations with respect to the final desired concentrations (*see* **Note 5**).
9. Dissolve the DiFMUP powder at 50 mM in 100% DMSO (5 mg vial dissolved in 342.3 μL of DMSO). Prepare 1.2 mL of an intermediate stock of DiFMUP at 2 mM in reaction buffer (1152 μL reaction buffer + 48 μL of DiFMUP) and keep it protected from light.

3.2 Filling the Plate

The assay is done in triplicates, which are pipetted on consecutive wells in the different plate rows (Fig. 1b). Thus, every three wells correspond to a different peptide concentration in triplicate. There is a blank control without enzyme for every peptide concentration (also done in triplicates). Protein-containing wells are pipetted into rows A, C, E, and G, while Blank wells are pipetted into rows B, D, F, and H.

1. Pipet 70 μL of reaction buffer into all wells of plate rows A, C, E, and G.
2. Pipet 80 μL of reaction buffer into all wells of plate rows B, D, F, and H.
3. Pipet 10 μL of peptide at the corresponding concentration in all the protein-containing wells and in all the blank-containing wells.
4. Pipet 10 μL of enzyme into all wells of rows A, C, E, and G.
5. Mix the content of the wells with a 12-channel micropipette. Avoid creating air bubbles.
6. Incubate the plate for 20 min at room temperature. Cover the plate with plastic or aluminum foil to avoid dust particles entering the wells.
7. In the meantime, prepare the plate reader and the acquisition protocol so it is ready to start the measurement.

3.3 Addition of DiFMUP

1. After the incubation time, remove the cover from the plate.
2. Transfer the DiFMUP solution into a reagent reservoir (*see* **Note 6**).
3. Pipet 10 μL of 2 mM DiFMUP into all the wells with a 12-channel micropipette.
4. Mix 3–4 times the well contents with the multichannel micropipette avoiding air bubbles (*see* **Note 7**).
5. Load the plate in the plate reader and start the kinetic measurement immediately.

3.4 Data Analysis

Plate reader data is usually exported as an Excel file.

1. In a separate spread sheet, copy the original raw data.
2. Average the blank triplicates for every peptide concentration.
3. Subtract the corresponding averaged blank to each of the triplicates for the corresponding peptide concentration measurement. Plot the results (as fluorescence signal versus time) as in Fig. 1c.
4. Calculate the slope values of the linear ranges for every peptide concentration.
5. Represent the slope values against the log of peptide concentration (Fig. 1d). Calculate the EC_{50} with the appropriate software (we use GraphPad Prism).
6. Repeat the experiment at least three times with values in triplicates.

4 Notes

1. DiFMUP is dissolved in 100% DMSO. We prepare an original stock at 50 mM and store aliquots of 100 μL in small glass vials at −20 °C.
2. The plate reader measurement must be optimized. One of the most important parameters to control is the gain of the detector. Overflowed (saturated) signal should be avoided, at least for the early time point measurements. Higher or lower gain adjustment will depend, for example, on the plate reader instrument, intrinsic activity of the phosphatase and its concentration in the assay or the concentration of substrate used. We suggest to do control measurements with the lowest and the highest activator concentration at different gains to find the optimal recorded fluorescence signal within the dynamic range. Also, the kinetic interval of the measurement should be short enough to allow recording the linear ranges of the enzymatic activity.

3. The protein of interest should be at highest purity. Original purified protein stock should be kept frozen at −80 °C in small volume aliquots for single use. Repeatedly cycles of freezing and defrosting of protein aliquots affect the activity of the protein.
4. The solubility of each peptide depends on its primary sequence and it is pH sensitive. If observing precipitates in the peptide solution, make sure that the pH of the buffer is correct and suitable for the peptide solubility. If the peptides or modulator molecules are only soluble in DMSO, it is desired not to exceed a final DMSO concentration of 5%.
5. The final modulator molecule concentration range needs to be empirically tested. It is important to have a sufficiently wide concentration range, so that when analyzing and fitting the results, plateaus are present at both ends of the curve when fitting the data for EC50 calculation. Also, make sure that the 10× serial dilutions of peptide or activator are properly mixed with the pipet.
6. The reaction will start as soon as the substrate is added, so it is important to be fast at this stage and to add the DiFMUP next to the plate reader. The use of a reagent reservoir helps with fast pipetting, but might need to use higher volume of DiFMUP solution to fill the cavity of the reservoir. Alternatively, 0.2 mL PCR tubes (placed for example into an empty pipet tip rack for 10 μL tips) can be filled with the necessary volume of DiFMUP. This saves unnecessary waste of DiFMUP and allows also a fast transfer of DiFMUP into the 96-well plate with a multichannel pipet.
7. Air bubbles due to mixing or to the presence of detergent in the reaction buffer interfere with the read and must be avoided. A fast way to remove them is using a plastic squeeze bottle half filled with 70% ethanol. If the spout is not in contact with the ethanol (it can be cut to shorten it out), vapors of ethanol are released upon squeezing the bottle (but don't squeeze too hard, otherwise ethanol might leak into the sample). When applied to the wells of the plate, the vapors of ethanol will remove the bubbles.

References

1. Neel BG, Gu H, Pao L (2003) The 'Shp'ing news: SH2 domain-containing tyrosine phosphatases in cell signaling. Trends Biochem Sci 28:284–293
2. Yang J, Liu L, He D et al (2003) Crystal structure of human protein-tyrosine phosphatase SHP-1. J Biol Chem 278:6516–6520
3. Wang W, Liu L, Song X et al (2011) Crystal structure of human protein tyrosine phosphatase SHP-1 in the open conformation. J Cell Biochem 112:2062–2071
4. Hof P, Pluskey S, Dhe-Paganon S et al (1998) Crystal structure of the tyrosine phosphatase SHP-2. Cell 92:441–450

5. Barford D, Neel BG (1998) Revealing mechanisms for SH2 domain mediated regulation of the protein tyrosine phosphatase SHP-2. Structure 6:249–254
6. Sharma Y, Bashir S, Bhardwaj P et al (2016) Protein tyrosine phosphatase SHP-1: resurgence as new drug target for human autoimmune disorders. Immunol Res 64:804–819
7. Varone A, Spano D, Corda D (2020) Shp1 in solid cancers and their therapy. Front Oncol 10:935
8. Watson HA, Wehenkel S, Matthews J, Ager A (2016) SHP-1: the next checkpoint target for cancer immunotherapy? Biochem Soc Trans 44:356–362
9. Dempke WCM, Uciechowski P, Fenchel K, Chevassut T (2018) Targeting SHP-1, 2 and SHIP pathways: a novel strategy for cancer treatment? Oncology 95:257–269
10. Matozaki T, Murata Y, Saito Y et al (2009) Protein tyrosine phosphatase SHP-2: a proto-oncogene product that promotes Ras activation. Cancer Sci 100:1786–1793
11. Bard-Chapeau EA, Li S, Ding J et al (2011) Ptpn11/Shp2 acts as a tumor suppressor in hepatocellular carcinogenesis. Cancer Cell 19: 629–639
12. Li S, Hsu DD, Wang H, Feng GS (2012) Dual faces of SH2-containing protein-tyrosine phosphatase Shp2/PTPN11 in tumorigenesis. Front Med 6:275–279
13. Butterworth S, Overduin M, Barr AJ (2014) Targeting protein tyrosine phosphatase SHP2 for therapeutic intervention. Future Med Chem 6:1423–1437. Erratum in: Future Med Chem (2014) 6:2132
14. Gelb BD, Tartaglia M (2006) Noonan syndrome and related disorders: dysregulated RAS-mitogen activated protein kinase signal transduction. Hum Mol Genet 15(Spec No 2):R220–R226
15. Yokosuka T, Takamatsu M, Kobayashi-Imanishi W et al (2012) Programmed cell death 1 forms negative costimulatory microclusters that directly inhibit T cell receptor signaling by recruiting phosphatase SHP2. J Exp Med 209:1201–1217
16. Marasco M, Berteotti A, Weyershaeuser J et al (2020) Molecular mechanism of SHP2 activation by PD-1 stimulation. Sci Adv 6:eaay4458
17. Chen YN, LaMarche MJ, Chan HM et al (2016) Allosteric inhibition of SHP2 phosphatase inhibits cancers driven by receptor tyrosine kinases. Nature 535:148–152
18. Yuan X, Bu H, Zhou J et al (2020) Recent advances of SHP2 inhibitors in cancer therapy: current development and clinical application. J Med Chem 63:11368–11396

Chapter 19

Peptides as Baits for the Coprecipitation of SH2 Domain-Containing Proteins

Azin Kiani, Pablo Rios, and Maja Köhn

Abstract

Phosphotyrosine (pTyr)-containing amino acid sequences have regulatory effects on proteins that contain pTyr recognition motifs, such as Src Homology 2 (SH2) domains. Using pTyr-containing peptides as a bait for coprecipitation, by immobilization of the synthesized phosphopeptides to beads and incubation with cell lysates, enables to study the binding preference of the SH2 domain for the specific pTyr-sequence obtained from a pTyr-containing protein in a complex biological environment. Using phosphopeptides allows to not only assess the wild-type sequence, but also peptides that can contain modified sequences which carry a nonhydrolyzable pTyr or other modifications varying the binding strength and selectivity, for example, to create strong SH2 domain binders to inhibit their interaction with pTyr-containing proteins. This pulldown experiment can be used as an assay to evaluate the ability of a peptide to bind to the protein of interest in the cell lysate or investigate the selectivity of the peptide. Therefore, immobilizing phosphopeptides and using them as a pulldown tool has a wide range of applications.

Key words Src Homology 2 (SH2) domains, pTyr, Immobilization, Pulldown, Selectivity

1 Introduction

Src Homology 2 (SH2) domains are protein domains of approximately 100 amino acids that interact with phosphotyrosine (pTyr)-containing motifs with binding affinity but without any catalytic activity on them. According to [1], in total, there are 121 SH2 domains and 111 proteins that contain either one or two of these domains. The broad range of proteins containing SH2 domains include those that function as kinases, adaptors, phosphatases, ubiquitin ligases, transcription factors, guanine nucleotide exchange factors, and phospholipid-based secondary signaling molecules [1]. Structurally, SH2 domains contain a β-sheet consisting of seven antiparallel strands (βA-βG) which is in between two α-helices, αA and αB. The *N*-terminal region of SH2 domains is conserved for binding to pTyr and forms a positively charged pocket from αA to βD. This region contains a conserved arginine

Teresa Carlomagno and Maja Köhn (eds.), *SH2 Domains: Functional Modules and Evolving Tools in Biology*, Methods in Molecular Biology, vol. 2705, https://doi.org/10.1007/978-1-0716-3393-9_19,

residue, which is essential for binding of this domain to pTyr through bi-dentate hydrogen binding. At the *C*-terminus, the hydrophobic pocket from βD to βG binds to the *C*-terminus of pTyr-containing motifs and provides specificity [2, 3].

For SH2 domains as pTyr recognition motifs, the binding energies related to the pTyr pocket and the specificity-determining region (provided by the *C*-terminus) are not equivalent, and the pTyr-dependent interaction dominates the interaction. However, the selectivity of an SH2 domain for its motif is not absolute because both *N*- and *C*-terminus of SH2 domains provide interactions with the pTyr-containing motifs and can have slight interaction with nonspecific pTyr peptides. The combination of these strong and weak bindings, can create even more complex signaling networks [4, 5].

SH2 domains function as key mediators of cellular processes. For example, they regulate protein functional states, serve as adaptors between a receptor and a protein to activate signaling cascades, or structurally stabilize transcription factor complexes. Of therapeutic significance, many diseases promoting intracellular signaling cascades contain key SH2 domain-mediated proteins. Therefore, as an effector in disease-related signaling cascades, the SH2 domains have been validated as an intracellular molecular target for inhibitors. It is well recognized that protein-selective SH2 domain binding compounds would be a significant benefit to the clinic [6]. Furthermore, detecting the natural ligands of these domains with peptides containing pTyr or its nonhydrolyzable mimetics can help with the understanding of their role and design active compounds to control their function in a therapeutic setting.

One efficient way to investigate SH2 domains that bind to a pTyr motif is using peptides as tools for pulling down proteins containing these domains (Fig. 1a). Covalent immobilization of the peptide to a solid phase (beads) provides the stable basis for incubation with a cell lysate or, if known, proteins of interest to evaluate efficiency and selectivity of binding through pulling down the protein(s) bound to the peptides on the beads.

The SulfoLink coupling resin is an efficient agarose support to immobilize peptides that contain a sulfhydryl group through thioether bond formation. In this reaction, beads containing iodoacetyl groups react specifically with free sulfhydryl groups provided by a coupled cysteine to the peptide sequence, to be applied for pulldown experiments that are described below in detail, including the peptide synthesis.

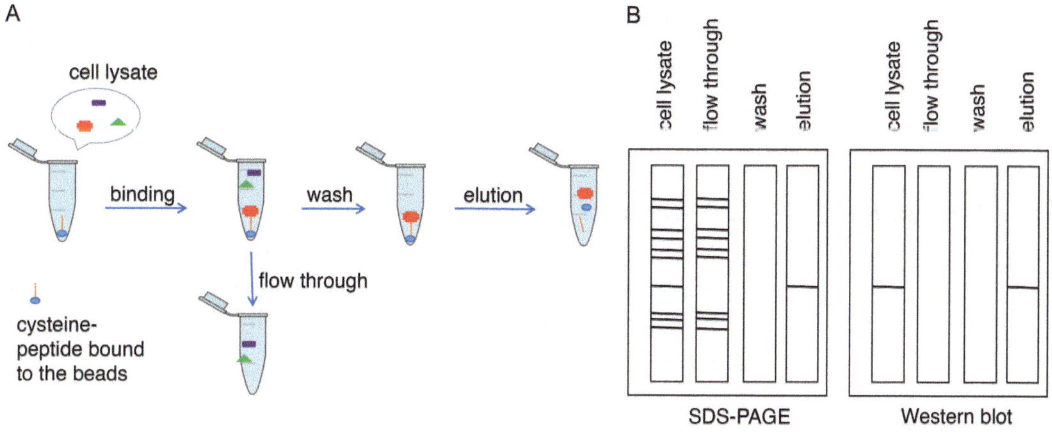

Fig. 1 Schematic overview of pulling down an SH2 domain-containing protein using an immobilized peptide: (**a**) Upon incubation of cell lysate with cysteine-peptide (light brown) that is immobilized to the beads (blue), SH2 domain-containing protein (red) binds to the peptide. The unbound proteins (green and violet) are separated in the flow through sample. After washing and boiling steps, the SH2 domain-containing protein dissolves in the elution sample. (**b**) The cell lysate sample contains all the endogenous proteins shown in the SDS-PAGE, and among them, the SH2 domain-containing protein is detected by the specific antibody in the Western blot. The flow through contains all the endogenous proteins of the cell lysate except the SH2 domain-containing protein (or to a much lesser extent than in the cell lysate sample if the pulldown was not complete), and in the Western blot, no protein (or much less than in the cell lysates if the pulldown was not complete) is detected. The wash sample contains no protein in both SDS-PAGE and Western blot. In the elution sample, the SH2 domain-containing protein is the only protein in the SDS-PAGE, unless other proteins bind to the peptide as well (*see* **Note 9**), and is detected by the specific antibody in the Western blot

2 Materials

2.1 Standard Fluorenylmethyloxy-carbonyl (Fmoc) Solid-Phase Peptide Synthesis (SPPS)

1. An automated peptide synthesis robot. We used a MultiPep RSi peptide synthesizer from Intavis.
2. Precise balance.
3. 5 mL syringes with polyethylene filter at the bottom and cap.
4. Rink Amide resin, standard Fmoc- and side chain protected amino acids. The side chain protecting groups are acid labile, whereas the Fmoc group is base labile.
5. *N*-hydroxybenzotriazole hydrate (HOBt), *N,N*-dimethylformamide (DMF), O-benzotriazole-*N,N,N′,N′*-tetramethyluronium hexafluorophosphate (HBTU), *N*-methylmorpholine (NMM), *N*-methyl-2-pyrrolidone (NMP), acetic anhydride (Ac_2O), 2,6-lutidine, piperidine, trifluoroacetic acid (TFA), triisopropylsilane (TIPS), diethyl ether.

2.2 Manual Amino Acid Coupling

1. A phosphotyrosine (pTyr) mimetic.
2. A linker molecule (we use (2-(2-(Fmoc-amino)ethoxy)ethoxy) acetic acid).

3. Fmoc-Cys(Trt)-OH (Cys).
4. Shaker.
5. Reagents from 2.1 (5.).
6. A 5 mL syringe with filter and plunger.

2.3 Purification

1. A megafuge 11R centrifuge.
2. MALDI (matrix assisted laser desorption/ionization)- or ESI (electrospray ionization)-mass spectrometer (MS) coupled to high-performance liquid chromatography (HPLC-ESI).
3. Preparative HPLC.

2.4 Immobilization of the Peptide

1. 1.5 mL tubes.
2. Resin bed (we use SulfoLink).
3. An orbital wheel.
4. Coupling buffer: 50 mM Tris–HCl, 5 mM EDTA-Na, pH 8.5.
5. Cell lysis buffer.
6. A microcentrifuge.

2.5 Cell Line and Lysis Buffer

1. Cell line of interest.
2. RIPA buffer for lysis: 25 mM Tris–HCl pH 7.6, 150 mM NaCl, 5 mM EDTA, 1% Triton X-100, 1% sodium deoxycholate, 0.1% SDS.

2.6 Western Blot Read-out

1. SDS sample buffer: 125 mM Tris–HCl (pH 6.8), 20% glycerol, 4% SDS, 0.1% bromophenol blue, 5% β-mercaptoethanol.
2. Sodium dodecylsulfate polyacrylamide gel electrophoresis (SDS-PAGE).
3. Polyvinylidene fluoride (PVDF) membrane.
4. Transfer buffer: 25 mM Tris–HCl, pH 8.3, 190 mM glycine, 0.1% (w/v) SDS, and 20% (v/v) methanol.
5. Antibody against SH2 domain-containing protein of interest.
6. Tris-buffered saline with 0.1% Tween (TBS-T).
7. Nonfat milk powder.
8. Shaker.
9. Enhanced chemiluminescence (ECL) Western blotting detection reagent.
10. Digital imaging system.

3 Methods

3.1 Synthesis of Peptide-Containing Phosphotyrosine

The synthesis in the scale of 50 µmol from *C*- to *N*-terminus, for an automated peptide synthesizer, is described below. The reaction steps of the peptide synthesis take place in a 5 mL syringe; however, the volumes might need to be adapted.

1. Weigh 50 µmol Fmoc-Rink Amide resin (this is 1 equivalent) in the reaction syringe and place it in and automated synthesizer.
2. Prior to the first coupling, add 1 mL DMF to the resin for 20 min.
3. Deprotect the Fmoc group protected resin by reacting twice with 800 µL of the 20% piperdine in DMF (the Fmoc group deprotection reagent) for 3 min and then 8 min.
4. Next, wash the resin with 2 mL DMF.
5. In a separate reaction tube, mix in 2 mL DMF, 3 equivalents (in mol) HBTU, 6 equivalents NMM, 6 equivalents NMP, and 3 equivalents of the Fmoc group protected amino acid with 3 equivalents HOBt for preactivation for 5 min. Add this coupling mixture into the syringe containing the resin. The reaction then takes place for 90 min. Afterward discard the coupling mixture, keep the resin in the syringe.
6. Repeat the previous step in order to increase the coupling efficiency.
7. Perform capping of the nonreacted free amine groups by using 2 mL of 5% acetic acid anhydrate and 5% 2,6-lutidine in DMF. Mix it with the resin for 5 min. Then remove the capping mixture from the syringe.
8. Wash the resin with 2 mL DMF.
9. Couple the next Fmoc group protected amino acid by repeating **steps 3–8**.
10. After the last amino acid is coupled, cleave the Fmoc group twice again with the Fmoc group deprotection reagent for 3 and 8 min.
11. Wash with 2 mL DMF.
12. Wash with 2 mL dichloromethane (DCM).
13. Wash with 2 mL DMF.
14. Remove the syringe from the synthesizer, insert a plunger into the barrel to seal the resin from the atmosphere and cap the tip. It can either be stored at −20 °C or used in Subheading 3.2, **step 1** or 3.2, **step 2**, depending on whether an unnatural pTyr mimetic should be incorporated.

3.2 Synthesis of Peptide with pTyr Mimetics

To synthesize the peptide with an unnatural pTyr mimetic, synthesize it from *C*-Terminus to the amino acid prior to the pTyr mimetic. The automatic synthesis must be done with open *N*-terminus (the Fmoc group cleaved and without any final capping step). The coupling of the unnatural amino acid must be done manually as instructed below and the rest of the synthesis after unnatural amino acid can be performed again in automated synthesizer (as described in Subheading 3.1):

1. In the reaction, syringe containing the protected peptide with open *N*-terminus (amount in mol, which equals 1 equivalent), coupled to the resin, add 1 mL of DMF and shake for 20 min.
2. In a beaker, prepare the reaction mixture by mixing 3 equivalents pTyr mimetic, 3 equivalents HOBT, 3 equivalents HBTU and 6 equivalents NMM in 2 mL DMF.
3. Discard the DMF from syringe, add the reaction mixture and cap the tip of the syringe.
4. Let the syringe shake for 4 h at room temperature, discard the reaction mixture but not the resin, wash the resin 3 times, each 5 min with 2 mL DMF.
5. Repeat **steps 2–4**, but let the syringe shake overnight at room temperature (*see* **Note 1**).
6. Prepare 2 mL of capping mixture as 5% acetic acid anhydrate and 5% 2,6-lutidine in DMF in order to acetylate the not reacted *N*-terminal amine groups.
7. Discard the solution but not the resin, wash the resin 3 times, each 2 mL DMF for 5 min.
8. Add capping mixture to the resin in the syringe, cap the syringe tip, and shake it for 20 min.
9. Discard the solution but not the resin, wash the resin 3 times, each time with 2 mL DMF for 5 min.
10. Repeat **steps 8** and **9**.
11. Put the syringe in the synthesizer and proceed the reaction of the rest of peptide sequence starting from **step 3** in Subheading 3.1.

3.3 Manual Coupling of Linker

In order to ensure enough space between peptide sequence and the cysteine that provides the sulfhydryl group, cysteine is coupled to the peptide after a linker molecule. The coupling of linker to the peptide is performed as described below:

1. In the reaction, syringe-containing peptide (amount in mol, which equals 1 equivalent) with protected side chains and unprotected *N*-terminus but coupled to the resin, add 1 mL of DMF, cap the syringe and shake for 20 min.

2. In a beaker, prepare the reaction mixture by mixing 3 equivalents linker, 3 equivalents HOBT, 3 equivalents HBTU, and 6 equivalents NMM in 2 mL DMF.
3. Discard the DMF from the syringe, add the reaction mixture and cap it.
4. Let the syringe shake for 4 h at room temperature, then discard the reaction mixture and wash the resin the syringe 3 times, each time 5 min with 2 mL DMF.
5. Repeat **steps 2–4**, but let the syringe shake overnight at room temperature.
6. Prepare 2 mL of the capping mixture as 5% acetic acid anhydrate and 5% 2,6-lutidine in DMF in order to acetylate the not reacted *N*-terminal amine groups.
7. Discard the solution in resin, wash 3 it times, each time using 2 mL DMF for 5 min.
8. Add the capping mixture to the resin, cap the syringe tip and shake it for 20 min.
9. Discard the solution but not the resin and wash the resin 3 times, each time with 2 mL DMF for 5 min.
10. Repeat **steps 8** and **9**.
11. Deprotect the Fmoc group of the linker by preparing 20% piperidine in 2 mL DMF and adding it to the syringe two times for 3 and 8 min. Wash the syringe as described in **step 9**, cap the tip and seal the syringe with paraffin from the atmosphere. It can either be stored at −20 °C or used in Subheading 3.4 (*see* **Note 2**).

3.4 Manual Coupling of Cysteine

This coupling procedure is the same as the coupling of the linker, described above in Subheading 3.3. In **step 2** of Subheading 3.3, replace the linker with Fmoc-Cys(Trt)-OH.

3.5 Peptide Deprotection and Purification

In order to make sure that the synthesis was successful after synthesis of the peptide with the synthesizer, coupling of linker and cysteine, a test cleavage can be done after each step as below:

3.5.1 Peptide Deprotection

1. Prepare 1 mL of the cleavage cocktail as 95% TFA, 2.5% TIPS, and 2.5% H_2O (all volume %). This will also cleave the side chain protecting groups of the amino acids used in standard Fmoc-SPPS.
2. Transfer a small amount of peptide-resin to a new syringe and add the cleavage cocktail carefully. Clean the outside of the syringe with a laboratory tissue to remove potential residual acid. The cleavage can take from 4 h to overnight depending on the protecting groups on the peptide side chains.

3. Pour the cleavage cocktail containing the deprotected and cleaved peptide in previously cooled at −20 °C diethyl ether and centrifuge for 5 min, at 4 °C and 2700 × *g*.
4. Discard the supernatant and validate the success of the synthesis by determining the mass of precipitate by standard MALDI or HPLC-ESI.
5. If the synthesis was successful, proceed with cleaving the whole peptide as described in **steps 1–4**. Start with 1 mL of cleavage cocktail, if the peptide does not cleave completely, increase time or volume of cleavage.
6. It is better to purify the peptide right after fully deprotecting it. However, if at this step storage is inevitable, store it at −20 °C.

3.5.2 Peptide Purification

1. Dissolve the peptide in solvents compatible with the solvents in the purification method (*see* **Notes 3** and **4**). For HPLC, this is usually a mixture of acetonitrile in water with 0.05% acid (we use TFA).
2. Purify the dissolved peptide by preparative high performance liquid chromatography.
3. Collect the fractions and check the mass using MALDI or HPLC-ESI.
4. Freeze dry the fraction with the correct mass.

3.6 Coupling of Peptide to the SulfoLink Resin

3.6.1 Preparing the Resin Bed

1. Ensure that SulfoLink coupling resin and all other reagents have room temperature.
2. Suspend the resin and take 500 μL (250 μL resin) using wide-bore pipette and transfer it to a 1.5 mL tube. Prepare two samples, one for coupling to the peptide and one as control (*see* **Note 5**).
3. Centrifuge the tubes 3 min at 1000 × *g* and discard the supernatant carefully.
4. Wash the resin by adding three times, each 5 min, 1 mL of coupling buffer and stirring at room temperature. Centrifuge the tubes 3 min at 1000 × *g* and discard the supernatant after each wash.

3.6.2 Immobilization of the Peptide

1. Dissolve 2 mg of peptide in 1 mL of coupling buffer and add it to the resin. For the control, add 1 mL of coupling buffer.
2. Depending on the stability of the peptide it can be stirred in rotator wheel with the beads in the tubes at room temperature up to one night (*see* **Note 6**).
3. The next day wash the resin 3 times with coupling buffer.

4. Prepare a solution of 50 mM L-Cysteine in coupling buffer and add 1 mL to the tubes to block not coupled resins. Let it stir for 1 h at room temperature.
5. Wash the resin 3 times with coupling buffer and then 3 times with lysis buffer.

3.7 Pull Down

In order to pull down proteins containing SH2 domains with affinity of binding to the peptide of interest, incubate immobilized peptide with a cell lysate. If the peptide contains pTyr, use protease and phosphatase inhibitors in the lysis buffer to prevent protein degradation and dephosphorylation of pTyr. For the peptide containing the mimetic, a phosphatase inhibitor is not needed.

1. Prepare the cell lysate. Use 500 μL of buffer for approximately nine million cells. Sonicate cells in buffer 3 times for 3 s. Incubate on ice for 30 min. Centrifuge for 10 min, 1000 × *g* at 4 °C.
2. Incubate the resin-peptide in the 1.5 mL tubes with 1 mL of a cell lysate overnight at 4 °C (*see* **Note 7**).
3. Centrifuge the sample 3 min, 1000 × *g* at 4 °C.
4. Collect the supernatant. This is the flow through and not bound mixture.
5. Wash the resin-peptide with lysis buffer 3 times, each 5 min at 4 °C and collect the supernatant.
6. Boil samples from **steps 4** and **5** in protein sample buffer for 3 min at 95 °C for later SDS-PAGE analysis.
7. Boil the Resin in sample buffer 3 min at 95 °C.
8. Centrifuge the sample 5 min at 1000 × *g* and collect the supernatant elution. This sample contains bound proteins to the resin-peptide.

These samples boiled in sample buffer can be stored at −20 °C or used to proceed (*see* **Note 8**).

3.8 Detection of the Pulldown Protein using Western Blot

If the proteins of interest are known, success of pulling down by the immobilized peptide can be detected by Western blot (*see* **Note 9**). The Western blot is for detecting if the protein of interest with the SH2 domain in the cell lysates binds to the peptide, so one can enrich or isolate the endogenous SH2 domain-containing protein from the cell lysate. In order to be accessible to immunodetections, proteins are required to be transferred to a membrane as described below:

1. Load 20 μL of each sample to an SDS-PAGE and run the gel (Fig. 1b). Store the rest of boiled samples at −20 °C.

2. Transfer the proteins from the SDS-PAGE gel to the transfer membrane for 1 h at 100 mA in transfer buffer.
3. Block the membrane with 5% nonfat milk in TBS-T for 1 h.
4. Prepare the antibody against SH2 domain-containing protein of interest at the suggested dilution by the manufacturer in 5% nonfat milk in TBS-T and add the mixture to membrane.
5. Incubate overnight at 4 °C. The next day, remove the primary antibody and wash the membrane 3 times with TBS-T each 10 min.
6. Add the diluted secondary antibody in 5% nonfat milk in TBS-T to the membrane and mix for 1 h at room temperature.
7. Remove the secondary antibody and wash the membrane 3 times with TBS-T, each 10 min at room temperature.
8. Incubate the membrane with ECL detection reagents for 1 min and image the membrane (Fig. 1b).

4 Notes

1. Although double coupling of unnatural amino acids to the peptide sequence increases the efficiency of synthesis, in case of lack of supplies, once overnight coupling can also be done.
2. The optimum way to store the peptide before purification, is before its cleavage and keeping it on resin with protecting side chains, washed with DCM and covered the syringe with Paraffin.
3. For dissolving the peptide before purification, take into account that traces of TFA after cleavage can still be present and make the pH acidic.
4. Depending on the sequence of the peptide of interest, the pH can influence solubility.
5. Less volume of the agarose beads can also be used, and accordingly, the respective volumes of the rest of materials should be adjusted.
6. If the peptide is oxidized, perform a reduction with tris(2-carboxyethyl)phosphine (TCEP). Add 0.1 mL of 25 mM TCEP to the 2 mL peptide solution in coupling buffer. Incubate the mixture at room temperature for 30 min.
7. Incubation time with the cell lysate can be optimized depending on affinity and stability of the binding. This step must be done at 4 °C.
8. If the proteins of interest are known, in order to optimize the assay conditions, pulling down of purified recombinant protein

with the immobilized peptide can also be done prior to performing the experiment with the cell lysate.

9. If the pulled down proteins are unknown or for selectivity, alternatively the sample from **step 8** in Subheading 3.7 can be analyzed by a proteomic read out [7].

References

1. Kaneko T, Joshi R, Feller S et al (2012) Phosphotyrosine recognition domains: the typical, the atypical and the versatile. Cell Commun Signal 10:1–20
2. Waksman G, Kuriyan J (2004) Structure and specificity of the SH2 domain. Cell 116:45–51
3. Kaneko T, Huang H, Zhao B et al (2010) Loops govern SH2 domain specificity by controlling access to binding pockets. Sci Signal 3:34–34
4. Zhou S, Shoelson SE, Chaudhuri M et al (1993) SH2 domains recognize specific phosphopeptide sequences. Cell 72:767–778
5. Machida K, Mayer BJ (2005) The SH2 domain: versatile signaling module and pharmaceutical target. BBA Proteins Proteomics 1747:1–25
6. Kraskouskaya D, Duodu E, Arpin CC et al (2013) Progress towards the development of SH2 domain inhibitors. Chem Soc Rev 42: 3337–3370
7. Eberl HC, Werner T, Reinhard FB et al (2019) Chemical proteomics reveals target selectivity of clinical Jak inhibitors in human primary cells. Sci Rep 9:1–14

References

Chapter 20

Biomolecular Condensation of SH2 Domain-Containing Proteins on Membranes

Longhui Zeng and Xiaolei Su

Abstract

The plasma membrane serves as an effective platform for signal transduction of membrane receptor pathways. Activation of the T-cell receptor (TCR) triggers the formation of membrane-associated condensates that are formed through liquid–liquid phase separation. These condensates are assembled by multivalent interactions between the tyrosine-phosphorylated receptor/adaptor and the SH2 domain-containing protein at membrane-proximal milieu. Here, we describe a biochemical reconstitution system that has been implemented to decipher the mechanisms of phospholipase PLCγ1-mediated LAT condensate formation. To characterize the interaction between specific phosphotyrosine–SH2 pair, we developed a total internal reflection fluorescence (TIRF) microscopy-based system to quantify the binding preference of each SH2 domain to specific tyrosine in the context of membranes. An assay to determine the condensate-mediated protection of phosphotyrosines from being dephosphorylated by phosphatase is also elaborated. These assays could be applied to study other transmembrane receptor pathway as well as condensates formed on endomembrane systems including the endoplasmic reticulum, mitochondrion, and Golgi apparatus.

Key words Phase separation, Condensate, Biochemical reconstitution, PLCγ1, Plasma membrane, TIRF, CD45

1 Introduction

Eukaryotic cells organize biochemical reactions by compartmentalization through the formation of both membrane-bound organelles such as mitochondrion and lysosome and membraneless compartments assembled by liquid–liquid phase separation [1–3]. These membraneless compartments include (a) 3D condensates such as stress granules in the cytoplasm, and promyelocytic leukemia protein (PML) bodies in the nucleus and (b) 2D membrane-associated condensates such as the T-cell microclusters and nephrin-slit diaphragm protein complexes on the plasma membrane [4–6].

Teresa Carlomagno and Maja Köhn (eds.), *SH2 Domains: Functional Modules and Evolving Tools in Biology*, Methods in Molecular Biology, vol. 2705, https://doi.org/10.1007/978-1-0716-3393-9_20,

Linker for activation of T cells (LAT) is a single-transmembrane adaptor protein that connects the upstream T-cell receptor stimulation to the downstream Ras, ERK, and calcium pathway that mediates T-cell activation [7, 8]. Our previous work demonstrated that LAT, through its phosphorylated tyrosines, forms multivalent interactions with Grb2 and Sos1 to form liquid-like condensates on the membrane [9]. These multivalent interactions include two types: (a) pY-SH2: phosphotyrosine of LAT with the SH2 domain of Grb2, (b) SH3-PRM: the SH3 domain of Grb2 with the proline-rich motif (PRM) of Sos1. Although the mechanism of LAT condensate formation has been elucidated, how LAT condensates are regulated remained less clear. Recently, we discovered that phospholipase PLCγ1, through its three SH domains (nSH2-cSH2-SH3), cooperates with Grb2 to regulate LAT condensation in a nonmonotonic manner: PLCγ1 promotes LAT clustering and increases cluster size at low concentrations but the effect is compromised at high concentrations [10].

Here, we describe a biochemical reconstitution assay for PLCγ1-regulated LAT condensation (Fig. 1a) based on a previously established system for reconstituting LAT microclusters

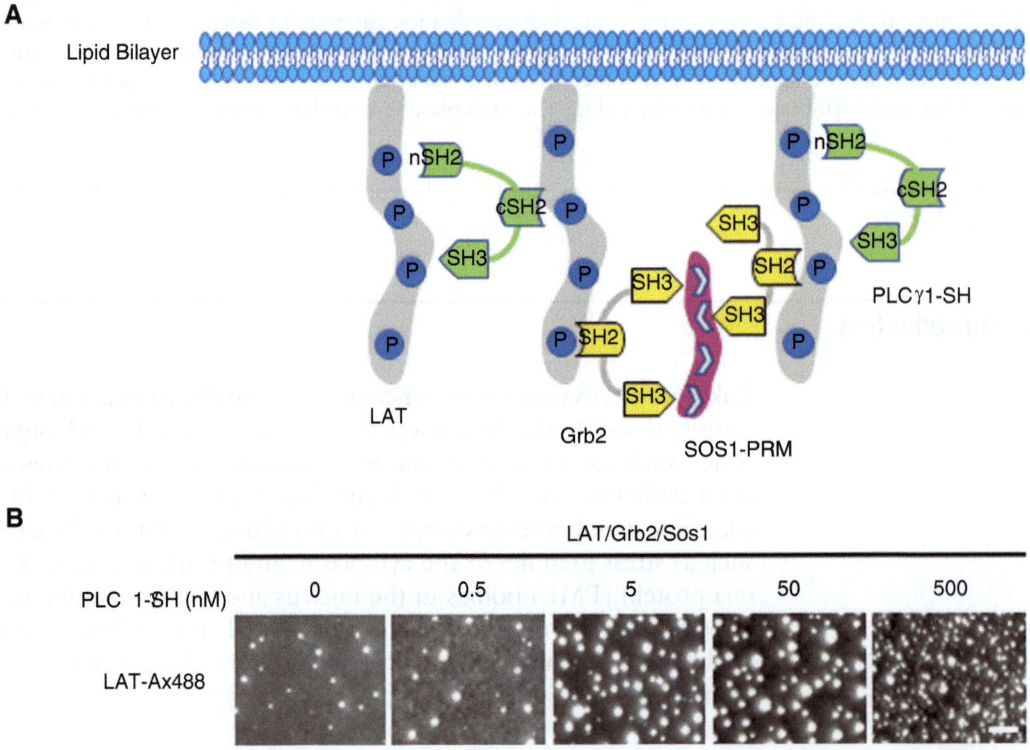

Fig. 1 Reconstituting PLCγ1-regulated LAT condensates. (**a**) Cartoon illustrating that PLCγ1 regulates LAT condensate formation. (**b**) PLCγ1 regulates LAT condensates formation nonmonotonically as imaged by TIRF microscopy. Scale bar: 5 μm

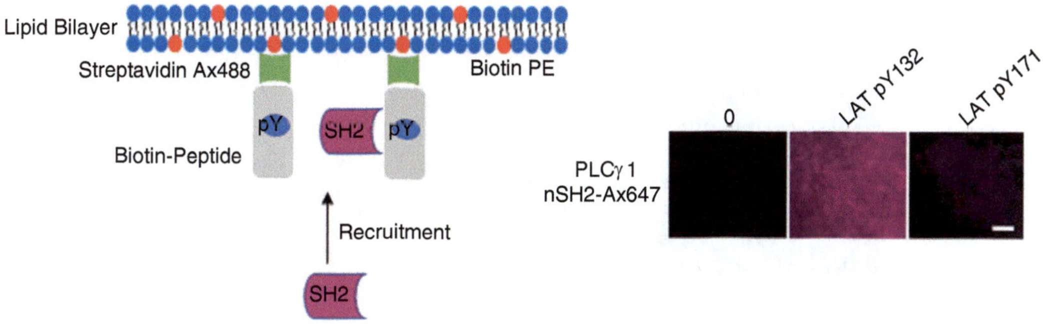

Fig. 2 Measuring the interaction preference of SH2 domain to phosphotyrosines. Left panel: Schematics of the assay. Right panel: TIRF microscopy revealed PLCγ1 nSH2 interacting with LAT pY132 peptide on membranes. Scale bar: 5 μm

[11]. Prephosphorylated LAT was fluorescently labeled on a C-terminal Cysteine residue and was attached on Ni-NTA lipid-containing bilayers through an N-terminal His8 tag. LAT formed microclusters on membranes with Grb2, Sos1, PLCγ1, as monitored by total internal reflection fluorescence (TIRF) microscopy. PLCγ1 increases LAT condensate formation only at or below physiologically relevant concentrations (~50 nM) (Fig. 1b).

The C-terminal four phosphotyrosines of LAT are essential for recruiting downstream SH2-containing proteins to transduce the TCR signaling [12, 13]. PLCγ1 contains two SH2 domains that can interact with some of these four phosphotyrosines and crosslink LAT. The binding affinity of pY to SH2 can be determined by classical methods such as surface plasmon resonance (SPR). However, SPR is usually performed in solution which lacks the membrane context of LAT. Complementing that, we developed an assay to determine the relative binding preference of each of these two SH2 domains to the four phosphotyrosines of LAT on membranes (Fig. 2). To elaborate, peptides containing individual phosphotyrosines were synthesized and modified with an N-terminal Biotin tag, which allows the attachment of peptides onto biotin-functionalized lipid bilayers through streptavidin. The PLCγ1 SH2 domains were purified, fluorescently labeled and incubated with individual peptides. TIRF microscopy was used to quantify the membrane-associated SH2 domain. Using this assay, we discovered that PLCγ1 nSH2 preferentially binds LAT pY132 whereas PLCγ1 cSH2 favorably interacts with LAT pY171 [10].

Previous study showed that LAT condensate formation is phosphorylation dependent [9]. CD45, a transmembrane phosphatase abundant on the T-cell surface, dephosphorylates LAT in vitro [9, 14, 15]. We found that PLCγ1 protects LAT from being dephosphorylated by CD45 [10]. CD45 is spatially excluded from the PLCγ1-mediated LAT clusters (Fig. 3). In this assay, LAT condensates were preformed with PLCγ1 on supported lipid bilayers. Then

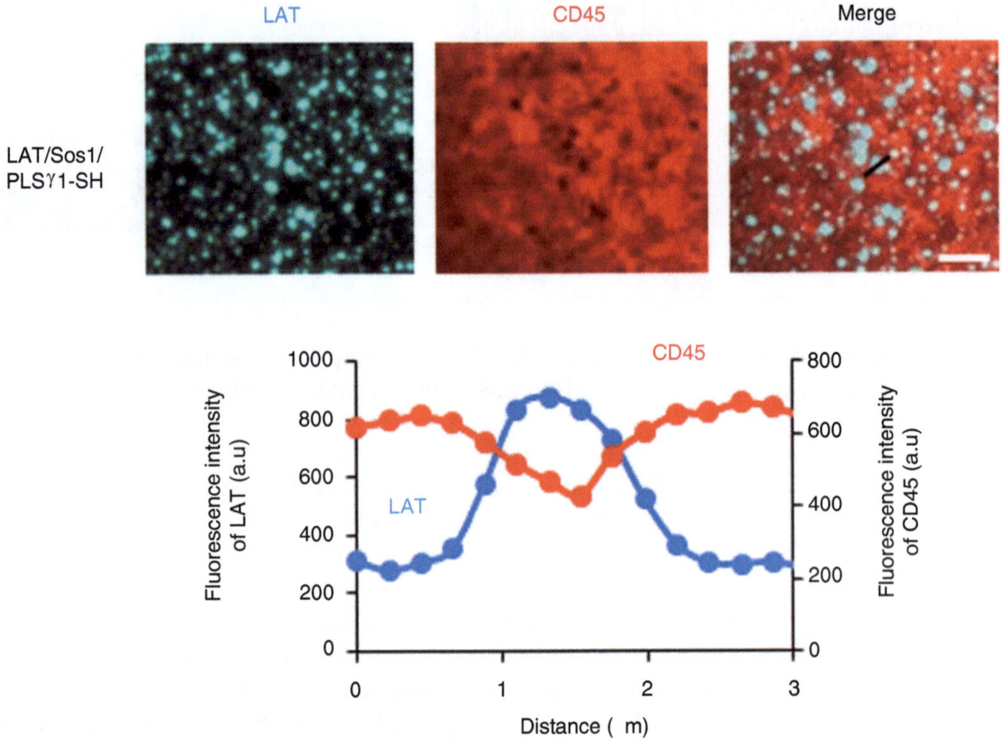

Fig. 3 PLCγ1 protects LAT from dephosphorylation by CD45. TIRF microscopy revealed the size-exclusion of CD45 by PLCγ1 driven LAT cluster. The fluorescence intensity of LAT and CD45 were quantified along the black line in the merged image. Scale bar: 5 μm

fluorescently labeled CD45 cytoplasmic domain with an N-terminal His10 tag was attached to the membrane. The localization of CD45 was imaged by TIRF microscopy. Samples were collected and applied to western blot analysis for LAT phosphorylation [10].

2 Materials

1. 96-well glass-bottom plate.
2. Glass vials.
3. Hellmanex III.
4. NaOH solution: 5 M NaOH.
5. Adhesive PCR sealing foil sheets.
6. 25 μL gastight syringe, 250 μL gastight syringe.
7. PD MiniTrap G25 column.
8. Vanadate: 2 mM vanadate.
9. Anti-LAT pY132 antibody.

10. Proteins: Streptavidin-AX488. Grb2, LAT, Sos1, PLCγ1 nSH2-cSH2-SH3, PLCγ1 nSH2, and CD45 were homemade as described [9, 10].
11. Synthesized phosphotyrosine peptides.
12. Basic lipid mix: 99.9% POPC, and 0.1% PEG5000 PE.
13. NTA lipid mix (for anchoring his-tagged protein): 97.9% POPC, 2% DOGS-NTA, and 0.1% PEG5000 PE.
14. Biotin lipid mix (for anchoring biotin-labeled protein): 99.8% POPC, 0.1% DSPE-PEG (2000) Biotin, and 0.1% PEG5000 PE.
15. Basic buffer: 50 mM HEPES, 150 mM NaCl, and 1 mM TCEP, pH 7.4.
16. Clustering buffer: 50 mM HEPES, 150 mM NaCl, 1 mM TCEP, and 1 mg/mL BSA, pH 7.4.
17. Oxygen scavenger solution: 0.2 mg/mL glucose oxidase, 0.035 mg/mL catalase, 70 mM β-mercaptoethanol, and 25 mM glucose in clustering buffer.

3 Methods

3.1 Generation of Small Unilamellar Vesicles (SUVs) and Supported Lipid Bilayers (Modified from Ref. [11])

1. Warm up lipid stocks to room temperature by placing them on bench for 15–30 min (*see* **Note 1**).
2. Wash and rinse glass vial with chloroform three times in fume hood. Preseed 1 mL of chloroform in each vial as a cushion.
3. Use glass syringes to transfer individual lipids into the glass vial. Prepare lipid mix up to 4 μmol per vial (*see* **Note 2**) following the composition described in Subheading 2.
4. Place the glass vial in ~45 °C water bath and dry the lipid mix with a stable flow of argon (*see* **Note 3**). The drying process is about 2–4 min for each vial and in the end, the lipid mix appears as multiple white circles that evenly decorate the bottom and lower side wall of the vial.
5. Further dry lipids for 3 h in a vacuum desiccator.
6. Add 1.5 mL of PBS to each vial and resuspend the lipid mix by vortex. Divide the resuspension into two 1.5 mL EP tubes.
7. A freeze-thaw method is used to break down lipid suspensions to SUVs: freeze the lipid suspension in liquid nitrogen, then thaw it in a water bath at room temperature. Repeat 30 cycles of freeze-thaw until the cloudy suspension turns clear (*see* **Note 4**).
8. Centrifuge the lipid suspension at 60,000 × *g* for 45 min at 4 ° C. Transfer the SUV-containing supernatant into a clean tube prefilled with argon. Store the SUVs at 4 °C.

9. Wash the glass-bottom plate with 5% Hellmanex III overnight, rinse with ddH_2O, and wash again with freshly made NaOH solution three times at 50 °C. (*see* **Note 5**). Rinse with ddH_2O and PBS to completely remove NaOH.
10. Add 200 μL of PBS into each well, and then 20 μL of SUV. Incubate the plate for 1 h at 37 °C to form supported lipid bilayers on the glass.
11. Remove 100 μL supernatant and wash out excessive SUVs with 500 μL basic buffer three times.
12. Add 100 μL clustering buffer to block the bilayers for 30 min at 37 °C.
13. Remove 100 μL supernatant and keep about 100 μL solution covering the supported lipid bilayers. This will be ready for further assays described below.

3.2 Reconstitution of PLCγ1-Regulated LAT Condensates on Supported Lipid Bilayers

1. Make 100 μL of 6 nM His8-phosphorylated LAT-Alexa488 (pLAT) in clustering buffer, add it onto NTA bilayers and incubate for 3 h at 37 °C. The final concentration of pLAT is 3 nM in a total of 200 μL solution.
2. Remove 100 μL supernatant and wash with 500 μL of clustering buffer three times. Remove 50 μL solution (*see* **Note 6**).
3. Prepare 50 μL oxygen scavenger solution (*see* **Note 7**) containing 6 μM Grb2, 0.6 μM Sos1 and PLCγ1 nSH2-cSH2-SH3 at 0, 1, 10, 100, or 1000 nM, add the mix into each well and incubate for 30 min at 37 °C.
4. The formation of LAT condensates was monitored by TIRF microscopy. Image J (Fiji) was used to analyze the clustering level [11].

3.3 Quantification of Binding Preference of SH2 Domains to Membrane-Associated Phosphopeptides

1. Add 100 μL 4 nM Streptavidin-Alexa488 in clustering buffer to the well coated with biotin bilayers and incubate for 30 min at 37 °C (*see* **Note 8**).
2. Remove 100 μL solution and wash with 500 μL clustering buffer three times. Remove 100 μL solution.
3. Add 100 μL 4 nM biotinylated phosphotyrosine peptides in clustering buffer and incubate for 30 min at 37 °C (*see* **Note 9**).
4. Remove 100 μL supernatant and wash with 500 μL clustering buffer three times. Remove 100 μL solution.
5. Add 100 μL oxygen scavenger solution containing 200 nM SH2-647 and incubate for 30 min at 37 °C. The SH2 domain was labeled with CoA 647 through an N-terminal ybbR tag [10] (*see* **Note 10**).

6. TIRF microscopy was used to determine the membrane-associated SH2 domain. The quality of bilayers can be evaluated by performing a FRAP assay on streptavidin-Alexa488. ImageJ was used to quantify the fluorescence intensity (*see* **Note 11**).

3.4 Protection of LAT Phosphorylation by PLCγ1 from Phosphatase

1. Add 100 μL 20 nM pLAT in clustering buffer onto NTA bilayers and incubate for 3 h at 37 °C.
2. Remove 100 μL supernatant and wash with 500 μL clustering buffer three times. Remove 50 μL solution.
3. Add 50 μL oxygen scavenger solution containing 2 μM Sos1 with or without 2 μM PLCγ1 fragment containing the nSH2-cSH2-SH3 domains and incubate for 30 min at 37 °C.
4. *Size-exclusion of CD45 monitored by TIRF microscopy*: Add 100 μL oxygen scavenger solution containing 8 nM His10-CD45-TMR protein and incubate for 1 h at 37 °C. LAT clusters and CD45 localization were imaged by TIRF microscopy. ImageJ was used to quantify the fluorescence intensity (*see* **Note 12**).
5. *LAT dephosphorylation detected by Western blot*: Following **step 3**, add 100 μL oxygen scavenger solution containing His10-CD45-TMR protein and incubate for 5 min at 37 °C. Remove all 200 μL supernatant, add 50 μL SDS-PAGE loading buffer with 2 mM vanadate to terminate the dephosphorylation reaction. Boil the sample for 5 min at 95 °C, and centrifuge for 5 min at 15,000 rpm. Apply 10 μL supernatant for SDS-PAGE and immunoblot analysis. LAT phosphorylation was detected by an antibody recognizing pY132 [10].

4 Notes

1. To reduce oxidization of stock lipids, we recommend ordering small volume of lipids by requesting bottling service from Avanti.
2. To prevent cross-contamination, glass syringe should be rinsed three times between taking different lipids. Rinse the syringe one more time with lipids to avoid dilution if transferring small amount of lipids (there is residual amount of chloroform retained in the syringe).
3. Nitrogen can be alternatively used to dry the lipid mix when argon is unavailable.
4. The freeze-thawed lipid suspension can be stored at −80 °C. Add an additional 5 cycles of freeze-thaw when using the frozen lipid suspension.

5. Sufficient cleaning of the glass is the key to achieve high quality of lipid bilayers. We suggest keeping the shelf life of Hellmanex III below 6 months once opened. 5 M NaOH solution should be made freshly.
6. Bilayers should always be covered with a minimal volume of 50 μL solution.
7. The oxygen scavenger solution should be prepared freshly. We recommend preparing a 200x stock of enzyme mix composed of glucose oxidase and catalase and store it in −80 °C freezers. Make fresh oxygen scavenger solutions by mixing the enzymes, glucose and β-mercaptoethanol in clustering buffer.
8. Instead of directly making 0.1% Biotin SUVs, 2% Biotin SUVs can be made and mixed with basic SUVs in 1:20.
9. Each peptide starts with 2 alanines and contains about 6 amino acids on each side of the phosphotyrosine.
10. The SH2 domain can be labeled by alternative methods including SNAP tag, cysteine-maleimide, or amine-NHS.
11. Only images with a similar fluorescence intensity of streptavidin-Alexa488 can be compared among groups.
12. The level of CD45 exclusion is generally correlated with the size of LAT clusters. The optimal size of LAT clusters can be achieved by adjusting stoichiometry of PLCγ1 to LAT.

Acknowledgment

Longhui Zeng is the recipient of a Cancer Research Institute/ Irvington Postdoctoral Fellowship (CRI3516). Xiaolei Su has received support from the Charles H. Hood Foundation Child Health Research Awards, the Andrew McDonough B+ Foundation Research Grant, the Gilead Sciences Research Scholars Program in Hematology/Oncology, the Rally Foundation and Bear Necessities Foundation a Collaborative Pediatric Cancer Research Awards Program, the Yale SPORE in skin cancer DRP Award, the Yale DeLuca Pilot Award, the NIH MIRA (R35) program (GM138299), the Human Frontier Science Program Early-Career Research Grant, and the American Cancer Society Research Scholar Grant.

References

1. Banani SF, Lee HO, Hyman AA et al (2017) Biomolecular condensates: organizers of cellular biochemistry. Nat Rev Mol Cell Biol 18: 285–298
2. Alberti S, Gladfelter A, Mittag T (2019) Considerations and Challenges in Studying Liquid-Liquid Phase Separation and Biomolecular Condensates. Cell 176:419–434
3. Shin Y, Brangwynne CP (2017) Liquid phase condensation in cell physiology and disease. Science 357:eaaf4382

4. Xiao Q, Mcatee CK, Su X (2021) Phase separation in immune signalling. Nat Rev Immunol 22:188
5. Jaqaman K, Ditlev JA (2021) Biomolecular condensates in membrane receptor signaling. Curr Opin Cell Biol 69:48–54
6. Zhao YG, Zhang H (2020) Phase Separation in membrane biology: the interplay between membrane-bound organelles and membraneless condensates. Dev Cell 55:30–44
7. Zhang W, Sloan-Lancaster J, Kitchen J et al (1998) LAT: the ZAP-70 tyrosine kinase substrate that links T cell receptor to cellular activation. Cell 92:83–92
8. Abraham RT, Weiss A (2004) Jurkat T cells and development of the T-cell receptor signalling paradigm. Nat Rev Immunol 4:301–308
9. Su X, Ditlev JA, Hui E et al (2016) Phase separation of signaling molecules promotes T cell receptor signal transduction. Science 352: 595–599
10. Zeng L, Palaia I, Šarić A et al (2021) PLCγ1 promotes phase separation of T cell signaling components. J Cell Biol 220:e202009154
11. Su X, Ditlev JA, Rosen MK et al (2017) Reconstitution of TCR Signaling Using Supported Lipid Bilayers. Methods Mol Biol 1584:65–76
12. Zhu M, Janssen E, Zhang W (2003) Minimal requirement of tyrosine residues of linker for activation of T cells in TCR signaling and thymocyte development. J Immunol 170:325–333
13. Zhang W, Trible RP, Zhu M et al (2000) Association of Grb2, Gads, and phospholipase C-gamma 1 with phosphorylated LAT tyrosine residues. Effect of LAT tyrosine mutations on T cell angigen receptor-mediated signaling. J Biol Chem 275:23355–23361
14. Charbonneau H, Tonks NK, Walsh KA et al (1988) The leukocyte common antigen (CD45): a putative receptor-linked protein tyrosine phosphatase. Proc Natl Acad Sci U S A 85:7182–7186
15. Courtney AH, Lo WL, Weiss A (2018) TCR signaling: mechanisms of initiation and propagation. Trends Biochem Sci 43:108–123

INDEX

Teresa Carlomagno and Maja Köhn (eds.), *SH2 Domains: Functional Modules and Evolving Tools in Biology*, Methods in Molecular Biology, vol. 2705, https://doi.org/10.1007/978-1-0716-3393-9,

Index

MIX
Papier aus verantwortungsvollen Quellen
Paper from responsible sources
FSC® C105338

If you have any concerns about our products,
you can contact us on
ProductSafety@springernature.com

In case Publisher is established outside the EU,
the EU authorized representative is:
Springer Nature Customer Service Center GmbH
Europaplatz 3, 69115 Heidelberg, Germany

Printed by Libri Plureos GmbH
in Hamburg, Germany